Computational Analysis and Conformational Modeling for Protein Structure and Interaction

Computational Analysis and Conformational Modeling for Protein Structure and Interaction

Guest Editors

Clement Agoni
Indrani Bera

MDPI

Basel • Beijing • Wuhan • Barcelona • Belgrade • Novi Sad • Cluj • Manchester

Guest Editors

Clement Agoni
UCD Conway Institute of Biomolecular and Biomedical Research
University College Dublin
Belfield
Ireland

Indrani Bera
UCD Conway Institute of Biomolecular and Biomedical Research
UCD
Dublin
Ireland

Editorial Office

MDPI AG
Grosspeteranlage 5
4052 Basel, Switzerland

This is a reprint of the Special Issue, published open access by the journal *Biomolecules* (ISSN 2218-273X), freely accessible at: https://www.mdpi.com/journal/biomolecules/special_issues/46A2O4AUY8.

For citation purposes, cite each article independently as indicated on the article page online and as indicated below:

Lastname, A.A.; Lastname, B.B. Article Title. *Journal Name* **Year**, *Volume Number*, Page Range.

ISBN 978-3-7258-5825-5 (Hbk)
ISBN 978-3-7258-5826-2 (PDF)
https://doi.org/10.3390/books978-3-7258-5826-2

Contents

About the Editors

Clement Agoni

Clement Agoni is a researcher in the School of Health Sciences at the University of KwaZulu-Natal, South Africa, where he was recently awarded a prestigious Y2 rating by the National Research Foundation (NRF). He also is a Research Associate with the National Institute for Theoretical and Computational Sciences (NITheCS) in South Africa. Dr. Agoni, until his recent role, served as a Marie Curie Career Fit Plus fellow at the Conway Institute in University College Dublin's school of medicine, Ireland. Dr. Agoni's research expertise lies in computational pharmacology and structural bioinformatics. He applies molecular modeling, AI-driven simulations, and high-performance computing to deepen the understanding of protein–ligand interactions and accelerate drug discovery for emerging infectious diseases, cancer therapeutics, and bioactive peptides.

Indrani Bera

Indrani Bera is a multidisciplinary researcher specialising in structural bioinformatics, proteomics, and cheminformatics. Her work integrates analysing large-scale proteomic data with structural and molecular modelling to elucidate protein function, interactions, and post-translational modifications. Her current research interests are the analysis of proteomics/peptidomics data for understanding how various protein/peptide motifs are expressed in different cellular conditions using computational biology techniques and studying how biological macromolecules interact and function using structural bioinformatics. Her previous studies focused on elucidating the structure, function, and dynamics of proteins and protein–ligand interactions.

Preface

This Reprint focuses on "Computational Analysis and Conformational Modeling for Protein Structure and Interaction", a rapidly advancing field at the intersection of computational biology, structural bioinformatics, and drug discovery. The scope of this collection encompasses methods and applications of molecular modeling, conformational dynamics, protein–ligand interactions, and artificial intelligence-driven methods that provide valuable insights into biological mechanisms and therapeutic development. The aim of this Reprint is to bring together contributions that highlight both methodological innovations and practical applications in understanding protein structure and function. By showcasing studies that span from fundamental computational approaches to translational insights, the collection seeks to illustrate how in silico tools can accelerate the discovery of new therapeutics. The motivation for curating this Reprint stems from the growing importance of computational strategies in addressing scientific and societal challenges in modern biomedical research. As experimental techniques generate vast amounts of structural and sequence data, computational methods have become indispensable in extracting meaningful knowledge and guiding experimental validation. This Reprint is addressed to researchers, educators, students, and professionals working in computational biology, structural bioinformatics, pharmaceutical sciences, and related disciplines. It is intended to serve as both a reference for specialists and an entry point for new researchers seeking to understand how computational tools can be harnessed to solve pressing biomedical problems.

Clement Agoni and Indrani Bera
Guest Editors

biomolecules

MDPI

Review

Molecular Modelling in Bioactive Peptide Discovery and Characterisation

Clement Agoni [1,2,3,*], Raúl Fernández-Díaz [1,4], Patrick Brendan Timmons [5], Alessandro Adelfio [5], Hansel Gómez [5] and Denis C. Shields [1,2,*]

1 School of Medicine, University College Dublin, D04 C1P1 Dublin, Ireland; raul.fernandezdiaz@ucdconnect.ie
2 Conway Institute of Biomolecular and Biomedical Science, University College Dublin, D04 C1P Dublin, Ireland
3 Discipline of Pharmaceutical Sciences, School of Health Sciences, University of KwaZulu-Natal, Durban 4000, South Africa
4 IBM Research, D15 HN66 Dublin, Ireland
5 Nuritas Ltd., Joshua Dawson House, D02 RY95 Dublin, Ireland; timmons.patrick@nuritas.com (P.B.T.); adelfio.alessandro@nuritas.com (A.A.); gomez.hansel@nuritas.com (H.G.)
* Correspondence: clement.agoni@ucd.ie (C.A.); denis.shields@ucd.ie (D.C.S.)

Abstract: Molecular modelling is a vital tool in the discovery and characterisation of bioactive peptides, providing insights into their structural properties and interactions with biological targets. Many models predicting bioactive peptide function or structure rely on their intrinsic properties, including the influence of amino acid composition, sequence, and chain length, which impact stability, folding, aggregation, and target interaction. Homology modelling predicts peptide structures based on known templates. Peptide–protein interactions can be explored using molecular docking techniques, but there are challenges related to the inherent flexibility of peptides, which can be addressed by more computationally intensive approaches that consider their movement over time, called molecular dynamics (MD). Virtual screening of many peptides, usually against a single target, enables rapid identification of potential bioactive peptides from large libraries, typically using docking approaches. The integration of artificial intelligence (AI) has transformed peptide discovery by leveraging large amounts of data. AlphaFold is a general protein structure prediction tool based on deep learning that has greatly improved the predictions of peptide conformations and interactions, in addition to providing estimates of model accuracy at each residue which greatly guide interpretation. Peptide function and structure prediction are being further enhanced using Protein Language Models (PLMs), which are large deep-learning-derived statistical models that learn computer representations useful to identify fundamental patterns of proteins. Recent methodological developments are discussed in the context of canonical peptides, as well as those with modifications and cyclisations. In designing potential peptide therapeutics, the main outstanding challenge for these methods is the incorporation of diverse non-canonical amino acids and cyclisations.

Keywords: molecular modelling; bioactive peptides; homology modelling; virtual screening; molecular docking; molecular dynamics simulation; AlphaFold; Protein Language Models

1. Introduction

Molecular modelling techniques are a group of computational approaches that can be used to understand the complex relationships between the molecular structure and function of small molecules and biological macromolecules [1,2]. These techniques have the potential to accelerate the design and discovery of new drugs with enhanced therapeutic properties by providing valuable insights into the structural characteristics, dynamics, and interactions at a molecular level [3–6]. In this review, we provide an introduction to the compositional, structural, and conformational aspects of peptides and of their protein targets and then go on to identify the main molecular modelling methods used in the course of bioactive peptide identification, spanning homology modelling, docking, virtual screening of peptide libraries, molecular dynamics analyses, and machine-learning prediction.

The increase in computational power and the growing availability of vast amounts of structural data have greatly improved molecular modelling techniques. At the same time, improvements in molecular modelling methods have significantly reduced the time and resources needed for hit identification, hit-to-lead optimization, and lead optimisation in drug discovery [7,8]. This progress has contributed to the discovery of a number of FDA-approved drugs, such as Viracept and Zanamivir, both of which were developed using structure-based drug design methods [1,5,7,9,10]. It has also led to the identification of new possible uses for known drugs (repurposing). For example, in silico molecular modelling identified Ouabain, Digitoxin, and Niclosamide as FDA-approved drugs with strong binding patterns to key viral proteins, suggesting they could be effective treatments for COVID-19 [11]. Similarly, through virtual screening and experimental evaluation, cepharanthine, clofazimine, metergoline, imatinib, and efloxate were identified as potential repurposable drugs with the ability to interfere with SARS-CoV-2 viral entry [12].

Bioactive peptides are promising therapeutic alternatives to small molecules due to their biosafety, medicinal properties, and nutritional and cosmetic benefits, as well as their known effectiveness at low concentrations. These peptides are typically short chains, with fewer than 50 amino acids, and exhibit a wide range of biological activities, including antimicrobial, antihypertensive, anti-inflammatory, anticancer, and antioxidant properties [13–17]. Biological systems already use peptides to address various physiological and pharmacological challenges [18,19]. Beyond natural hormone analogues based on peptides, peptide drug candidates are also utilised to interfere with protein–protein interactions and to target or inhibit intracellular molecules [20,21]. Several peptide-based drugs, such as dulaglutide [22], albiglutide [23], liraglutide (Victoza) [24], and glucagon-like peptide-1 (GLP-1) [25], have been approved for clinical use in treating metabolic disorders. Additionally, peptide-based hormone analogues like plecanatide [26], leuprolide (Lupron) [27], goserelin (Zoladex) [28], octreotide [29], and lanreotide [30] are also available for clinical applications. Collectively, these drugs have generated billions of dollars in sales, fuelling the growing interest in bioactive peptides [31].

The traditional discovery and characterisation of bioactive peptides is, however, challenged by their unique inherent structural diversity and dynamic nature, as well as their poor bioavailability [17,19,32,33]. Experimental methods for identifying and characterising bioactive peptides are typically time-consuming and resource-intensive, highlighting the need for more efficient approaches [34,35]. However, these traditional methods can be significantly accelerated by implementing molecular modelling techniques or integrating them with artificial intelligence (AI) solutions. Such methods can predict peptide structures, explore peptide–protein interactions, investigate the structural dynamics of peptides, and repurpose bioactive peptides with known activities [36–40]. Molecular modelling techniques can, therefore, aid in the rational design of peptide-based drugs to improve their activity, facilitate virtual screening of peptide libraries to identify potential drug candidates, and support structure–activity relationship studies to optimize peptides, and possibly also repurpose peptides with known bioactivities [37,41–43]. The application of molecular modelling has been particularly noticeable in discovering potential milk-derived bioactive peptides, where techniques like molecular docking and molecular dynamics (MD) simulations have shown great potential in enhancing the efficiency of generating and discovering bioactive peptides [44–46].

While molecular modelling methods have been largely successful in the discovery of small-molecule therapeutics, it remains uncertain whether this success can be replicated in the design and discovery of bioactive peptides [37,44,47,48]. Evidence suggests that directly applying small-molecule or protein design strategies is not well-suited for peptide drug design [38,49–52]. This is primarily due to peptide conformational flexibility, which limits the effectiveness of small-molecule computational methods. Also, the incorporation of nonstandard amino acids and the use of diverse cyclisation chemistries in peptide drug design pose difficulties for traditional protein-modelling software [38]. As a result, customised computational approaches specifically tailored to peptides are needed and are gradually gaining more and more interest.

The recent integration of molecular modelling techniques with artificial intelligence (AI) and machine-learning methods has led to innovative advancements such as AlphaFold's protein structure prediction and Protein Language Models, which could significantly enhance the bioactive peptide discovery process [44,53–60]. The use of molecular modelling techniques must, however, always be in tandem with rigorous experimental validations to guarantee effectiveness.

2. Structural Characteristics of Bioactive Peptides

The biological activity of any given peptide and its ability to interact with molecular targets are determined by its structural characteristics, such as amino acid composition, sequence, secondary structure, molecular weight, chain length, and the presence of disulfide bonds [16,17,61] (Figure 1). The specific arrangement of amino acids, especially aromatic and charged residues, plays a significant role in determining properties like solubility, binding affinity, and reactivity. These relationships are often elucidated through structure–activity relationship (SAR) studies, which highlight how variations in peptide structure can affect their functional outcomes [62–64].

Bioactive peptides typically adopt secondary structures such as α-helices and β-sheets, which can be stabilised through strategies like peptide cyclisation, the introduction of D-amino acids, and covalent cross-linking [16,62,65]. Shorter peptides (2–20 amino acids) with molecular weights between 180 and 1000 Da often exhibit greater bioactivity compared to longer chains, with approximately 42% of the most active peptides having molecular weights below 1000 Da [19,66]. Shorter peptides tend to exhibit higher bioavailability due to their enhanced absorption and reduced susceptibility to proteolytic degradation and overall favourable pharmacokinetic properties [66]. Larger peptides, on the other hand, may exhibit more specific and potent effects compared to smaller peptides or traditional small molecules. This specificity results from the increased number of interactions they can form with target sites, such as through hydrogen bonding and hydrophobic interactions, leading to fewer off-target effects and reduced side effects overall [18,19]. There may be a complex relationship between optimal size and bioactivity, with tumours being modelled to retain larger peptides better and smaller peptides needing to have a higher affinity to be retained in the tumour [67]. Disulfide bonds formed between cysteine residues help stabilise specific secondary structures and maintain the overall conformation of many peptides [62,68,69]. The size of peptides also influences their ability to traverse biological barriers, with smaller peptides generally exhibiting higher permeability, often diffusing passively through lipid bilayers [70,71]. Studies have shown an inverse relationship between peptide size and permeability, with smaller peptides demonstrating higher flux rates across barriers like the Caco-2 cell monolayer [70,72]. As size increases, passive diffusion becomes less effective, and peptides rely more on active transport mechanisms [70]. Medium-sized peptides may utilise specific transporters and/or receptor-mediated transcytosis or function as cell-penetration peptides [70,73]. Larger peptides face significant challenges but can employ strategies such as transcytosis, macropinocytosis, or targeted delivery systems to enhance their transport [74,75]. PepT1 and PepT2, oligopeptide transporters, facilitate the uptake of di- and tripeptides with diverse chemical properties [76]. These oligopeptide transporters also transport neutral tripeptide-like β-lactam antibiotics and other peptide-mimetic drugs [77]. As peptide size increases, the impact of charge on permeability decreases [19,78]. Positively charged peptides generally permeate more efficiently due to favourable electrostatic interactions with negatively charged lipid head groups in membranes. However, as peptide size increases, these interactions become less significant compared to other factors like steric hindrance and hydrophobicity [72,79]. Ongoing research continues to develop innovative strategies to improve the transport of larger peptides, expanding their potential as drug delivery vehicles [80,81].

2.1. Amino Acid Composition: The Building Blocks of Bioactivity

Peptide amino acid composition plays a pivotal role in determining its structure and function [82]. Collectively, the composition of amino acids in a peptide influences the peptide's ability to form secondary structures, such as alpha helices and beta sheets [83,84]. Different amino acids possess varying side chains that can be hydrophobic, hydrophilic, charged, or neutral [83,85–87]. The presence of specific amino acids, such as aromatic residues (e.g., tryptophan and phenylalanine) or charged residues (e.g., lysine and glutamic acid), can significantly affect a peptide's solubility, binding affinity, and reactivity with target molecules [88,89]. Hydrophobic (Val, Ile, and Pro) and positively charged (His, Arg, and Lys) amino acids are enriched with anti-inflammatory peptides [66].

Hydrophobic interactions between amino acids contribute to peptide structure and stability, which can have an important effect on their bioavailability [90,91]. This is particularly important in the gastrointestinal tract, where peptides with a higher proportion of hydrophobic amino acids tend to be more resistant to proteolysis, increasing their bioavailability and allowing them to reach their target sites [92–95]. On the other hand, polarity determines solubility and the ability of any given peptide to cross biological membranes [96–98]. Peptides with a high net charge (positive or negative) tend to interact preferentially with other charged molecules with the opposite sign. The net charge of a peptide, determined by its proportion of positively and negatively charged amino acids, can also influence its interaction with other molecular targets, like extracellular receptors or enzymes [99,100].

In computational analysis, amino acid composition can be quantified to assess how variations in sequence impact biological activity [101,102]. This is particularly important in machine-learning approaches to analyse large datasets of peptide sequences to identify patterns that correlate specific amino acid compositions with desired activities, such as antimicrobial or anti-inflammatory effects [103].

2.2. Amino Acid Sequence and Stereochemistry

The order in which amino acids are arranged along a peptide's backbone plays a significant role in determining its functional properties [16,104,105]. Specific amino acid sequences can confer unique binding affinities and specificities to receptors or enzymes, allowing the peptide to interact with its target molecules and exert its biological effects [106–108]. Also, tripeptides with tryptophan and tyrosine at their C-terminus have been shown to exhibit strong radical-scavenging activity, while different combinations of amino acids in tripeptide chains have also shown different antioxidant activities [109,110].

The stereochemistry of a peptide plays a crucial role in determining its stability and resistance to proteolysis. In a study by Lu et al., a modified version of the cationic antimicrobial peptide Pep05 (KRLFKKLLKYLRKF) was examined. In this derivative, all L-lysine and L-arginine residues were substituted with their D-amino acid counterparts. This alteration resulted in significantly enhanced stability and reduced toxicity in vitro. However, the peptide exhibited only minimal activity and caused severe toxicity in vivo [111]. The cruciality of peptide chirality is also established in the report by Chen et al., where the substitution of L-histidine with D-histidine in a peptide reduced the antioxidant activity [112]. Understanding the interplay between amino acid sequence, chain length, and bioactivity is pivotal for the rational design of therapeutic peptides. By elucidating these relationships through molecular modelling, it is possible to design peptides that maintain structural integrity while exhibiting the desired biological activities.

2.3. Cyclisation of Bioactive Peptides

The cyclisation of some bioactive peptides, characterised by closed-loop structures, commonly enhances their stability and binding affinity compared to linear peptides [113,114]. Cyclic peptides are typically formed through various cyclisation methods, such as head-to-tail cyclisation, lactamisation, and disulfide bond formation, with the additional structural constraint shown to increase their rigidity and resistance to enzymatic degradation [115–117]. Despite this rigidity, many cyclic peptides can adopt multiple conformations in solution, allowing for effective interaction with diverse biological targets [118,119]. Many cyclic peptides are approximately planar, and when they are amphipathic with one hydrophobic face and one hydrophilic or cationic face, this can facilitate interactions with lipid membranes and membrane permeability [120–122]. These structural features contribute to the diverse biological activities, including antibacterial and immunosuppressive effects [116,123]. Cyclic peptides are often non-ribosomally synthesised by enzymes encoded in complex bacterial gene clusters, with non-canonical amino acids and linkages that differ extensively from those found in ribosomally encoded peptides [115,124]. mRNA display technology has been harnessed to create screening combinatorial libraries of cyclic peptides that can, in vitro, ribosomally incorporate non-natural amino acids, generating a new space of cyclic peptides to be experimentally screened [125].

2.4. Structural Stability of Bioactive Peptides

The ability of a peptide to adopt a stable conformation is essential for its proper interaction with other biomolecules, such as proteins or nucleic acids. A structurally stable peptide is more likely to exhibit predictable and consistent functional properties [126]. It allows for specific binding interactions, enzymatic activity, and efficient targeting of cellular processes [90,127,128]. For example, bioactive peptides intended for food applications or as oral drugs must withstand the harsh acidic and proteolytic conditions of the gastrointestinal tract in order to maintain their structure and exert their biological effects [44,94].

Factors such as secondary structure formation, disulphide bonds, and glycosylation contribute to peptide stability, increasing their efficacy in vivo [129–132]. The stability of peptides is, therefore, crucial in the pharmaceutical design and development of novel peptides. For example, the more stable GLP-1 peptide analogue semaglutide exhibits a lower degradation rate by DPP4 protease and increased binding affinity to albumin [133]. Several reports have extensively reviewed the available strategies to improve the stability of bioactive peptides [90,91,134]. While molecular modelling is essential for understanding the structural stability of bioactive peptides and offers valuable insights into their dynamics and interactions, accurately predicting peptide stability remains challenging. This difficulty arises from the peptides' inherent structural diversity and sensitivity to environmental conditions [91,135]. Machine-learning approaches have been applied to predict peptide stability versus gastrointestinal digestion using a training dataset of 109 peptides with known data from in vitro experimentally simulated gastrointestinal digestions [136].

2.5. Folding, Aggregation, and Peptide Conformation

The tendency of bioactive peptides to fold into specific conformations or aggregate can significantly impact their bioavailability and biological activity [17,137]. Aggregation can reduce the peptide's solubility and decrease its ability to reach its target sites, potentially altering its functional properties [138–140]. Protein misfolding is associated with numerous disorders, including the neurodegenerative diseases Alzheimer's and Parkinson's [141–143]. By studying the principles of peptide stability, insights into the mechanisms underlying protein misfolding and aggregation can be gained. Furthermore, understanding the interplay between peptide folding and misfolding provides valuable information for designing strategies to modulate protein conformation and restore proper folding [144,145].

Conformation changes can expose or hide specific functional groups, altering the peptide's ability to interact with other molecules [146]. Therefore, understanding the peptide's three-dimensional structure is crucial for predicting and optimising its biological activity. Molecular modelling methods present advanced platforms for understanding folding mechanisms, aggregation tendencies, and conformational dynamics [147,148]. Notable molecular methods such as molecular dynamics (MD) simulations have been extensively employed to investigate peptide folding, revealing insights into the energy landscape that governs this process, with enhanced sampling MD methods allowing for a more comprehensive exploration of folding pathways compared to conventional MD simulations [149]. MD simulations can provide detailed Gibbs free energy landscapes that highlight the stability of native conformations and the kinetics of folding events [150,151].

Overall, even small structural modifications could influence peptide bioactivity [104]. Molecular modelling tools can provide insights into how structural characteristics influence bioactivity by considering the effects of any changes at a molecular level [99,152,153].

2.6. Peptide Structure Determination: Experimental Characterisation

Structure determination is crucial for thoroughly understanding the functions and bioactivity of peptides. Several experimental techniques have been at the centre of providing essential structural features of peptides, many of which rely primarily on spectroscopic techniques, notably, Nuclear Magnetic Resonance (NMR) spectroscopy [154–156]. NMR enables the direct study of peptide and small protein structures in solution or solid states, eliminating the need for crystallization. This provides a significant advantage over X-ray crystallography methods, particularly for peptides and small proteins that are typically difficult to crystallise. The ability of NMR to predict the structure of peptide structures under various environmental conditions also allows for the characterisation of the structural dynamics of peptides in different temperature, pH, and lipid composition conditions. Nonetheless, environmental variability can sometimes lead to NMR-derived peptide models diverging from their functional conformations. Other methods such as Circular Dichroism (CD) and Fourier Transform Infrared (FTIR) spectroscopy methods, often employed in combination, have also been employed to determine and characterise the secondary structure of peptides [157,158]. However, obvious issues with the small size and flexibility of peptides remain a challenge for the structural determination of peptides, often necessitating a multi-technique approach. In addition, experimental methods are also usually time-consuming and resource-intensive, making computational methods attractive.

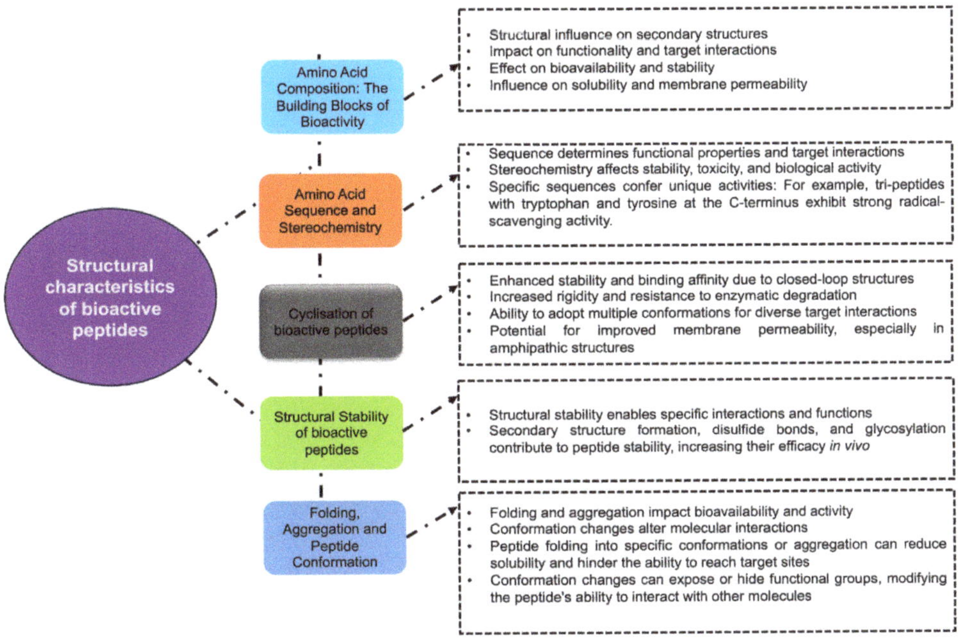

Figure 1. Structural characteristics influencing the bioactivity of peptides. This figure illustrates the key structural features of bioactive peptides and their impact on various functional properties.

3. Current Molecular Modelling Methods in Bioactive Peptide Discovery

Integrating molecular modelling methods into peptide discovery pipelines has allowed for the prediction and analysis of the folding, stability, interactions, and binding affinities of peptides with target proteins [4,36,39,159–161]. Conventional molecular modelling methods that have augmented traditional peptide design and discovery processes over the years include homology modelling, molecular docking, virtual screening, and molecular dynamics simulations [4,36,160,162,163]. Figure 2 describes a simple decision criterion to select the appropriate molecular modelling method for the discovery of bioactive peptides.

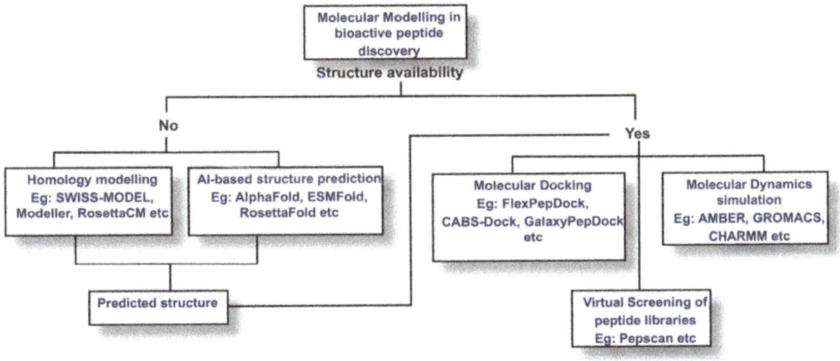

Figure 2. Molecular modelling methods for the discovery of bioactive peptides. The process begins with determining the availability of a peptide structure. If unavailable, structure prediction is performed using homology modelling or AI-based tools. For available structures, molecular docking and molecular dynamics simulations are employed. Virtual screening of peptide libraries can be integrated to identify potential candidates.

3.1. Homology Modelling in Bioactive Peptide Discovery

Homology modelling is a technique that enables the prediction of the 3D structure of any given peptide, under the assumption that peptides with similar sequences will fold in similar ways [164,165]. It has become an indispensable tool in the discovery and characterisation of bioactive peptides, particularly when experimental structural data are lacking. The historically important methods are reviewed below to outline some of the important principles; however, the current most powerful methods use AlphaFold (see below, Section 4.1) with appropriate template homologues. Overall, the process of homology modelling involves three steps: template identification, sequence–template alignment, and model construction [165–168]. The accuracy of the predicted peptide structures from homology modelling, therefore, largely depends on the quality of the template employed, the similarity between the target and template proteins, and the refinement methods used [169]. It is generally preferred that a high sequence identity, with a minimum threshold of 30%, exists between the query structure and the template structure, as this is shown to translate to greater confidence in the homology [170,171]. Homology modelling is particularly useful when experimental methods, such as X-ray crystallography or nuclear magnetic resonance (NMR) spectroscopy, are either not feasible or too time-consuming [169,172,173].

Homology modelling has been employed in several efforts to discover bioactive peptides. For example, a recent discovery of a novel 37-residue antimicrobial peptide used a support vector machine (SVM) pipeline, with the peptide's structure predicted by I-TASSER [174,175], a template-based method for protein structure and function prediction. The predicted structure was subsequently characterised for its anti-tumoural properties through further experimental evaluations [174]. Structural parameters from homology modelling in E. coli proteins and peptide binding data were also integrated to develop an algorithm that significantly outperformed either method alone in predicting peptide binding sites of the molecular chaperone DnaK [176]. The algorithm also performed well on a blind test set, effectively identifying binders and non-binders with no prior information. In another report, several trypsin-inhibitor peptides from the Cucurbitaceae family, characterised by conserved disulfide bonds and an arginine-rich active centre, were modelled using homology modelling based on the X-ray structure of MCTI-II from bitter gourd [177]. The peptides were analysed in both native and trypsin-bound states, with energy minimisation supporting their binding behaviour. The study concluded that evolutionary mutations in non-conserved regions modulate the inhibitory function of the peptides without altering their conserved structural fold or binding interactions. Homology modelling has also been previously employed to elucidate the structural properties that account for the antimicrobial activity of the protegrin family of antimicrobial peptides. Sixty-two (62) protegrin- and protegrin-analogue peptides were studied using homology modelling, and molecular descriptors were correlated with experimental activity against six (6) microbial species, as well as hemolytic and cytotoxic effects, revealing that structural characteristics, rather than sequence similarity alone, are key determinants of antimicrobial activity [178]. This provides valuable insights into optimising peptide design for enhanced therapeutic efficacy.

Despite its advantages, homology modelling faces challenges, such as the availability of suitable templates and difficulties in accurately modelling flexible regions like loops, as these often exhibit significant variability among homologous proteins [168,179]. In structures with greater flexibility, the established correlation between sequence and structural divergence is less clear due to structural changes occurring without corresponding sequence variations [180]. Also, the reliability of generated models is contingent upon rigorous validation techniques that assess their accuracy against experimental structures when available [181].

3.2. Molecular Docking

Molecular docking facilitates the prediction of how peptides interact with their biological targets [48,182]. Molecular docking simulates the binding of peptides to specific receptor sites, providing insights into the ability of a peptide to bind to receptors and into the specific potential interactions [160,183–185]. Molecular docking involves fitting a peptide or ligand into a binding site of a target protein, which may include the active site or other interaction regions, while predicting how well it binds, as indicated by an associated numerical score [40,186,187]. The process typically includes sampling multiple docking poses, which are then ranked based on their predicted binding energies [188–190]. Molecular docking simulations often incorporate the concept of multipose binding, thus recognising that certain interactions in certain protein–ligand complexes are characterised by multiple orientations or conformations rather than single, fixed poses [191]. Multipose binding also provides a more comprehensive and realistic representation of ligand–protein interactions, potentially leading to more accurate predictions and better-informed drug-design strategies [191].

Various strategies exist for peptide–protein docking, including template-based [192,193], local [194], and global docking methods, each with different levels of accuracy and computational cost, depending on the quality of input data [195]. Unlike docking algorithms for small molecules, which require minimal computational resources, docking algorithms for peptides are generally more computationally intensive, particularly in virtual screening experiments [38,49]. Moreover, docking programs do not provide a one-size-fits-all solution for peptide ligands. Traditional molecular docking programs such as AutoDock Vina and MOE-Dock [196–198] often struggle with ligands longer than three amino acids [38,199]. Therefore, using peptide-specific docking protocols is essential for achieving greater accuracy. These docking tools offer unique features to handle the complexities of peptide flexibility and protein surface interactions [200,201].

The ability of molecular docking to unravel structural mechanisms of peptide–protein interaction and thus guide the identification and design of novel potential bioactive peptides has been extensively explored. In very few instances, these reports are accompanied by experimental evaluations. For instance, two potential antioxidant peptides, Tyr-Gly-Arg-Asp-Glu-Ile-Ser-Val (YGRDEISV) and Leu-Asp-Leu-Val-Lys-Pro-Gln (LDLVKPQ), identified using liquid chromatography–tandem mass spectrometry, were shown through molecular docking to block the entrance to the active site of myeloperoxidase as a proposed mechanism of action [202]. Similarly, the binding mode of Phe-Asp-Gly-Asp-Phe (FDGDF) to superoxide dismutase (SOD) was unravelled via molecular docking to augment experiment efforts in characterising the binding mechanism. The docking analysis provided structural insights into the experimentally established antioxidant properties, highlighting crucial interactions and hotspot residues [203].

Molecular docking can also be followed by MD simulations to provide a more comprehensive understanding of peptide behaviour within biological systems. While molecular docking offers initial insights into binding interactions, molecular dynamics can elucidate the stability and dynamics of peptide–protein complexes over time [39]. The growing number of peptide–protein structures has enabled detailed characterisation, driving advances in peptide docking through diverse sampling techniques, which result in models that enhance predictions of binding specificity and aid inhibitor design [204].

Considering the vital role of peptide-mediated interactions in regulatory processes that involve dynamic protein interactions [205–207], modelling these interactions could unravel structural insights to guide the discovery and design of targeted inhibitors. The inhibitory mechanism of two peptides derived from camel and goat milk, IMEQQQTEDE-QQDK and DQHQKAMKPWTQPK, identified using liquid chromatography–quadrupole time-of-flight mass spectrometry as potent inhibitors of pancreatic α-amylase (PAA), was characterized through molecular docking [208]. The analysis revealed that these peptides exhibit a strong affinity for the catalytic site of PAA, indicating their potential as effective inhibitors. In another study, five novel peptides were identified through LC-MS/MS and evaluated for their ACE-inhibitory activity [209]. The peptides, including Ala-Leu-Gly-Pro-Gln-Phe-Tyr and others, exhibited IC50 values ranging from 0.012 mM to 1.680 mM. Crucially, molecular docking studies played a key role in confirming the ACE-inhibitory effects, revealing that the peptides primarily interact with ACE through hydrogen bonding and electrostatic forces. Several molecular docking tools have been developed specially for peptide–protein docking, as highlighted in Table 1.

Table 1. List of some prominent peptide–protein docking tools.

Tool	Synopsis	Server
HPEPDOCK 2.0 [210]	-Web server for blind peptide–protein docking that uses a hierarchical algorithm to efficiently sample peptide conformations and perform global docking across the protein surface.-Generates peptide structures with MODPEP and has demonstrated competitive performance in benchmarks, achieving high success rates in both global and local docking.-No licence required.-Available only as a web server.	http://huanglab.phys.hust.edu.cn/hpepdock/ (accessed on 12 December 2024)
CABS-Dock [211]	-Emphasises conformational flexibility for peptides and proteins using coarse-grained modelling, making it ideal for flexible docking scenarios.-Is available as a user-friendly web server and stand-alone tool, enabling efficient exploration of binding sites and poses.-Standalone executable available.-Free for academic and non-profit use under the MIT licence.	https://biocomp.chem.uw.edu.pl/CABSdock (accessed on 30 November 2024)
AutoDock CrankPep (ADCP) v1.1 [212]	-Tailored for peptide docking, offering flexible sampling of peptide conformations and achieving strong performance in predicting peptide–protein interactions.-Its emphasis on flexibility makes ADCP a valuable tool for exploring diverse binding modes in peptide–protein complexes.-Available under the GNU LGPL v2.0 OpenSource licence.-ADCP is part of the ADFR suite, which can be downloaded as a standalone package.-Not available as a standalone executable.	https://ccsb.scripps.edu/adcp/ (accessed on 30 November 2024)
GalaxyPepDock [213]	-Specialises in predicting peptide–protein interactions by refining docking poses using molecular dynamics.-Excels when target protein structures are known, providing reliable binding affinity predictions.-Freely accessible and does not require licence for use.-The Galaxy source code, which GalaxyPepDock is part of, is licenced under the Academic Free Licence version 3.0.-Standalone executable available.	https://galaxy.seoklab.org/cgi-bin/submit.cgi?type=PEPDOCK (accessed on 30 November 2024)

Table 1. *Cont.*

Tool	Synopsis	Server
Rosetta FlexPepDock [214]	-High-resolution local docking algorithm that refines peptide binding modes within predefined binding sites.-Ideal for accurate predictions when detailed structural information about the binding site is available.-Rosetta licence required.-Command line options available.	http://flexpepdock.furmanlab.cs.huji.ac.il/overview.php (accessed on 30 November 2024)
PepCrawler [215]	-Refines peptide–protein interactions using a Rapidly Exploring Random Tree (RRT) algorithm.-Predicts and optimises peptide binding conformation and estimates binding affinity.-Useful for designing peptide inhibitors by refining docking solutions and improving peptide–protein binding.	Not available
HADDOCK2.2 [216]	-Suitable for peptide–protein docking, especially when binding site or interaction data is available.-Integrates biochemical and biophysical information to improve docking accuracy.-Standalone executable available.-Requires licence for non-commercial agreement.	https://www.bonvinlab.org/software/haddock2.2/ (accessed on 30 November 2024)
pepATTRACT [217]	-Flexible protein–peptide docking algorithm that performs efficient, coarse-grained global searches on the protein surface.-Enables rapid identification of potential binding modes for peptide–protein interactions.-Freely available.	https://bioserv.rpbs.univ-paris-diderot.fr/services/pepATTRACT/ (accessed on 30 November 2024)
PIPER-FlexPepDock [218]	-Fragment-based high-resolution peptide–protein docking protocol, integrating Rosetta's fragment picker, PIPER for rigid-body docking, and FlexPepDock for flexible refinement.-Approach achieves accurate global peptide–protein docking, validated against X-ray crystallography data, enabling high-resolution modelling of peptide–protein interactions.-Requires a Rosetta licence for the FlexPepDock component.-PIPER component requires a separate commercial licence.	http://piperfpd.furmanlab.cs.huji.ac.il/ (accessed on 30 November 2024)
ClusPro PeptiDock 2.0 [219]	-An efficient method for docking peptide motifs to free receptor structures by conducting a motif-based search to retrieve structural fragments from the Protein Data Bank (PDB) that closely resemble the peptide's bound conformation.-Utilises a Fast Fourier Transform (FFT)-based docking approach to perform global rigid-body docking of fragments to the receptor.-Freely available. -No standalone executable.	https://cluspro.org/peptide/index.php (accessed on 30 November 2024)

Table 1. *Cont.*

Tool	Synopsis	Server
DynaDock [220]	-For docking peptides into flexible receptors, utilising the following two-step procedure: scanning the protein–peptide conformational space to identify approximate ligand poses, followed by refinement using optimised potential molecular dynamics (OPMD).-OPMD method employs soft-core potentials for protein–peptide interactions and a novel optimisation scheme, demonstrating significant improvements in sampling capability compared to conventional molecular dynamics and soft-core scaling methods.	Not available
AnchorDock [221]	-Automatically targets the docking search to the most relevant parts of the conformational space by precomputing the free peptide's structure and by computationally identifying anchoring spots on the protein surface.-Free peptide conformation undergoes anchor-driven simulated annealing molecular dynamics simulations around the predicted anchoring spots.	Not available
MDockPep [222]	-User-friendly server for global docking of flexible peptides to protein receptors, starting from a peptide sequence and protein structure.-Docking results are scored using ITScorePeP, a statistical potential-based function, and validated with the peptiDB benchmark, showing high accuracy in both bound and unbound cases.-No licence required.-No standalone executable.	https://zougrouptoolkit.missouri.edu/mdockpep/ (accessed on 30 November 2024)
PEPFOLD3 [223]	-Predicting 3D structures of linear peptides (5–50 amino acids) in aqueous solution and peptide–protein interactions.-Supports both de novo and biased predictions, generating native-like peptide conformations when the interaction site is known.-No licence required.-No standalone executable.	http://bioserv.rpbs.univ-paris-diderot.fr/services/PEP-FOLD3 (accessed on 30 November 2024)
DINC 2.0 [224]	-Meta-docking strategy that overcomes this by incrementally docking ligand fragments, enabling accurate prediction of peptide-based inhibitors.-Improves upon the original by allowing docking of larger ligands (over 25 flexible bonds). -No licence required.-No standalone executable.	https://dinc-ensemble.kavrakilab.rice.edu/ (accessed on 30 November 2024)
GOLD [225]	-Automated program that predicts small-molecule binding to macromolecules by using a genetic algorithm.-Accounts for ligand flexibility and partial protein flexibility while ensuring displacement of loosely bound water molecules during binding.-Licence required. -No standalone executable.	Not available

Table 1. *Cont.*

Tool	Synopsis	Server
Rosetta FlexPepDock ab initio [226]	-Allows for simultaneous docking and de novo folding of flexible peptides starting from an approximate peptide binding site specification.-Utilises the Rosetta fragments library and a coarse-grained structural model to effectively sample peptide conformations and rigid-body orientations on the receptor surface, followed by all-atom refinement to accurately model side-chain interactions.-Rosetta licence required.-Command line options available.-No standalone executable.	Not available
PaFlexPepDock [227]	-Parallel docking approach combining ab initio peptide folding, peptide docking, and flexible receptor refinement.-Showed improved interface modelling and energy funnel construction by refining receptor flexibility during docking.-Rosetta licence required.	Not available
PepComposer [228]	-Requires only the target protein structure and a rough definition of the binding site to start the design process.-Identifies peptide scaffolds with similar backbone arrangements and optimises sequences to best fit the target binding site. -Freely available.	Not available

3.2.1. Challenges of Peptide–Protein Docking

Unlike traditional protein–ligand or protein–protein docking, peptide–protein docking faces distinct difficulties primarily due to the inherent flexibility of peptides and the often unknown binding sites on target proteins [52,127,194,221,229,230]. This flexibility complicates the identification of favourable interactions, as conventional docking algorithms for small molecules often assume a fixed ligand conformation [210,231]. The challenge becomes even more complex when receptor flexibility is considered, as peptide–protein binding can induce conformational changes in the protein. These changes may involve side chain reorganisation and/or possible backbone rearrangement [194,232,233]. The lack of stable conformation of peptides prior to binding makes peptide–protein docking computationally expensive to fully sample the conformational space [127,229]. This limits the practical application of existing docking algorithms for high-throughput screening in drug discovery. The classical rigid docking approach of traditional molecular docking often fails to capture key structural adaptations, necessitating the incorporation of binding models such as induced fit and conformational selection in some docking approaches [234]. The concept of induced fit suggests that ligand binding triggers structural rearrangements in the protein, requiring flexible docking techniques to accommodate these changes [235,236]. Induced fit docking (IFD) can account for significant side-chain conformational changes and small backbone relaxation, and the approach has shown success, with some methods, like Schrödinger's IFD-MD approach successfully predicting 85% of 258 protein–ligand pairs correctly [237]. On the other hand, the conformational selection model proposes that proteins exist in multiple pre-formed conformations, with ligands selectively binding to the most favourable state, highlighting the need for ensemble docking strategies [238]. Since many biological systems exhibit aspects of both models, integrating hybrid approaches that consider induced fit effects while leveraging conformational ensembles enhances the predictive power of docking algorithms and improves reliability in drug discovery applications [239–243].

In addressing the flexibility challenge, some docking tools, particularly global docking protocols, typically employ rigid-body docking by ignoring receptor flexibility, with an added advantage of low computational cost [194]. However, these tools would often allow side-chain flexibility in later modelling steps. Some advanced tools like Gold [244], which uses ensemble docking, and DynaDock [220], which uses soft potentials, attempt to mimic flexibility. Coarse-grained protein models, like CABS-Flex [245] and Rosetta FlexPepDock ab initio [226], can also be applied to capture large-scale backbone rearrangements, especially in disordered or loop regions near binding sites. For instance, Rosetta FlexPepDock ab initio generates coarse-grained representations of both the peptide and the receptor while simultaneously exploring rigid body orientations around the peptide-binding surface of the target.

Current scoring functions used in peptide–protein docking may also not adequately account for all relevant interactions between peptides and their targets. Many of these functions are optimised for small molecules or larger proteins, potentially leading to inaccurate predictions of peptide binding affinities [52,200,246]. Most peptide docking tools employ specific scoring functions to handle the complexities of peptides. For instance, Rosetta FlexPepDock uses the Rosetta energy function, MM-PBSA is used in HADDOCK and BiPPred [247], semi-empirical physicochemical scoring in pyDockWEB [248], and MD-derived binding energy in pepATTRACT. Other methods, such as CABS-dock, employ knowledge-based scoring functions. Structural clustering, the use of coevolutionary information, mutagenesis data, sequence-based predictions, and template-based comparisons have also been employed to improve model selection [193,249–252]. Hybrid approaches that combine mixed scoring functions with biological insights have been shown to yield the most accurate peptide–protein docking results [193,194,252].

Another challenge of peptide–protein docking is the instance of unknown or improperly defined peptide binding sites. This contrasts with protein–ligand docking, where the binding sites are usually characterised. Consequently, peptide–protein docking often requires a global search across the entire protein surface to identify potential binding modes, which increases computational complexity and time [210,253,254].

Peptide-mediated interactions also tend to be weaker and more transient than those involving larger ligands or proteins [255–257]. Many of these binding interactions are subtle regulators and appear evolutionarily optimised for low rather than high affinity [258]. The transient nature of these interactions also complicates experimental validation, as traditional methods like X-ray crystallography may not capture fleeting complexes effectively. The smaller interface between peptides and their protein partners results in lower binding affinities, making it difficult to accurately model these interactions using standard scoring functions [127,194,259].

3.2.2. Clustering in Molecular Docking

Clustering enhances the efficiency and accuracy of molecular docking, particularly with flexible receptors/peptides and large datasets [260–262]. The clustering technique reduces computational complexity by condensing vast conformational ensembles, such as those from MD simulations, into a manageable set of representative structures. This allows for virtual screening against a diverse range of receptor/peptide states without docking to every single conformation. By grouping similar conformations and selecting representative structures, clustering helps capture essential receptor/peptide flexibility and improves docking efficiency, often leading to increased hit rates [250]. Peptide docking tools such as CABS-dock [263], HADDOCK [216], Cluspro PeptiDock [219], and FlexPepDock [226] incorporate clustering to enhance their accuracy. These cluster-based docking methods apply clustering algorithms to the ensemble of docked poses based on root mean square deviations of the structures [264]. Furthermore, clustering aids in the interpretation of results by grouping similar binding modes and identifying key conformational states, facilitating the analysis of ligand-receptor interactions. Overall, clustering is essential for streamlining computational demands, representing receptor flexibility, and improving the effectiveness of structure-based drug design approaches.

3.3. Virtual Screening of Peptide Libraries

Physical screening of peptide libraries is an important contributor to peptide development [265]. Virtual screening (VS) can search libraries of peptides, usually with docking software, to identify or up-rank the most promising peptides with high affinity for a protein target [43,266,267]. VS can also allow for the repurposing of peptides with known functions for new functions [268].

VS are generally divided into two main categories, namely structure-based virtual screening (SBVS) and ligand-based virtual screening (LBVS). SBVS utilises the three-dimensional structure of biological targets, obtained through X-ray crystallography, NMR spectroscopy, cryo-electron microscopy (cryo-EM), or computational modelling, to evaluate large libraries of peptide compounds [43,269–271]. SBVS allows for the prediction of binding affinities and poses of potential peptide ligands within the target's binding site, effectively narrowing down candidates for experimental testing. SBVS can handle flexible binding sites and predict novel binding modes, which is particularly valuable for peptides with complex conformations [272,273]. On the other hand, LBVS methods, typically combined with docking and modelled on a known structure of a lead molecule complexed with a target protein, include similarity searching, pharmacophore modelling, machine-learning approaches, and shape-based screening. They leverage information from known active compounds to identify potential new bioactive peptides [274–276]. LBVS techniques are efficient in early-stage discovery, applicable to diverse targets (including those with unavailable experimentally defined 3D structures), and capable of handling flexible binding sites [277].

The efficiency and accuracy of virtual screening largely depend on how well the receptor's binding pocket is characterised, with the binding pocket typically defined through molecular docking, based on the crystal structure of the proteins bound to ligands [278]. Although VS has become increasingly popular for identifying potential hits in small-molecule drug discovery, its application in the screening of peptide libraries has yet to gain momentum since the screening of peptides is relatively computationally intensive and requires 3D-structure of the protein and the appropriate scoring and search algorithms [267,268].

Several studies have successfully utilised VS to discover potential bioactive peptides, yielding some promising results. For example, one virtual screening approach using molecular docking screened over 1400 peptides, leading to the identification of macrocyclic peptide 22, which showed improved target affinity and biological activity [279]. These findings were validated through X-ray crystallography, further expanding the accessible chemical space for drug design. In another report by Amarasinghe et al., VS was employed to explore a vast library of nearly 380,000 non-natural amino acids, facilitating the identification of mutations at specific "sweet spots" on the peptide that could enhance its affinity for Keap1 [270]. Despite screening only a subset (less than 3%) of the full virtual library, the method successfully pinpointed mutations with the potential to improve protein–peptide binding. The study highlights the power of VS in rapidly evaluating large chemical spaces and postulates that scaling up the process could reveal even more effective peptide variants. In another report, a hierarchical structure-based virtual screening approach was employed to identify novel peptide inhibitors of human α-glucosidase [280]. Small di- and tripeptides, naturally present in the digestive system, were screened, with MD simulations used to evaluate their stability compared to the known inhibitor, acarbose. Four promising peptides with low RMSD variability were synthesised, and one (compound 2) demonstrated significant inhibitory activity in vitro. VS, combined with MD simulations, also played a crucial role in identifying and refining potential peptide scaffolds that closely resembled the three-dimensional shape and electrical properties of ATP-like compounds. These structures, upon further refinement and optimisation, could potentially serve as potent inhibitors of eIF4A1, thereby exhibiting promising antitumour effects. [281,282]. VS also played a pivotal role in identifying novel inhibitors of pancreatic lipase (PL), a key enzyme involved in triglyceride metabolism and linked to acute pancreatitis [283]. By screening a tripeptide library, the method successfully prioritised the top nine peptides for synthesis and in vitro testing, including two found to have strong inhibitory effects. Li et al. also identified 17 novel potential anti-inflammatory peptides from broccoli fermented by Lactobacillus strains through VS [284]. Augmented by in vitro assessments, it was further revealed that one of the peptides, SIWYGPDRP, showed the strongest inhibitory effect on NO release from inflammatory cells. Additionally, two other peptides, RFR and KASFAFAGL, effectively inhibited the secretion of TNF-α and IL-6, respectively. VS, combined with experimental validation, has also been applied in the discovery of a cyclic peptide inhibitor of thrombin from a library of 108,659 multiconformer cyclic peptides [285].

As represented in Figure 3, VS for bioactive peptides is a systematic approach that begins with the retrieval of peptides from peptide libraries, e.g, PepBank [286], PlantPepDB [287], Peptipedia [288], BIOPEP-UWM [289], Antimicrobial Peptide Database [290], StraPep [291], and CyclicPepedia [292]. Customised libraries can also be designed. This is followed by molecular docking simulations to evaluate binding affinities with target proteins. The peptides are then scored and ranked based on their predicted interactions. Additional filtering may be applied using extensive modelling techniques such as MD simulation and binding free energy calculations. Physicochemical properties may also be assessed to further refine candidate selection. Afterwards, the most promising candidates are synthesised and experimentally evaluated for biological activity. This streamlined process enhances the efficiency of identifying potential bioactive peptides while minimising resource expenditure.

Virtual screening thus provides a relevant computational approach to discovering bioactive peptides, enabling the rapid identification of potential candidates from vast peptide libraries while significantly reducing the time and resources required for experimental validation.

3.4. Molecular Dynamics Simulation

Molecular dynamics (MD) simulation is a computational technique that can be used to study the physical movements of atoms and molecules over time to provide atomistic-level structural dynamics and thermodynamic properties of various systems, including biomolecules such as proteins and peptides [293–296]. MD simulation involves various steps, including initialisation, energy minimisation, equilibration, and production runs, as highlighted in Figure 4. This is followed by an analysis of structural stability, dynamics, and interactions [295].

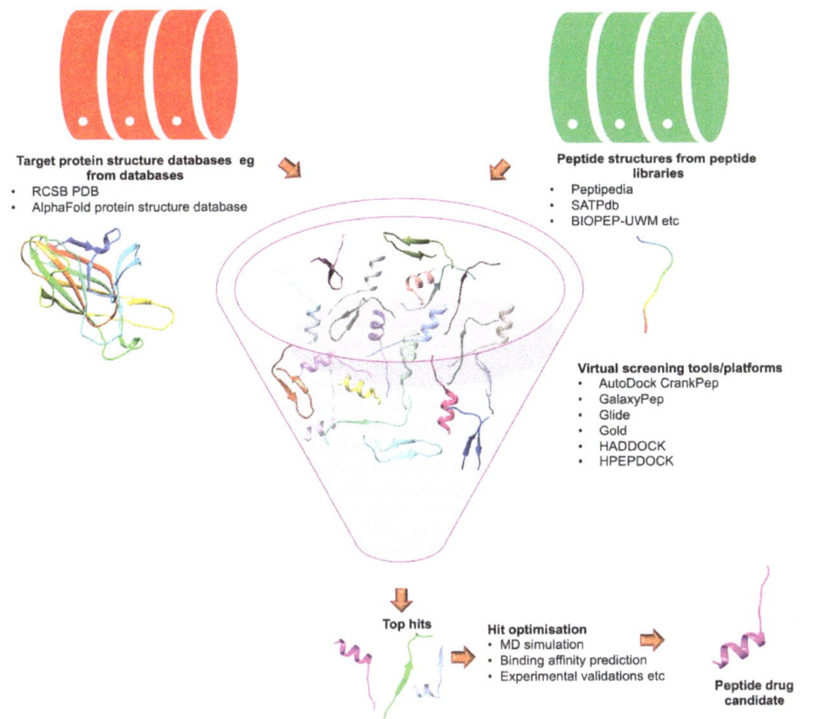

Figure 3. Schematic representation of the virtual screening-based discovery of bioactive peptides.

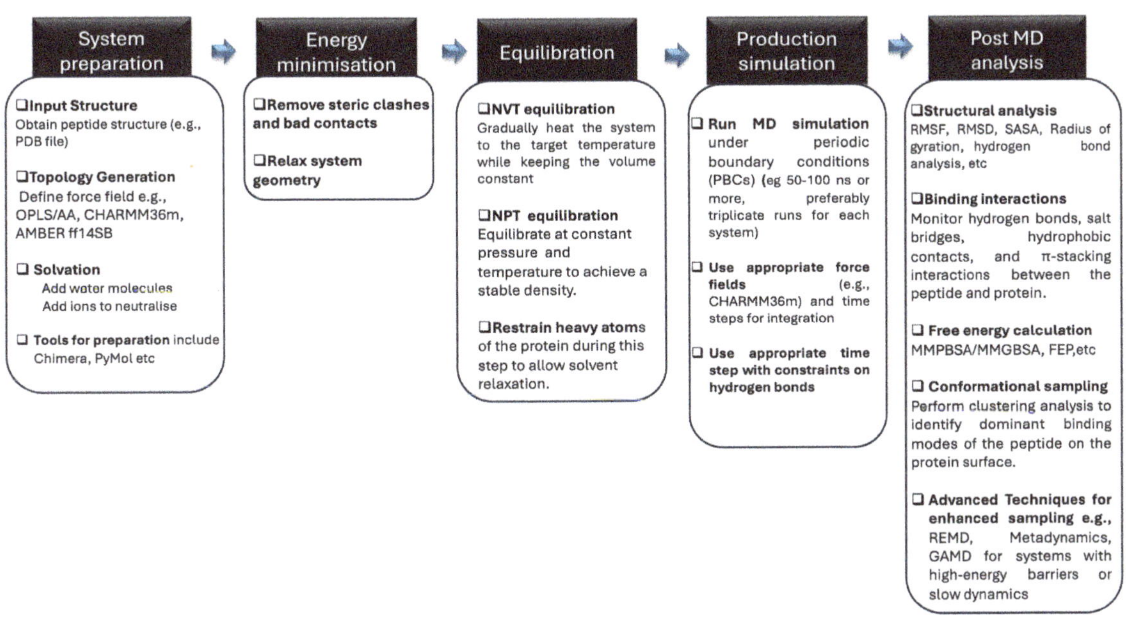

Figure 4. Overview of the MD simulation process for peptides.

In comparison to molecular docking, MD simulations provide a dynamic picture of protein–ligand interactions, offering several advantages over static molecular docking approaches. MD accounts for full atomistic flexibility, captures induced fit effects, includes explicit solvation, and provides insights into binding kinetics and conformational sampling [297–299]. In terms of scoring, MD enables a more reliable estimation of thermodynamic parameters, allows for ensemble averaging, incorporates entropic contributions, and can refine docking poses [185,294]. MD also captures time-dependent interactions and typically uses more advanced force fields [300,301]. While computationally more expensive, MD's ability to provide a more realistic and comprehensive view of the binding process makes it an increasingly valuable tool in drug discovery, offering improved accuracy in predicting binding affinities and identifying potential drug candidates.

To perform MD simulations, several tools have been developed over the years, notably GROMACS [302], AMBER [303], CHARMM [304], NAMD [305], and OpenMM [306]. These tools, along with the use of graphic processing units (GPUs) on high-performance computing (HPC) systems, have significantly improved the speed and accuracy of simulations [307–309]. Some of these platforms allow for all-atom MD simulations, where every atom in a peptide and its environment are modelled, typically employing force fields such as AMBER or CHARMM and also enabling precise predictions of protein structures, binding modes of protein–ligand or protein–protein interactions [310,311].

An essential aspect of MD simulations is the accurate representation of the solvent environment. Solvent modelling plays a crucial role in the molecular modelling of peptides, influencing their folding, stability, and interactions [312–315]. Two primary approaches exist: explicit solvent models, such as TIP3P and TIP4P etc., which accurately capture hydrogen bonding and hydration effects but are computationally expensive [316,317], and implicit solvent models, like Generalized Born, Poisson–Boltzmann, and SASA, which offer faster calculations but lack explicit solvation details [318–320]. Key challenges include the trade-off between accuracy and computational efficiency, force field limitations, and the difficulty in capturing long-range electrostatic interactions [321]. Recent advances, such as hybrid solvation approaches, polarisable water models, and AI-driven techniques, aim to improve accuracy while maintaining computational feasibility [322–325]. Selecting an appropriate solvent model requires balancing computational cost with the level of detail needed for peptide simulations.

Recent advancements in computational power have enabled the MD simulation of large systems, reaching hundreds of thousands of atoms and millisecond-long time scales, thus enhancing the ability to study the complex conformational space of peptides and proteins [163,326,327]. In the study of peptides, MD simulation has also been very pivotal in several efforts towards the design and development of bioactive peptides. MD simulation has allowed for the investigation of conformational changes, flexibility, and interactions of peptides at an atomic level over time, aiding in the design of more potent and stable bioactive peptides [328–330]. This is particularly important in the context of peptide-based drug design, where small changes in sequence can lead to significant differences in biological activity [331]. For instance, in the study of antimicrobial peptides (AMPs), MD has elucidated the mechanisms of action of AMPs, revealing how they interact with bacterial membranes as well as the physicochemical properties that dictate their efficacy [332–335]. Simulations have also shown how AMPs can disrupt membrane integrity, leading to cell lysis, thus providing insights into the structural features that enhance their activity against resistant strains [336–340]. Using MD simulations, the stability and conformational rigidity of 12 stereo-engineered amphipathic peptides were tested against Gram-positive, Gram-negative, and antibiotic-resistant bacteria [341]. The results showed that syndiotactic peptides retained their designed electrostatic environment, exhibiting bactericidal activity and underscoring the potential of stereo-engineered peptides for therapeutic use [341]. MD simulations have also complemented experimental results to advance the Deno design of cell-penetrating peptides and further provide insights into mechanisms of CPP internalisation, membrane interactions, and peptide-induced membrane changes [342–345].

MD simulations have also furthered the atomistic-level understanding of protein–peptide interactions, enabling the exploration of their binding affinities and conformational changes upon interaction with target proteins [48,163], although they are often computationally expensive. To address this challenge of computational intensiveness, reports such as that of Chen et al. have proposed a new method that utilises high-temperature MD simulations with the RSFF2C force field to simultaneously predict peptide binding sites and poses [342]. By sampling thousands of binding events during microseconds of high-T MD, this approach accurately predicts peptide-receptor structures through density-based clustering analysis, achieving root-mean-square deviation (RMSD) values below 2.5 Å. This method was shown to also significantly improve the precision of protein–peptide docking, surpassing traditional approaches.

MD simulations have also been useful in modelling cyclic peptides, particularly in elucidating their solution structures, a crucial step toward quantitatively understanding the properties of cyclic peptides, using enhanced sampling techniques [346]. The reliability and accuracy of MD for simulating and understanding biomolecular systems, including peptides, are largely dependent on the conformational sampling methods (see Section 3.4.2 below) and the choice of force field (see Section 3.4.1 below) employed.

The integration of MD simulations and machine-learning (ML) methods has also sought to improve the accuracy of modelling peptides while aiding in the discovery of bioactive peptides [160,347–349]. Traditional MD methods, which typically rely on classical force fields and standard thermodynamic conditions, often struggle to capture complex, high-dimensional molecular interactions, but ML could offer flexible, unbiased solutions [350–352]. By learning from existing trajectories, ML accelerates convergence, guides simulations to relevant regions of configurational space, and improves exploration of rare events [348,353]. Additionally, ML enables the extraction of potential energy surfaces (PES) from quantum mechanical calculations, enhancing MD accuracy [354]. Its integration into Markov State Models (MSMs) further advances the analysis of MD trajectories, which can lead to a wider application in the study of the functional conformational changes of biological molecules [355].

MD simulation has, therefore, been useful in elucidating the dynamic behaviour and binding mechanisms of peptides, providing insights that can inform the design of peptide-based therapeutics and enhance our understanding of protein–protein interactions.

3.4.1. Force Fields for MD Simulation Towards Bioactive Peptide Discovery

Selecting an appropriate MD force field is crucial for accurately modelling peptide behaviour, as it directly influences the reliability of simulations in capturing peptide dynamics, stability, and interactions with other biomolecules. Force fields determine the potential energy of a system based on the positions of its constituent atoms, using mathematical functions and parameters that describe various bonded and non-bonded interactions [301]. The accuracy and reliability of MD simulations are largely dependent on the choice of force field employed to accurately sample all the relevant conformations of the system [301]. Choosing the right force field is crucial in bioactive peptide discovery as it enables accurate prediction of peptide conformations and dynamics under physiological conditions, helps assess binding affinities between peptides and target proteins, and provides insights into peptide stability and folding mechanisms, key factors for their therapeutic effectiveness [301,356,357].

Although several force fields exist for MD simulations, most are not specifically tailored for peptides. Commonly used force fields include the AMBER force field [358,359], known for its detailed handling of proteins and nucleic acids; the CHARMM force field [360], which offers comprehensive coverage of biomolecules; GROMOS [361], designed for efficient simulations of biomolecules; and OPLS [362], primarily focused on small molecules but applicable to peptides. Recent advancements have introduced polarisable force fields like the Drude Polarisable Force Field and AMOEBA, which account for electronic polarisation effects that significantly influence molecular interactions [363–365]. Some of these force fields have been employed to study peptide systems; however, their efficiency in accurately parameterising peptides remains unclear. For instance, four different force fields—Amber ff99SB-disp [366], Amber ff99SB-ILDN [367], CHARM36IDPSFF [368], and CHARMM36m [369]—were comparatively investigated for their ability to capture the secondary structure element of five peptides previously suggested to possess polyproline II (PPII) structure [356]. Findings suggested that while all these force fields can generate conformational ensembles with polyproline II (PPII) structures, they differ significantly in their propensity to sample other secondary structures like β-sheets and random coils [356].

A few force fields have, however, been reported to consider peptides specifically. For instance, the ECEPP-05 force field parameterises both organic molecules and peptides where torsional parameters for peptides were derived by fitting molecular mechanics energies to quantum-mechanical energy maps of model peptides like Ac-Ala-NMe and Ac-Gly-NMe [370]. The reliability of ECEPP-05 was validated through simulations of small peptide crystal structures, which showed high accuracy with minimal deviations from experimental X-ray data, confirming its effectiveness in parameterising peptides [370]. The residue-specific force field (RSFF2) reported by Li et al. was shown to significantly improve agreement with experimental data in an MD simulation investigation of 256 peptides [371]. RSFF2 exhibited superiority over other derivatives of the Amber ff99SB force field in accurately modelling peptides by comparing their performance in reproducing experimental 3JHNHα scalar coupling constants for two-residue peptides. Although RSFF2 performed better, all force fields demonstrated similar limitations in capturing neighbouring residue effects (NREs), indicating potential challenges in their treatment of nonbonded interactions between adjacent side chains or terminal groups. RSFF2 and RSFF1 have also been shown to accurately identify the crystal-like conformations of cyclic peptides, with RSFF2 successfully identifying the most stable conformations for over half of the tested cyclic peptides [372]. In another study evaluating the ability of seven advanced force fields to reproduce experimental NMR data for 12 benchmark cyclic peptides, RSFF2+TIP3P, RSFF2C+TIP3P, and Amber14SB+TIP3P successfully captured the structure of 10 peptides, while Amber19SB+OPC demonstrated accuracy for eight peptides [373]. Regardless of these efforts, the challenge remains in parameterisation for non-standard amino acids and the computational demands of polarisable models [374]. The emergence of machine-learning force fields could address some of these challenges and consequently improve the accuracy of peptide parameterisation [375].

3.4.2. Enhanced Sampling Methods

Enhanced sampling methods in molecular dynamics (MD) simulations address the limitations of conventional MD by overcoming energy barriers and exploring complex conformational landscapes [376,377]. Traditional MD often struggles with insufficient sampling due to rough energy landscapes with numerous local minima [378,379].

Notable enhanced sampling techniques like umbrella sampling [380], replica-exchange molecular dynamics (REMD) [381], metadynamics [382,383], accelerated molecular dynamics (aMD) [384], simulated annealing [385], and Gaussian accelerated molecular dynamics

(GaMD) [386] amongst others, as shown in Table 2, are known to overcome the challenges associated with traditional MD [387]

Some of these enhanced methods have been employed to study peptide systems. For instance, Turner et al. employed REMD to investigate how amyloid-β (Aβ42) peptides interact with a platinum–phenanthroline complex (Pt(phen)) [388]. The REMD simulations revealed that the peptide adopts more compact conformations with increased helical content and reduced β-strand structure when bound to the platinum complex. REMD allowed for a thorough sampling of the diverse conformational states of the peptide, providing crucial insights into how metal complexes influence peptide structure and stability. In another study that introduced a method for calculating peptide dimer formation and dissociation rates, REMD simulations were employed [389]. By analysing continuous trajectories from replicas at various temperatures, the study accurately determined transition rates and validated the employed methods against alternative estimation techniques. The findings showed that REMD offers more than double the sampling efficiency compared to standard MD at lower temperatures. Clayton et al. also presented a hierarchical approach for exploring peptide conformational space by combining aMD and metadynamics [390]. Initially, aMD quickly scans accessible peptide states using an implicit solvent model, providing a qualitative view of conformations. Collective variables identified from the aMD were then used in well-tempered metadynamics with explicit solvent to accurately quantify free energy landscapes, despite the slower simulation speed. This combined method leveraged the strengths of both rapid exploration from aMD and precise free energy estimation from metadynamics, more efficiently and thoroughly characterising peptide states than conventional MD, which is often limited by slower timescales and insufficient sampling. Multiple simulated annealing–molecular dynamics (MSA-MD) has also been employed to improve peptide and miniprotein structure predictions [391]. MSA-MD was shown to significantly outperform MD and conventional simulated annealing–MD (SA-MD) in reaching native structures. The method successfully generated a cluster of near-native structures, demonstrating its enhanced effectiveness in structural prediction compared to traditional MD techniques. Wang et al. also incorporated GaMD to design a computational method called Pep-GaMD to overcome the limitations of conventional MD in simulating peptide binding and unbinding, which are hindered by long biological timescales and peptide flexibility [392]. Based on the GaMD technique, Pep-GaMD selectively boosts the peptide's potential energy to better model its flexibility, while a dual-boost algorithm enhances the entire system's potential energy. This approach enables the capture of repetitive binding and dissociation events, allowing accurate calculation of binding free energies and kinetics that align well with experimental data. Pep-GaMD significantly improves simulation efficiency and provides deeper insights into peptide–protein binding mechanisms compared to conventional MD. GaMD allowed for more efficient sampling of the conformational space to analyse large cyclic peptide libraries, to investigate how solvent affects the free energy landscape of cyclic lariat peptides, and showing that permeability is significantly influenced by N-methylation and solvent exposure [393].

Table 2. Table of enhanced sampling methods.

Method	Purpose and Application	Reference
Metadynamics	Accelerates the exploration of free energy landscapes by applying a history-dependent bias potential that discourages the system from revisiting previously explored states, pushing it out of energy wells to sample higher-energy states. This approach is useful in protein folding, ligand binding, and phase transition.	[394]

Table 2. *Cont.*

Method	Purpose and Application	Reference
Umbrella Sampling	Enhances sampling in areas with high energy barriers by dividing the system into smaller, more manageable "windows" for easier sampling. Biasing forces are applied within each window, and the results are combined using the weighted histogram analysis method (WHAM) to reconstruct the free energy landscape. It is commonly applied in studying protein–ligand interactions, membrane fusion, and reaction mechanisms.	[380]
Replica Exchange Molecular Dynamics	Enhances sampling by running multiple simulations at different temperatures, periodically swapping configurations. Higher temperatures help the system overcome energy barriers, and the exchanges allow low-temperature replicas to benefit from broader sampling. This method is used in applications such as protein conformational sampling, studies of thermodynamic properties, and the exploration of phase transitions.	[395]
Adaptive Biasing Force (ABF)	Efficiently computes free energy profiles along a selected reaction coordinate by applying an adaptive biasing force to flatten energy barriers, promoting the exploration of rare conformational states. Over time, the force adjusts as the simulation progresses, making it useful for studying diffusion processes, chemical reactions, and molecular conformations.	[396]
Steered Molecular Dynamics (SMD)	Mimics experiments like atomic force microscopy (AFM) by applying an external force to specific parts of a system, driving it along a reaction pathway. It is used to study processes such as protein–ligand unbinding and the mechanical properties of proteins.	[397,398]
Accelerated Molecular Dynamics (aMD)	Lowers energy barriers by adding a boost potential to flatten energy wells, enabling faster sampling of conformations and helping the system escape local minima more easily. It is useful for studying large-scale conformational changes in proteins, molecular motors, and membrane simulations.	[399,400]
Path Sampling Methods	Transition Path Sampling (TPS) and Forward Flux Sampling (FFS) are methods designed to simulate rare events by focusing on sampling transition pathways directly, rather than the entire phase space. These techniques gather trajectories connecting initial and final states, particularly during rare transitions between stable states, and are applied in areas such as chemical reactions, protein folding, and nucleation processes.	[401]
Variationally Enhanced Sampling (VES)	Optimises biasing potential using a variational principle to enhance sampling in targeted regions. It constructs a free-energy landscape adaptively and applies an optimised bias potential for efficient exploration. This approach is particularly useful in applications such as protein folding, phase transitions, and complex chemical systems.	[402]

Table 2. *Cont.*

Method	Purpose and Application	Reference
Gaussian Accelerated Molecular Dynamics (GaMD)	An extension of aMD that enhances conformational sampling while preserving the Gaussian distribution of the biased potential. It applies a boost potential similar to aMD but incorporates the constraint of maintaining Gaussian statistics, enabling accurate reweighting. This approach is useful for studying protein–ligand interactions and large biomolecular systems.	[386]

4. Machine Learning and Structural Modelling in Bioactive Peptide Discovery

Integrating artificial intelligence (AI) and structural modelling techniques can significantly accelerate the discovery of bioactive peptides. AI methods have enhanced the efficiency of predicting, designing, and validating peptides with specific biological functions using tools developed to specifically support bioactive peptide discovery [17,57,403].

AI approaches are able to utilise available datasets of known peptides and their biological activities to train models that can predict the structure, function, and efficacy of new peptide sequences and consequently aid in the optimisation of potential peptide-based therapeutics [404,405]. While these datasets provide valuable insights, their size is relatively modest compared to the extensive datasets typically used in AI applications, which may limit the scalability and robustness of predictions. Despite this, such approaches are instrumental in optimising potential peptide-based therapeutics. Using either simple machine-learning models, deep neural networks, or, more recently, protein large language models (pLLM), AI methods offer valuable predictions and hypothesis-generation capabilities that can significantly advance structural modelling efforts for bioactive peptide discovery [405–410].

Predicting and generating bioactive peptides computationally typically involves classifier methods, predictive systems, and deep-generative models like generative adversarial networks (GANs) and variational autoencoders (VAEs), and, more recently, diffusion-based models [411,412]. However, the successful application of these approaches requires the optimisation of processing workflows and the rigorous validation of predictive models to ensure accuracy and reliability [404,405].

The recent ability of AI methods to successfully predict the three-dimensional structures of proteins using primary sequences presents an opportunity for advancement in the structural modelling research into amino acid-based structures [404,413,414]. Several AI-based tools have been developed over the years to guide amino acid-based structure prediction, such as Raptor X [415], I-TASSER-MTD [416], trRosetta [417], and APPTEST [418], and more recently, RosettaFold [419], ESMFold [420], and AlphaFold [55]. Some of the recently developed AI-based tools (Table 3) have demonstrated reliable accuracy in predicting the structures of peptides, with many being AlphaFold-based implementations.

Table 3. Common AI-based methods that use sequence information to predict 3D structures of peptides.

Tool	Synopsis	Availability
AlphaFold [55,421,422]	-Alignment-based deep learning model that predicts protein structures with high accuracy, including peptide conformations.-Adapted for cyclic peptides and can predict structures based on single sequences.	-AlphaFold Server is freely available for non-commercial research.-The code for AlphaFold3 is now downloadable for academic use.-Training weights on request for scientists with academic affiliations.
ColabFoldv1.5.5 [423]	-User-friendly interface implementing AlphaFold technology.-Allows batch predictions of structures in a single session.	-Web server-like interface implemented as a Google Colab notebook.-Can be installed and run locally.-Code is open-source and shared on GitHub at https://github.com/sokrypton/ColabFold (accessed on 30 November 2024).-Uses AlphaFold2 models, which typically include shared model weights for local deployment.-Can be run on high-performance computing (HPC) clusters.
Highfold v1.0 [424]	-AlphaFold implementation specifically designed for predicting cyclic peptide structures and their complexes.-Incorporates constraints like head-to-tail connections and disulfide bridges.	-The code for HighFold is shared on GitHub at https://github.com/hongliangduan/HighFold (accessed on 30 November 2024).-Uses AlphaFold2 models, which typically include shared model weights for local deployment.-https://github.com/sokrypton/ColabDesign/blob/main/af/examples/af_cyc_design.ipynb (accessed on).
AfCycDesign [425]	-AlphaFold implementation tailored for the rapid and accurate prediction and design of cyclic peptide monomers.-Addresses limitations related to insufficient training data.	-Available as an online web server through Neurosnap.-Implemented within the ColabDesign framework, which utilises AlphaFold for structure prediction and design.-Accessible via Google Colab, allowing users to run it without local installation.-Source code available at https://github.com/sokrypton/ColabDesign/blob/main/af/examples/af_cyc_design.ipynb (accessed on 30 November 2024).
RoseTTAFold [419]	-Deep-learning approach that excels in protein structure prediction and can be applied to peptide structures.-Provides high accuracy compared to traditional methods.	-Accessible via the Robetta web server.-Source code for RoseTTAFold is available on https://github.com/RosettaCommons/RoseTTAFold (accessed on 30 November 2024).-Model weights available and shared along with the code.
OmegaFold [426]	-Alignment-based deep learning model alternative to AlphaFold.	-Available as a standalone package.-Source code available on GitHub (OmegaFold GitHub Repository).-Model weights available and downloadable during execution.
ESMFold-2 [420,427]	-Deep learning built on Meta AI's ESM-2 PLM.-Predicts 3D structures directly from amino acid sequences without relying on extensive homology (i.e., no alignments used in training or prediction).	-Available as a standalone package.-Open-source and does not require a commercial licence.-Code and pre-trained weights available on GitHub.-API available through BioLM.ai.
PepFlow [428]	-Deep-learning model trained on peptides of 15 or fewer residues to predict a wide range of dynamic folding patterns in peptides based on their energy landscapes.-Accurately predicts peptide structures and effectively recapitulates experimental peptide ensembles at a fraction of the running time of traditional approaches.	-Available as a GitLab repository at https://gitlab.com/oabdin/pepflow with code and documentation (accessed on 30 November 2024).

4.1. AlphaFold-Based Modelling and Bioactive Peptide Discovery

The introduction of AlphaFold, particularly its iterations AlphaFold 2 and the recently released AlphaFold 3, has marked a transformative phase in the modelling of protein structures and interactions, significantly impacting bioactive peptide discovery [429]. These deep learning-driven models have enhanced our ability to predict the three-dimensional structures of proteins and their complexes, facilitating the rational design of therapeutic peptides.

AlphaFold employs a two-step approach to predicting protein structures. It starts with a search of big sequence databases, including metagenomic sequences (e.g., Big Fantastic Database), to collect homologous sequences and build a Multiple Sequence Alignment (MSA) that provides evolutionary context and helps to identify conserved residues and potential structural features [55]. The core predictive power of AlphaFold lies in its advanced neural network architecture, which incorporates attention mechanisms that focus on relevant parts of the input data. This model predicts distances between pairs of amino acids and their relative orientations, ultimately generating a three-dimensional representation of the protein structure. ESMFold [420] is an alternative algorithm for modelling peptide structures. Instead of relying on MSA-based input, it uses hidden states derived from inference on a Protein Language Model (PLM). This PLM was pre-trained on a subset of UniRef-50 with a masked token prediction objective, enabling it to learn the conditional probability of a residue occupying a certain position given the context of the rest of the sequence. This method enables ESMFold to predict peptide structures with an accuracy slightly lower than AlphaFold2 but with a 60-fold speed-up. For peptide modelling with an interacting protein, a key benefit of AlphaFold2 over ESMFold is that it can model multiple interacting protein chains.

The application of AlphaFold in bioactive peptide discovery spans several areas [430,431]. One major application is predictive modelling, where AlphaFold accurately models peptide structures, allowing the prediction of how peptides fold and interact with target proteins [432]. This capability is particularly valuable for discovering therapeutic peptides with specific biological activities. For instance, AlphaFold was incorporated into a pipeline that accurately predicted peptide interactions with class I and class II MHC molecules, outperforming existing tools [432].

AlphaFold can accurately and rapidly model peptide–protein complexes without requiring multiple sequence alignment information for the peptide partner, as shown in an early implementation of AlphaFold monomer modelling (before the multimer version was available) by connecting the peptide to the receptor with a polyglycine linker [422]. AlphaFold demonstrated high-accuracy structure predictions for peptides ranging from 10 to 40 amino acids in length, particularly modelling α-helical, β-hairpin, and disulfide-rich peptides with accuracy comparable to or better than methods specifically designed for peptide structure prediction [433]. Similarly, Johansson-Åkhe et al. established that, aside from AlphaFold-Multimer's ability to predict peptide–protein complexes with high quality exhibiting DockQ scores greater than 0.8, it could also determine whether a peptide and a protein could interact [434].

A number of AlphaFold implementations have been developed specifically for peptides. For example, HighFold [424] is an AlphaFold-derived model that incorporates specific structural features like head-to-tail circularity and disulfide bridges to accurately predict cyclic peptide structures. AfCycDesign is proposed to be able to predict cyclic peptide structures with an RMSD of less than 1.5 Å in many cases and to design a library of 10,000 macrocyclic 7-13mer peptides with high-confidence predicted structures [425]. They adapted the AlphaFold code to define an "offset" as the relationship of termini of the supplied linear peptide to be one residue apart, and this approach was further explored to design cyclic peptides for the MDM2/p53 complex and 12 other protein–peptide complexes [435]. Additionally, the integration of AlphaFold with generative models for peptide design has enabled the creation of novel peptides through deep generative models, such as GANs and VAEs, which explore vast chemical spaces to identify peptides with optimal binding affinities and enhanced stability for targeted drug delivery [436,437]. A generative model, such as a variational autoencoder or transformer, could be trained on existing peptide sequences to create novel candidates with desired properties. These sequences are then input into AlphaFold, which predicts their 3D structures, providing crucial structural information. This process allows for iterative optimisation, where structural feedback refines the peptide sequences generated by the model. The approach could enable rapid screening and ranking of candidate peptides based on both sequence and predicted structural features, ultimately leading to experimental validation of the top candidates. However, while this integration offers a faster and more cost-effective method for exploring peptide diversity, challenges include the computational cost of AlphaFold predictions and the accuracy of its predictions for short peptides.

AlphaFold's computational efficiency also supports high-throughput screening, allowing for virtual screening of peptide libraries against target proteins. By predicting the binding affinities of thousands of peptide candidates, the most promising candidates for experimental validation can be efficiently prioritised [438–440].

Efforts to rank peptides towards determining which peptide would bind to a given protein target amongst a set of peptides has also recently been explored by new AlphaFold-derived methods such as Automated Pairwise Peptide-Receptor Analysis for Screening Engineered proteins (APPRAISE), a method that bridges the gap between protein structure prediction and binding prioritisation [441]. By utilising AlphaFold-Multimer and ESMFold for structural modelling, APPRAISE predicts the target-binding propensity of engineered proteins. It effectively classifies various engineered proteins, including adeno-associated viral vectors and peptides targeting SARS-CoV-2 and other receptors. Similarly, Chang et al. introduced a novel computational protocol that also leveraged AlphaFold's capabilities to accurately predict protein structures and their interactions with peptides [442]. The assay enabled the identification of the highest affinity binder among a set of candidate peptides by modelling receptor structures in the presence of two peptides. The protocol effectively captured the higher affinity peptide in a bound state while depicting the other peptide in an unbound form when their individual structures are well predicted. Testing on six protein receptors with known experimental binding affinities confirms the assay's efficacy, particularly for medium to strong peptide binders that adopt stable secondary structures upon binding.

AlphaFold has some limitations. Models showed consistently worse performance compared to native PDB structures when compared to four docking programs and two consensus techniques in a benchmark of 22 structurally known targets with known small-molecule ligands [443]. While AlphaFold-3 [444] permits a number of post-translational modifications of amino acids that are common in the PDB training dataset, more diverse noncanonical amino acids and unconventional cyclisation present challenges [430]. Thus, physics-based approaches like docking and molecular dynamics (MD) simulations remain essential in peptide drug discovery, additionally offering insights into binding mechanisms and the thermodynamic and kinetic properties of complexes. AlphaFold structures with the lowest RMSD do not always correspond to those with the highest pLDDT scores, and there are limitations in predicting Φ/Ψ angles and disulfide bond patterns [433], a limitation often ameliorated by relaxing the predicted structures with molecular modelling methods. Considering these limitations, synergistic integration of AI, physics-based methods coupled with experimental validation has the potential to transform peptide-based drug discovery. Overall, AlphaFold's multifaceted applications are revolutionising the field of peptide discovery and could accelerate the design and identification of peptide-based therapeutics.

4.2. Structure-Aware Machine-Learning Methods for Predicting Peptide Bioactivity

Apart from the direct modelling of protein–peptide interactions, new machine-learning advances can leverage structural information for directly predicting bioactivity. These models can then be used to predict different relevant properties for the computational design of bioactive peptides with a series of desired physico-chemical properties.

These new methods take as input a peptide or protein structure, either experimental or predicted. They use that information to predict a target property of the peptide, such as antimicrobial activity. The methods represent the structures as undirected graphs where the nodes are the residues (or, in some cases, the atoms), and the edges represent the contact between those residues. Usually, an edge is defined when the distance between two given residues is lower than a predefined threshold. Both the nodes and edges can have different properties, describing them from naive one-hot encodings to more complex PLM representations. A framework for this, Esm-AxP-GDL, [427] has three basic steps: (1) prediction of the folding conformation of the peptide using ESMFold, (2) the folded conformation is translated into a graph representation where each residue constitutes a node and is represented by an embedding vector extracted from ESM and where the edges are defined between nodes within a given distance, and (3) a GNN is then used to learn to predict whether a peptide is antimicrobial or not. The main drawback of these approaches is that they require fairly large datasets, with the Esm-AxP-GDL framework having been trained on ~25,000 predicted antimicrobial peptide structures, most of which were less than 100 amino acids. It is not clear whether such use of structural information will enhance peptide bioactivity prediction since the performance of the method proposed did not outperform sequence-based methods for smaller peptides [427].

5. Conclusions

The field of peptide computational modelling and virtual screening is rapidly changing in response to developments in AI (AlphaFold and Protein Language Models). Such models are clearly important in accelerating scientific research regarding molecular interactions involving peptides. Nevertheless, the prospect of computational methods contributing in a major way to the clinical development of new peptide therapeutics is unclear. To date, most clinically used peptides are either microbial noncanonical peptides of non-ribosomal origin or human peptides that have undergone extensive chemical modifications to improve their stability. Thus, the key to the design of clinically useful peptides may lie in not only identifying and modelling bioactivity but also predicting peptides that have a reasonable bioavailability in the human body. Therefore, an outstanding challenge in this field is to improve the structural modelling of peptides with diverse chemical modifications and cyclisations.

The integration of molecular modelling techniques in bioactive peptide characterisation offers significant potential for advancing peptide therapeutics. Combining computational approaches such as molecular docking, molecular dynamics simulations, and QM/MM hybrid methods can enhance predictive accuracy, while AI-driven techniques improve structure–activity relationship predictions. Hybrid in silico-experimental workflows can streamline peptide screening, reducing trial and error. Additionally, enhanced sampling and multi-scale modelling provide deeper insights into peptide conformations and interactions. Future research should focus on optimising these methods for scalability, precision, and adaptability to non-traditional peptide structures, ultimately accelerating the discovery of next-generation bioactive peptides.

While the computational predictions presented in this review have, in many cases, been followed up by experimental analyses that appear to validate the computational method, this is a very piecemeal approach to methodology validation and does not comprise a benchmark for the field. Ideally, blinded benchmarking datasets would be provided prior to their publication so that the research community can rigorously assess the efficiency of different prediction methods, in a similar manner to the CASP competitions used to assess protein structure prediction [445]. However, designing or selecting such a validation dataset to have sufficient generalisability to diverse applications may represent a substantial challenge. The ultimate validation is whether such bioactive peptide discoveries have led to useful reagents, either to advance our understanding of molecular signalling or to act as lead therapeutic compounds. In general, peptides face some hurdles in translating into the clinic. For example, there is a vast treasure trove of ribosomally synthesised antimicrobial peptides, but these have not translated directly into the clinic. A challenge is how to build on the insights gained from the mechanism of action of novel bioactive peptides (such as dynobactin A, [446]) to develop lead compounds that act by the same mechanism but are more drug-like [446–448].

Author Contributions: C.A.: Writing—original draft, review and editing, visualisation, conceptualisation. R.F.-D.: writing—original draft, editing, P.B.T.: review and editing, A.A.: review and editing, H.G.: review and editing, D.C.S.: review and editing, supervision. All authors have read and agreed to the published version of the manuscript.

Funding: C.A. was supported by Enterprise Ireland and received funding from the European Union Horizon 2020 research and innovation programme under the Marie Skłodowska-Curie [grant agreement number 847402. R.F.D. was supported by Science Foundation Ireland through the SFI Centre for Research Training in Genomics Data Science [18/CRT/6214].

Institutional Review Board Statement: Not applicable.

Informed Consent Statement: Not applicable.

Data Availability Statement: Not applicable.

Conflicts of Interest: Author Raúl Fernández-Díaz was employed by the company IBM Research. Authors Patrick Brendan Timmons, Alessandro Adelfio and Hansel Gómez were employed by the company Nuritas Ltd. The remaining authors declare that the research was conducted in the absence of any commercial or financial relationships that could be construed as a potential conflict of interest.

References

1. Adelusi, T.I.; Oyedele, A.-Q.K.; Boyenle, I.D.; Ogunlana, A.T.; Adeyemi, R.O.; Ukachi, C.D.; Idris, M.O.; Olaoba, O.T.; Adedotun, I.O.; Kolawole, O.E.; et al. Molecular Modeling in Drug Discovery. *Inform. Med. Unlocked* **2022**, *29*, 100880. [CrossRef]
2. Forster, M.J. Molecular Modelling in Structural Biology. *Micron* **2002**, *33*, 365–384. [CrossRef] [PubMed]
3. Bagchi, A. Molecular Modeling Techniques and In-Silico Drug Discovery. In *Reverse Engineering of Regulatory Networks*; Mandal, S., Ed.; Springer: New York, NY, USA, 2024; pp. 1–11. ISBN 978-1-07-163461-5.
4. Kazmirchuk, T.D.D.; Bradbury-Jost, C.; Withey, T.A.; Gessese, T.; Azad, T.; Samanfar, B.; Dehne, F.; Golshani, A. Peptides of a Feather: How Computation Is Taking Peptide Therapeutics under Its Wing. *Genes* **2023**, *14*, 1194. [CrossRef]
5. Leelananda, S.P.; Lindert, S. Computational Methods in Drug Discovery. *Beilstein J. Org. Chem.* **2016**, *12*, 2694–2718. [CrossRef]
6. Peluso, P.; Chankvetadze, B. Recent Developments in Molecular Modeling Tools and Applications Related to Pharmaceutical and Biomedical Research. *J. Pharm. Biomed. Anal.* **2024**, *238*, 115836. [CrossRef]
7. Aminpour, M.; Montemagno, C.; Tuszynski, J.A. An Overview of Molecular Modeling for Drug Discovery with Specific Illustrative Examples of Applications. *Molecules* **2019**, *24*, 1693. [CrossRef] [PubMed]
8. Naithani, U.; Guleria, V. Integrative Computational Approaches for Discovery and Evaluation of Lead Compound for Drug Design. *Front. Drug Discov.* **2024**, *4*, 1362456. [CrossRef]
9. Colman, P. Zanamivir: An Influenza Virus Neuraminidase Inhibitor. *Expert. Rev. Anti-Infect. Ther.* **2005**, *3*, 191–199. [CrossRef]
10. Kaldor, S.W.; Kalish, V.J.; Davies, J.F.; Shetty, B.V.; Fritz, J.E.; Appelt, K.; Burgess, J.A.; Campanale, K.M.; Chirgadze, N.Y.; Clawson, D.K.; et al. Viracept (Nelfinavir Mesylate, AG1343): A Potent, Orally Bioavailable Inhibitor of HIV-1 Protease. *J. Med. Chem.* **1997**, *40*, 3979–3985. [CrossRef]
11. Qayed, W.S.; Ferreira, R.S.; Silva, J.R.A. In Silico Study towards Repositioning of FDA-Approved Drug Candidates for Anti-coronaviral Therapy: Molecular Docking, Molecular Dynamics and Binding Free Energy Calculations. *Molecules* **2022**, *27*, 5988. [CrossRef]
12. Ginex, T.; Garaigorta, U.; Ramírez, D.; Castro, V.; Nozal, V.; Maestro, I.; García-Cárceles, J.; Campillo, N.E.; Martinez, A.; Gastaminza, P.; et al. Host-Directed FDA-Approved Drugs with Antiviral Activity against SARS-CoV-2 Identified by Hierarchical In Silico/In Vitro Screening Methods. *Pharmaceuticals* **2021**, *14*, 332. [CrossRef] [PubMed]
13. Akbarian, M.; Khani, A.; Eghbalpour, S.; Uversky, V.N. Bioactive Peptides: Synthesis, Sources, Applications, and Proposed Mechanisms of Action. *Int. J. Mol. Sci.* **2022**, *23*, 1445. [CrossRef] [PubMed]
14. Chauhan, V.; Kanwar, S.S. Chapter 4—Bioactive Peptides: Synthesis, Functions and Biotechnological Applications. In *Biotechnological Production of Bioactive Compounds*; Verma, M.L., Chandel, A.K., Eds.; Elsevier: Amsterdam, The Netherlands, 2020; pp. 107–137, ISBN 978-0-444-64323-0.
15. Du, Z.; Li, Y. Review and Perspective on Bioactive Peptides: A Roadmap for Research, Development, and Future Opportunities. *J. Agric. Food Res.* **2022**, *9*, 100353. [CrossRef]

16. Karami, Z.; Akbari-adergani, B. Bioactive Food Derived Peptides: A Review on Correlation between Structure of Bioactive Peptides and Their Functional Properties. *J. Food Sci. Technol.* **2019**, *56*, 535–547. [CrossRef]
17. Purohit, K.; Reddy, N.; Sunna, A. Exploring the Potential of Bioactive Peptides: From Natural Sources to Therapeutics. *Int. J. Mol. Sci.* **2024**, *25*, 1391. [CrossRef]
18. Rossino, G.; Marchese, E.; Galli, G.; Verde, F.; Finizio, M.; Serra, M.; Linciano, P.; Collina, S. Peptides as Therapeutic Agents: Challenges and Opportunities in the Green Transition Era. *Molecules* **2023**, *28*, 7165. [CrossRef]
19. Wang, L.; Wang, N.; Zhang, W.; Cheng, X.; Yan, Z.; Shao, G.; Wang, X.; Wang, R.; Fu, C. Therapeutic Peptides: Current Applications and Future Directions. *Sig. Transduct. Target Ther.* **2022**, *7*, 48. [CrossRef]
20. Birk, A.V.; Liu, S.; Soong, Y.; Mills, W.; Singh, P.; Warren, J.D.; Seshan, S.V.; Pardee, J.D.; Szeto, H.H. The Mitochondrial-Targeted Compound SS-31 Re-Energizes Ischemic Mitochondria by Interacting with Cardiolipin. *J. Am. Soc. Nephrol.* **2013**, *24*, 1250. [CrossRef]
21. Chang, Y.S.; Graves, B.; Guerlavais, V.; Tovar, C.; Packman, K.; To, K.-H.; Olson, K.A.; Kesavan, K.; Gangurde, P.; Mukherjee, A.; et al. Stapled A−helical Peptide Drug Development: A Potent Dual Inhibitor of MDM2 and MDMX for P53-Dependent Cancer Therapy. *Proc. Natl. Acad. Sci. USA* **2013**, *110*, E3445–E3454. [CrossRef]
22. Scott, L.J. Dulaglutide: A Review in Type 2 Diabetes. *Drugs* **2020**, *80*, 197–208. [CrossRef]
23. Trujillo, J.M.; Nuffer, W. Albiglutide: A New GLP-1 Receptor Agonist for the Treatment of Type 2 Diabetes. *Ann. Pharmacother.* **2014**, *48*, 1494–1501. [CrossRef] [PubMed]
24. Jackson, S.H.; Martin, T.S.; Jones, J.D.; Seal, D.; Emanuel, F. Liraglutide (Victoza). *Pharm. Ther.* **2010**, *35*, 498–529.
25. Holst, J.J. The Physiology of Glucagon-like Peptide 1. *Physiol. Rev.* **2007**, *87*, 1409–1439. [CrossRef]
26. Al-Salama, Z.T.; Syed, Y.Y. Plecanatide: First Global Approval. *Drugs* **2017**, *77*, 593–598. [CrossRef]
27. Wilson, A.C.; Vadakkadath Meethal, S.; Bowen, R.L.; Atwood, C.S. Leuprolide Acetate: A Drug of Diverse Clinical Applications. *Expert Opin. Investig. Drugs* **2007**, *16*, 1851–1863. [CrossRef]
28. Jonat, W. Goserelin (Zoladex TM)—Its Role in Early Breast Cancer in Pre- and Perimenopausal Women. *Br. J. Cancer* **2001**, *85*, 1–5. [CrossRef] [PubMed]
29. Lamberts, S.W.J.; van der Lely, A.-J.; de Herder, W.W.; Hofland, L.J. Octreotide. *N. Engl. J. Med.* **1996**, *334*, 246–254. [CrossRef]
30. Caplin, M.E.; Pavel, M.; Ćwikła, J.B.; Phan, A.T.; Raderer, M.; Sedláčková, E.; Cadiot, G.; Wolin, E.M.; Capdevila, J.; Wall, L.; et al. Lanreotide in Metastatic Enteropancreatic Neuroendocrine Tumors. *N. Engl. J. Med.* **2014**, *371*, 224–233. [CrossRef]
31. Anand, U.; Bandyopadhyay, A.; Jha, N.K.; Pérez de la Lastra, J.M.; Dey, A. Translational Aspect in Peptide Drug Discovery and Development: An Emerging Therapeutic Candidate. *Bio. Factors* **2023**, *49*, 251–269. [CrossRef]
32. Drucker, D.J. Advances in Oral Peptide Therapeutics. *Nat. Rev. Drug Discov.* **2020**, *19*, 277–289. [CrossRef]
33. Romanova, E.V.; Sweedler, J.V. Peptidomics for the Discovery and Characterization of Neuropeptides and Hormones. *Trends Pharmacol. Sci.* **2015**, *36*, 579–586. [CrossRef]
34. Ferreira, R.; Amado, F.; Vitorino, R. Empowering Peptidomics: Utilizing Computational Tools and Approaches. *Bioanalysis* **2023**, *15*, 1315–1325. [CrossRef] [PubMed]
35. Köcher, T.; Superti-Furga, G. Mass Spectrometry–Based Functional Proteomics: From Molecular Machines to Protein Networks. *Nat. Methods* **2007**, *4*, 807–815. [CrossRef] [PubMed]
36. Hashemi, Z.S.; Zarei, M.; Fath, M.K.; Ganji, M.; Farahani, M.S.; Afsharnouri, F.; Pourzardosht, N.; Khalesi, B.; Jahangiri, A.; Rahbar, M.R.; et al. In Silico Approaches for the Design and Optimization of Interfering Peptides Against Protein–Protein Interactions. *Front. Mol. Bio. Sci.* **2021**, *8*, 669431. [CrossRef]
37. Kuczera, K. Molecular Modeling of Peptides. In *Computational Peptidology*; Zhou, P., Huang, J., Eds.; Springer: New York, NY, USA, 2015; pp. 15–41, ISBN 978-1-4939-2285-7.
38. Redkar, A.; Ramakrishnan, V. Chapter 2—Modeling and Simulation of Peptides. In *De Novo Peptide Design*; Ramakrishnan, V., Patel, K., Goyal, R., Eds.; Academic Press: Cambridge, MA, USA, 2023; pp. 35–56, ISBN 978-0-323-99917-5.
39. Vidal-Limon, A.; Aguilar-Toalá, J.E.; Liceaga, A.M. Integration of Molecular Docking Analysis and Molecular Dynamics Simulations for Studying Food Proteins and Bioactive Peptides. *J. Agric. Food Chem.* **2022**, *70*, 934–943. [CrossRef] [PubMed]
40. Wang, Y.; Stebe, K.J.; De La Fuente-Nunez, C.; Radhakrishnan, R. Computational Design of Peptides for Biomaterials Applications. *ACS Appl. Bio. Mater.* **2024**, *7*, 617–625. [CrossRef]
41. Cardoso, M.H.; Orozco, R.Q.; Rezende, S.B.; Rodrigues, G.; Oshiro, K.G.N.; Cândido, E.S.; Franco, O.L. Computer-Aided Design of Antimicrobial Peptides: Are We Generating Effective Drug Candidates? *Front. Microbiol.* **2020**, *10*, 3097. [CrossRef]
42. Chang, L.; Mondal, A.; Perez, A. Towards Rational Computational Peptide Design. *Front Bio. Inform* **2022**, *2*, 1046493. [CrossRef]
43. Vincenzi, M.; Mercurio, F.A.; Leone, M. Virtual Screening of Peptide Libraries: The Search for Peptide-Based Therapeutics Using Computational Tools. *Int. J. Mol. Sci.* **2024**, *25*, 1798. [CrossRef]
44. FitzGerald, R.J.; Cermeño, M.; Khalesi, M.; Kleekayai, T.; Amigo-Benavent, M. Application of in Silico Approaches for the Generation of Milk Protein-Derived Bioactive Peptides. *J. Funct. Foods* **2020**, *64*, 103636. [CrossRef]

45. Nongonierma, A.B.; Dellafiora, L.; Paolella, S.; Galaverna, G.; Cozzini, P.; FitzGerald, R.J. In Silico Approaches Applied to the Study of Peptide Analogs of Ile-Pro-Ile in Relation to Their Dipeptidyl Peptidase IV Inhibitory Properties. *Front. Endocrinol.* **2018**, *9*, 329. [CrossRef]
46. Vukic, V.R.; Vukic, D.V.; Milanovic, S.D.; Ilicic, M.D.; Kanuric, K.G.; Johnson, M.S. In Silico Identification of Milk Antihypertensive Di- and Tripeptides Involved in Angiotensin I–Converting Enzyme Inhibitory Activity. *Nutr. Res.* **2017**, *46*, 22–30. [CrossRef]
47. Olğaç, A.; Orhan, I.E.; Banoglu, E. The Potential Role of In Silico Approaches to Identify Novel Bioactive Molecules from Natural Resources. *Future Med. Chem.* **2017**, *9*, 1665–1686. [CrossRef]
48. Shuli, Z.; Linlin, L.; Li, G.; Yinghu, Z.; Nan, S.; Haibin, W.; Hongyu, X. Bioinformatics and Computer Simulation Approaches to the Discovery and Analysis of Bioactive Peptides. *Curr. Pharm. Biotechnol.* **2022**, *23*, 1541–1555. [CrossRef] [PubMed]
49. Audie, J.; Swanson, J. Recent Work In The Development And Application Of Protein–Peptide Docking. *Future Med. Chem.* **2012**, *4*, 1619–1644. [CrossRef]
50. Das, R. Four Small Puzzles That Rosetta Doesn't Solve. *PLoS ONE* **2011**, *6*, e20044. [CrossRef]
51. Pike, D.H.; Nanda, V. Empirical Estimation of Local Dielectric Constants: Toward Atomistic Design of Collagen Mimetic Peptides. *Pept. Sci.* **2015**, *104*, 360–370. [CrossRef]
52. Rentzsch, R.; Renard, B.Y. Docking Small Peptides Remains a Great Challenge: An Assessment Using AutoDock Vina. *Brief. Bio. Inform.* **2015**, *16*, 1045–1056. [CrossRef]
53. Agüero-Chapin, G.; Galpert-Cañizares, D.; Domínguez-Pérez, D.; Marrero-Ponce, Y.; Pérez-Machado, G.; Teijeira, M.; Antunes, A. Emerging Computational Approaches for Antimicrobial Peptide Discovery. *Antibiotics* **2022**, *11*, 936. [CrossRef]
54. Doherty, A.; Wall, A.; Khaldi, N.; Kussmann, M. Artificial Intelligence in Functional Food Ingredient Discovery and Characterisation: A Focus on Bioactive Plant and Food Peptides. *Front. Genet.* **2021**, *12*, 768979. [CrossRef]
55. Jumper, J.; Evans, R.; Pritzel, A.; Green, T.; Figurnov, M.; Ronneberger, O.; Tunyasuvunakool, K.; Bates, R.; Žídek, A.; Potapenko, A.; et al. Highly Accurate Protein Structure Prediction with AlphaFold. *Nature* **2021**, *596*, 583–589. [CrossRef]
56. Lin, Z.; Akin, H.; Rao, R.; Hie, B.; Zhu, Z.; Lu, W.; dos Santos Costa, A.; Fazel-Zarandi, M.; Sercu, T.; Candido, S.; et al. Language Models of Protein Sequences at the Scale of Evolution Enable Accurate Structure Prediction. *bioRxiv* **2022**, 500902.
57. Perpetuo, L.; Klein, J.; Ferreira, R.; Guedes, S.; Amado, F.; Leite-Moreira, A.; Silva, A.M.S.; Thongboonkerd, V.; Vitorino, R. How Can Artificial Intelligence Be Used for Peptidomics? *Expert Rev. Proteom.* **2021**, *18*, 527–556. [CrossRef]
58. Rein, D.; Ternes, P.; Demin, R.; Gierke, J.; Helgason, T.; Schön, C. Artificial Intelligence Identified Peptides Modulate Inflammation in Healthy Adults. *Food Funct.* **2019**, *10*, 6030–6041. [CrossRef]
59. Zhang, H.; Saravanan, K.M.; Wei, Y.; Jiao, Y.; Yang, Y.; Pan, Y.; Wu, X.; Zhang, J.Z.H. Deep-Learning Based Bioactive Therapeutic Peptides Generation and Screening. *J. Chem. Inf. Model.* **2023**, *63*, 835–845. [CrossRef]
60. Zhang, Y.; Li, P.; Ma, Y.; Wang, J.; Chen, Y. Artificial Intelligence Accelerates the Mining of Bioactive Small Molecules from Human Microbiome. *Clin. Transl. Med.* **2022**, *12*, e1011. [CrossRef]
61. Alzaydi, A.; Barbhuiya, R.I.; Routray, W.; Elsayed, A.; Singh, A. Bioactive Peptides: Synthesis, Applications, and Associated Challenges. *Food Bioeng.* **2023**, *2*, 273–290. [CrossRef]
62. Hou, H.; Wang, J.; Wang, J.; Tang, W.; Shaikh, A.S.; Li, Y.; Fu, J.; Lu, L.; Wang, F.; Sun, F.; et al. A Review of Bioactive Peptides: Chemical Modification, Structural Characterization and Therapeutic Applications. *J. Biomed. Nanotechnol.* **2020**, *16*, 1687–1718. [CrossRef]
63. Li, Y.-W.; Li, B.; He, J.; Qian, P. Quantitative Structure–Activity Relationship Study of Antioxidative Peptide by Using Different Sets of Amino Acids Descriptors. *J. Mol. Struct.* **2011**, *998*, 53–61. [CrossRef]
64. Merritt, H.I.; Sawyer, N.; Arora, P.S. Bent Into Shape: Folded Peptides to Mimic Protein Structure and Modulate Protein Function. *Pept. Sci.* **2020**, *112*, e24145. [CrossRef]
65. Hong, S.Y.; Oh, J.E.; Lee, K.-H. Effect of D-Amino Acid Substitution on the Stability, the Secondary Structure, and the Activity of Membrane-Active Peptide. *Biochem. Pharmacol.* **1999**, *58*, 1775–1780. [CrossRef]
66. Rivera-Jiménez, J.; Berraquero-García, C.; Pérez-Gálvez, R.; García-Moreno, P.J.; Espejo-Carpio, F.J.; Guadix, A.; Guadix, E.M. Peptides and Protein Hydrolysates Exhibiting Anti-Inflammatory Activity: Sources, Structural FeatuRes. and Modulation Mechanisms. *Food Funct.* **2022**, *13*, 12510–12540. [CrossRef]
67. Schmidt, M.M.; Wittrup, K.D. A Modeling Analysis of the Effects of Molecular Size and Binding Affinity on Tumor Targeting. *Mol. Cancer Ther.* **2009**, *8*, 2861–2871. [CrossRef]
68. Chen, Y.; Yang, C.; Li, T.; Zhang, M.; Liu, Y.; Gauthier, M.A.; Zhao, Y.; Wu, C. The Interplay of Disulfide Bonds, α-Helicity, and Hydrophobic Interactions Leads to Ultrahigh Proteolytic Stability of Peptides. *Biomacromolecules* **2015**, *16*, 2347–2355. [CrossRef]
69. Góngora-Benítez, M.; Tulla-Puche, J.; Albericio, F. Multifaceted Roles of Disulfide Bonds. *Pept. Therapeutics. Chem. Rev.* **2014**, *114*, 901–926. [CrossRef]
70. Yang, N.J.; Hinner, M.J. Getting Across the Cell Membrane: An Overview for Small Molecules, Peptides, and Proteins. *Methods Mol. Biol.* **2015**, *1266*, 29–53. [CrossRef]

71. Pf, A.; Tp, B.; Ls, G.; Rw, H.; Dr, T. Evidence for a Polarized Efflux System in CACO-2 Cells Capable of Modulating Cyclosporin A Transport. *Biochem. Biophys. Res. Commun.* **1993**, *197*, 360–365. [CrossRef]
72. Pauletti, G.M.; Okumu, F.W.; Borchardt, R.T. Effect of Size and Charge on the Passive Diffusion of Peptides across Caco-2 Cell Monolayers via the Paracellular Pathway. *Pharm. Res.* **1997**, *14*, 164–168. [CrossRef]
73. Sánchez-Navarro, M.; Teixidó, M.; Giralt, E. Jumping Hurdles: Peptides Able To Overcome Biological Barriers. *Acc. Chem. Res.* **2017**, *50*, 1847–1854. [CrossRef]
74. Bhavsar, J.; Kasture, K.; Salvi, B.V.; Shende, P. Strategies for Transportation of Peptides across the Skin for Treatment of Multiple Diseases. *Ther. Deliv.* **2025**, *16*, 63–86. [CrossRef]
75. Kurabi, A.; Pak, K.; Chavez, E.; Doan, J.; Ryan, A.F. A Transcytotic Transport Mechanism across the Tympanic Membrane. *Sci. Rep.* **2022**, *12*, 984. [CrossRef]
76. Stauffer, M.; Jeckelmann, J.-M.; Ilgü, H.; Ucurum, Z.; Boggavarapu, R.; Fotiadis, D. Peptide Transporter Structure Reveals Binding and Action Mechanism of a Potent PEPT1 and PEPT2 Inhibitor. *Commun. Chem.* **2022**, *5*, 23. [CrossRef]
77. Smith, D.E.; Clémençon, B.; Hediger, M.A. Proton-Coupled Oligopeptide Transporter Family SLC15: Physiological, Pharmacological and Pathological Implications. *Mol. Asp. Med.* **2013**, *34*, 323–336. [CrossRef]
78. Khandelia, H.; Ipsen, J.H.; Mouritsen, O.G. The Impact of Peptides on Lipid Membranes. *Biochim. Biophys. Acta (BBA)—Biomembr.* **2008**, *1778*, 1528–1536. [CrossRef]
79. Cardenas, A.E.; Shrestha, R.; Webb, L.J.; Elber, R. Membrane Permeation of a Peptide: It Is Better to Be Positive. *J. Phys. Chem. B* **2015**, *119*, 6412. [CrossRef]
80. Wang, X.; Yang, Z.; Zhang, W.; Xing, L.; Luo, R.; Cao, S. Obstacles, Research Progress, and Prospects of Oral Delivery of Bioactive Peptides: A Comprehensive Review. *Front. Nutr.* **2024**, *11*, 1496706. [CrossRef]
81. Xiao, W.; Jiang, W.; Chen, Z.; Huang, Y.; Mao, J.; Zheng, W.; Hu, Y.; Shi, J. Advance in Peptide-Based Drug Development: Delivery Platforms, Therapeutics and Vaccines. *Sig. Transduct. Target Ther.* **2025**, *10*, 74. [CrossRef]
82. Lopez, M.J.; Mohiuddin, S.S. Biochemistry, Essential Amino Acids. In *StatPearls*; StatPearls Publishing: Treasure Island, FL, USA, 2024.
83. Alberts, B.; Johnson, A.; Lewis, J.; Raff, M.; Roberts, K.; Walter, P. The Shape and Structure of Proteins. In *Molecular Biology of the Cell*, 4th ed.; Garland Science: New York, NY, USA, 2002.
84. Fujiwara, K.; Toda, H.; Ikeguchi, M. Dependence of α-Helical and β-Sheet Amino Acid Propensities on the Overall Protein Fold Type. *BMC Struct. Biol.* **2012**, *12*, 18. [CrossRef]
85. Damodaran, S.; Parkin, K.L.; Fennema, O.R. *Fennema's Food Chemistry*; CRC Press: Boca Raton, FL, USA, 2007; ISBN 978-1-4200-2052-6.
86. Maloy, S. Amino Acids. In *Brenner's Encyclopedia of Genetics*, Maloy, S., Hughes, K., Eds.; 2nd ed.; Academic Press: San Diego, CA, USA, 2013; pp. 108–110, ISBN 978-0-08-096156-9.
87. Stollar, E.J.; Smith, D.P. Uncovering Protein Structure. *Essays Biochem.* **2020**, *64*, 649–680. [CrossRef]
88. Dietzen, D.J. Proteins. In *Principles and Applications of Molecular Diagnostics*; Rifai, N., Horvath, A.R., Wittwer, C.T., Eds.; Elsevier: Amsterdam, The Netherlands, 2018; pp. 345–380, ISBN 978-0-12-816061-9.
89. Ouellette, R.J.; Rawn, J.D. Proteins. In *Organic Chemistry*, Ouellette, R.J., Rawn, J.D., Eds.; 2nd ed.; Academic Press: Cambridge, MA, USA, 2018; pp. 929–971, ISBN 978-0-12-812838-1.
90. Al Musaimi, O.; Lombardi, L.; Williams, D.R.; Albericio, F. Strategies for Improving Peptide Stability and Delivery. *Pharmaceuticals* **2022**, *15*, 1283. [CrossRef]
91. Pei, J.; Gao, X.; Pan, D.; Hua, Y.; He, J.; Liu, Z.; Dang, Y. Advances in the Stability Challenges of Bioactive Peptides and Improvement Strategies. *Curr. Res. Food Sci.* **2022**, *5*, 2162–2170. [CrossRef] [PubMed]
92. Acquah, C.; Stefano, E.D.; Udenigwe, C.C. Role of Hydrophobicity in Food Peptide Functionality and Bioactivity. *J. Food Bio. Act.* **2018**, *4*, 88–98. [CrossRef]
93. Ahmed, T.; Sun, X.; Udenigwe, C.C. Role of Structural Properties of Bioactive Peptides in Their Stability during Simulated Gastrointestinal Digestion: A Systematic Review. *Trends Food Sci. Technol.* **2022**, *120*, 265–273. [CrossRef]
94. Amigo, L.; Hernández-Ledesma, B. Current Evidence on the Bioavailability of Food Bioactive Peptides. *Molecules* **2020**, *25*, 4479. [CrossRef]
95. Xie, N.; Wang, B.; Jiang, L.; Liu, C.; Li, B. Hydrophobicity Exerts Different Effects on Bioavailability and Stability of Antioxidant Peptide Fractions from Casein during Simulated Gastrointestinal Digestion and Caco-2 Cell Absorption. *Food Res. Int.* **2015**, *76*, 518–526. [CrossRef]
96. Avci, F.G.; Sariyar Akbulut, B.; Ozkirimli, E. Membrane Active Peptides and Their Biophysical Characterization. *Biomolecules* **2018**, *8*, 77. [CrossRef] [PubMed]
97. Qing, R.; Hao, S.; Smorodina, E.; Jin, D.; Zalevsky, A.; Zhang, S. Protein Design: From the Aspect of Water Solubility and Stability. *Chem. Rev.* **2022**, *122*, 14085–14179. [CrossRef]
98. Sikora, K.; Jaśkiewicz, M.; Neubauer, D.; Migoń, D.; Kamysz, W. The Role of Counter-Ions in Peptides—An Overview. *Pharmaceuticals* **2020**, *13*, 442. [CrossRef]

99. Vascon, F.; Gasparotto, M.; Giacomello, M.; Cendron, L.; Bergantino, E.; Filippini, F.; Righetto, I. Protein Electrostatics: From Computational and Structural Analysis to Discovery of Functional Fingerprints and Biotechnological Design. *Comput. Struct. Biotechnol. J.* **2020**, *18*, 1774–1789. [CrossRef]
100. Zhou, H.-X.; Pang, X. Electrostatic Interactions in Protein Structure, Folding, Binding, and Condensation. *Chem. Rev.* **2018**, *118*, 1691–1741. [CrossRef]
101. Choi, Y.; Sims, G.E.; Murphy, S.; Miller, J.R.; Chan, A.P. Predicting the Functional Effect of Amino Acid Substitutions and Indels. *PLoS ONE* **2012**, *7*, e46688. [CrossRef] [PubMed]
102. Wang, X.; Mishra, B.; Lushnikova, T.; Narayana, J.L.; Wang, G. Amino Acid Composition Determines Peptide Activity Spectrum and Hot-Spot-Based Design of Merecidin. *Adv. Biosyst.* **2018**, *2*, 1700259. [CrossRef] [PubMed]
103. Aguilera-Puga, M.D.C.; Plisson, F. Structure-Aware Machine Learning Strategies for Antimicrobial Peptide Discovery. *Sci. Rep.* **2024**, *14*, 11995. [CrossRef]
104. Ma, J.; Yan, L.; Yang, J.; He, Y.; Wu, L. Effect of Modification Strategies on the Biological Activity of Peptides/Proteins. *ChemBiochem* **2024**, *25*, e202300481. [CrossRef] [PubMed]
105. Wang, J.; Eck, M.J. Chapter 6—Protein Architecture: Relationship of Form and Function. In *Hematology*, Hoffman, R., Benz, E.J., Silberstein, L.E., Heslop, H.E., Weitz, J.I., Anastasi, J., Salama, M.E., Abutalib, S.A., Eds.; 7th ed.; Elsevier: Amsterdam, The Netherlands, 2018; pp. 59–67, ISBN 978-0-323-35762-3.
106. Apostolopoulos, V.; Bojarska, J.; Chai, T.-T.; Elnagdy, S.; Kaczmarek, K.; Matsoukas, J.; New, R.; Parang, K.; Lopez, O.P.; Parhiz, H.; et al. A Global Review on Short Peptides: Frontiers and Perspectives. *Molecules* **2021**, *26*, 430. [CrossRef]
107. Davenport, A.P.; Scully, C.C.G.; de Graaf, C.; Brown, A.J.H.; Maguire, J.J. Advances in Therapeutic Peptides Targeting G Protein-Coupled Receptors. *Nat. Rev. Drug Discov.* **2020**, *19*, 389–413. [CrossRef]
108. Saw, P.E.; Xu, X.; Kim, S.; Jon, S. Biomedical Applications of a Novel Class of High-Affinity Peptides. *Acc. Chem. Res.* **2021**, *54*, 3576–3592. [CrossRef]
109. Ohashi, Y.; Onuma, R.; Naganuma, T.; Ogawa, T.; Naude, R.; Nokihara, K.; Muramoto, K. Antioxidant Properties of Tripeptides Revealed by a Comparison of Six Different Assays. *Food Sci. Technol. Res.* **2015**, *21*, 695–704. [CrossRef]
110. Saito, K.; Jin, D.-H.; Ogawa, T.; Muramoto, K.; Hatakeyama, E.; Yasuhara, T.; Nokihara, K. Antioxidative Properties of Tripeptide Libraries Prepared by the Combinatorial Chemistry. *J. Agric. Food Chem.* **2003**, *51*, 3668–3674. [CrossRef]
111. Lu, J.; Xu, H.; Xia, J.; Ma, J.; Xu, J.; Li, Y.; Feng, J. D- and Unnatural Amino Acid Substituted Antimicrobial Peptides With Improved Proteolytic Resistance and Their Proteolytic Degradation Characteristics. *Front. Microbiol.* **2020**, *11*, 563030. [CrossRef]
112. Chen, H.-M.; Muramoto, K.; Yamauchi, F. Structural Analysis of Antioxidative Peptides from Soybean. *Beta.-Conglycinin. J. Agric. Food Chem.* **1995**, *43*, 574–578. [CrossRef]
113. Chan, L.Y.; Zhang, V.M.; Huang, Y.; Waters, N.C.; Bansal, P.S.; Craik, D.J.; Daly, N.L. Cyclization of the Antimicrobial Peptide Gomesin with Native Chemical Ligation: Influences on Stability and Bioactivity. *ChemBiochem* **2013**, *14*, 617–624. [CrossRef]
114. Costa, L.; Sousa, E.; Fernandes, C. Cyclic Peptides in Pipeline: What Future for These Great Molecules? *Pharmaceuticals* **2023**, *16*, 996. [CrossRef]
115. Ji, X.; Nielsen, A.L.; Heinis, C. Cyclic Peptides for Drug Development. *Angew. Chem. Int. Ed.* **2024**, *63*, e202308251. [CrossRef]
116. Joo, S.H. Cyclic Peptides as Therapeutic Agents and Biochemical Tools. *Biomol. Ther.* **2012**, *20*, 19–26. [CrossRef]
117. Zhang, R.-Y.; Thapa, P.; Espiritu, M.J.; Menon, V.; Bingham, J.-P. From Nature to Creation: Going around in Circles, the Art of Peptide Cyclization. *Bioorg. Med. Chem.* **2018**, *26*, 1135–1150. [CrossRef] [PubMed]
118. Bozovičar, K.; Bratkovič, T. Small and Simple, yet Sturdy: Conformationally Constrained Peptides with Remarkable Properties. *Int. J. Mol. Sci.* **2021**, *22*, 1611. [CrossRef]
119. Huang, H.; Damjanovic, J.; Miao, J.; Lin, Y.-S. Cyclic Peptides: Backbone Rigidification and Capability of Mimicking Motifs at Protein–Protein Interfaces. *Phys. Chem. Chem. Phys.* **2021**, *23*, 607–616. [CrossRef]
120. Dougherty, P.G.; Sahni, A.; Pei, D. Understanding Cell Penetration of Cyclic Peptides. *Chem. Rev.* **2019**, *119*, 10241–10287. [CrossRef]
121. Kalafatovic, D.; Giralt, E. Cell-Penetrating Peptides: Design Strategies beyond Primary Structure and Amphipathicity. *Molecules* **2017**, *22*, 1929. [CrossRef] [PubMed]
122. Mohammed, E.H.M.; Lohan, S.; Ghaffari, T.; Gupta, S.; Tiwari, R.K.; Parang, K. Membrane-Active Cyclic Amphiphilic Peptides: Broad-Spectrum Antibacterial Activity Alone and in Combination with Antibiotics. *J. Med. Chem.* **2022**, *65*, 15819–15839. [CrossRef] [PubMed]
123. Abdalla, M.A.; McGaw, L.J. Natural Cyclic Peptides as an Attractive Modality for Therapeutics: A Mini Review. *Molecules* **2018**, *23*, 2080. [CrossRef] [PubMed]
124. Singh, R.K.; Singh, P.; Sharma, R.; Mohapatra, T.M. Nonribosomal Peptide Synthesis in Microbes. *Recent Adv. Microbiol.* **2012**, *1*, 183.
125. Goto, Y.; Suga, H. The RaPID Platform for the Discovery of Pseudo-Natural Macrocyclic Peptides. *Acc. Chem. Res.* **2021**, *54*, 3604–3617. [CrossRef]

126. Saadi, S.; Ghazali, H.M.; Saari, N.; Abdulkarim, S.M. The Structural Reconformation of Peptides in Enhancing Functional and Therapeutic Properties: Insights into Their Solid State Crystallizations. *Biophys. Chem.* **2021**, *273*, 106565. [CrossRef]
127. London, N.; Movshovitz-Attias, D.; Schueler-Furman, O. The Structural Basis of Peptide-Protein Binding Strategies. *Structure* **2010**, *18*, 188–199. [CrossRef]
128. Shi, S.M.; Di, L. Strategies to Optimize Peptide Stability and Prolong Half-Life. In *Peptide Therapeutics: Fundamentals of Design, Development, and Delivery*; Jois, S.D., Ed.; Springer International Publishing: Cham, Switzerland, 2022; pp. 163–182, ISBN 978-3-031-04544-8.
129. Bellavita, R.; Braccia, S.; Galdiero, S.; Falanga, A. Glycosylation and Lipidation Strategies: Approaches for Improving Antimicrobial Peptide Efficacy. *Pharmaceuticals* **2023**, *16*, 439. [CrossRef]
130. Ma, Y.; Lee, C.-J.; Park, J.-S. Strategies for Optimizing the Production of Proteins and Peptides with Multiple Disulfide Bonds. *Antibiotics* **2020**, *9*, 541. [CrossRef]
131. SOLÁ, R.J.; GRIEBENOW, K. Effects of Glycosylation on the Stability of Protein Pharmaceuticals. *J. Pharm. Sci.* **2009**, *98*, 1223–1245. [CrossRef]
132. Zhang, L.; Chou, C.; Moo-Young, M. Disulfide Bond Formation and Its Impact on the Biological Activity and Stability of Recombinant Therapeutic Proteins Produced by Escherichia Coli Expression System. *Biotechnol. Adv.* **2011**, *29*, 923–929. [CrossRef]
133. Lau, J.; Bloch, P.; Schäffer, L.; Pettersson, I.; Spetzler, J.; Kofoed, J.; Madsen, K.; Knudsen, L.B.; McGuire, J.; Steensgaard, D.B.; et al. Discovery of the Once-Weekly Glucagon-Like Peptide-1 (GLP-1) Analogue Semaglutide. *J. Med. Chem.* **2015**, *58*, 7370–7380. [CrossRef]
134. Zheng, X.; Zhu, L.; Li, T.; Xu, W.; Liu, D.; Sheng, J.; Cao, H.; Shi, Y.; Wang, F. Improve Stability of Bioactive Peptides by Enzymatic Modular Synthesis of Peptides with *O*-Linked Sialyl Lewis x. *ACS Catal.* **2021**, *11*, 8042–8048. [CrossRef]
135. Gonzalez, N.A.; Li, B.A.; McCully, M.E. The Stability and Dynamics of Computationally Designed Proteins. *Protein Eng. Des. Sel.* **2022**, *35*, gzac001. [CrossRef]
136. Wang, F.; Sangfuang, N.; McCoubrey, L.E.; Yadav, V.; Elbadawi, M.; Orlu, M.; Gaisford, S.; Basit, A.W. Advancing Oral Delivery of Biologics: Machine Learning Predicts Peptide Stability in the Gastrointestinal Tract. *Int. J. Pharm.* **2023**, *634*, 122643. [CrossRef]
137. Nie, C.; Zou, Y.; Liao, S.; Gao, Q.; Li, Q. Peptides as Carriers of Active Ingredients: A Review. *Curr. Res. Food Sci.* **2023**, *7*, 100592. [CrossRef]
138. Rahban, M.; Ahmad, F.; Piatyszek, M.A.; Haertlé, T.; Saso, L.; Saboury, A.A. Stabilization Challenges and Aggregation in Protein-Based Therapeutics in the Pharmaceutical Industry. *RSC Adv.* **2023**, *13*, 35947–35963. [CrossRef]
139. Rajan, R.; Ahmed, S.; Sharma, N.; Kumar, N.; Debas, A.; Matsumura, K. Review of the Current State of Protein Aggregation Inhibition from a Materials Chemistry Perspective: Special Focus on Polymeric Materials. *Mater. Adv.* **2021**, *2*, 1139–1176. [CrossRef]
140. Zapadka, K.L.; Becher, F.J.; Gomes dos Santos, A.L.; Jackson, S.E. Factors Affecting the Physical Stability (Aggregation) of Peptide Therapeutics. *Interface Focus* **2017**, *7*, 20170030. [CrossRef] [PubMed]
141. Allen, S.G.; Meade, R.M.; White Stenner, L.L.; Mason, J.M. Peptide-Based Approaches to Directly Target Alpha-Synuclein in Parkinson's Disease. *Mol. Neurodegener.* **2023**, *18*, 80. [CrossRef] [PubMed]
142. Soto, C.; Pritzkow, S. Protein Misfolding, Aggregation, and Conformational Strains in Neurodegenerative Diseases. *Nat. Neurosci.* **2018**, *21*, 1332–1340. [CrossRef] [PubMed]
143. Sweeney, P.; Park, H.; Baumann, M.; Dunlop, J.; Frydman, J.; Kopito, R.; McCampbell, A.; Leblanc, G.; Venkateswaran, A.; Nurmi, A.; et al. Protein Misfolding in Neurodegenerative Diseases: Implications and Strategies. *Transl. Neurodegener.* **2017**, *6*, 6. [CrossRef] [PubMed]
144. Díaz-Villanueva, J.F.; Díaz-Molina, R.; García-González, V. Protein Folding and Mechanisms of Proteostasis. *Int. J. Mol Sci.* **2015**, *16*, 17193–17230. [CrossRef]
145. Khan, A.N.; Khan, R.H. Protein Misfolding and Related Human Diseases: A Comprehensive Review of Toxicity, Proteins Involved, and Current Therapeutic Strategies. *Int. J. Biol. Macromol.* **2022**, *223*, 143–160. [CrossRef]
146. Ha, J.-H.; Loh, S.N. Protein Conformational Switches: From Nature to Design. *Chemistry* **2012**, *18*, 7984–7999. [CrossRef]
147. Shea, J.-E.; Friedel, M.R.; Baumketner, A. Simulations of Protein Folding. In *Reviews in Computational Chemistry*; John Wiley & Sons, Ltd.: Hoboken, NJ, USA, 2006; pp. 169–228, ISBN 978-0-471-78036-6.
148. Zhang, J.; Li, W.; Wang, J.; Qin, M.; Wu, L.; Yan, Z.; Xu, W.; Zuo, G.; Wang, W. Protein Folding Simulations: From Coarse-Grained Model to All-Atom Model. *IUBMB Life* **2009**, *61*, 627–643. [CrossRef]
149. Jahn, T.R.; Radford, S.E. Folding versus Aggregation: Polypeptide Conformations on Competing Pathways. *Arch. Biochem. Biophys.* **2008**, *469*, 100–117. [CrossRef] [PubMed]
150. Galano-Frutos, J.J.; Nerín-Fonz, F.; Sancho, J. Calculation of Protein Folding Thermodynamics Using Molecular Dynamics Simulations. *J. Chem. Inf. Model.* **2023**, *63*, 7791–7806. [CrossRef]

151. Seibert, M.M.; Patriksson, A.; Hess, B.; van der Spoel, D. Reproducible Polypeptide Folding and Structure Prediction Using Molecular Dynamics Simulations. *J. Mol. Biol.* **2005**, *354*, 173–183. [CrossRef]
152. Díaz-Gómez, J.L.; Neundorf, I.; López-Castillo, L.-M.; Castorena-Torres, F.; Serna-Saldívar, S.O.; García-Lara, S. In Silico Analysis and In Vitro Characterization of the Bioactive Profile of Three Novel Peptides Identified from 19 kDa α-Zein Sequences of Maize. *Molecules* **2020**, *25*, 5405. [CrossRef] [PubMed]
153. Prakash Nirmal, N.; Singh Rajput, M.; Bhojraj Rathod, N.; Mudgil, P.; Pati, S.; Bono, G.; Nalinanon, S.; Li, L.; Maqsood, S. Structural Characteristic and Molecular Docking Simulation of Fish Protein-Derived Peptides: Recent Updates on Antioxidant, Anti-Hypertensive and Anti-Diabetic Peptides. *Food Chem.* **2023**, *405*, 134737. [CrossRef]
154. Luca, S.; Heise, H.; Baldus, M. High-Resolution Solid-State NMR Applied to Polypeptides and Membrane Proteins. *Acc. Chem. Res.* **2003**, *36*, 858–865. [CrossRef] [PubMed]
155. Rienstra, C.M.; Tucker-Kellogg, L.; Jaroniec, C.P.; Hohwy, M.; Reif, B.; McMahon, M.T.; Tidor, B.; Lozano-Pérez, T.; Griffin, R.G. De Novo Determination of Peptide Structure with Solid-State Magic-Angle Spinning NMR Spectroscopy. *Proc. Natl. Acad. Sci. USA* **2002**, *99*, 10260–10265. [CrossRef]
156. Schroeder, C.I.; Rosengren, K.J. Three-Dimensional Structure Determination of Peptides Using Solution Nuclear Magnetic Resonance Spectroscopy. In *Snake and Spider Toxins: Methods and Protocols*; Priel, A., Ed.; Springer: New York, NY, USA, 2020; pp. 129–162, ISBN 978-1-4939-9845-6.
157. Andrushchenko, V.V.; Vogel, H.J.; Prenner, E.J. Solvent-Dependent Structure of Two Tryptophan-Rich Antimicrobial Peptides and Their Analogs Studied by FTIR and CD Spectroscopy. *Biochim. Biophys. Acta (BBA) Biomembr.* **2006**, *1758*, 1596–1608. [CrossRef] [PubMed]
158. Bagińska, K.; Makowska, J.; Wiczk, W.; Kasprzykowski, F.; ChmurzyńSKI, L. Conformational Studies of Alanine-Rich Peptide Using CD and FTIR Spectroscopy. *J. Pept. Sci.* **2008**, *14*, 283–289. [CrossRef]
159. Allison, J.R. Computational Methods for Exploring Protein Conformations. *Biochem. Soc. Trans* **2020**, *48*, 1707–1724. [CrossRef]
160. Mondal, A.; Chang, L.; Perez, A. Modelling Peptide–Protein Complexes: Docking, Simulations and Machine Learning. *QRB Discov.* **2022**, *3*, e17. [CrossRef]
161. Zhou, P.; Wang, C.; Ren, Y.; Yang, C.; Tian, F. Computational Peptidology: A New and Promising Approach to Therapeutic Peptide Design. *Curr. Med. Chem.* **2013**, *20*, 1985–1996. [CrossRef]
162. Du, Z.; Comer, J.; Li, Y. Bioinformatics Approaches to Discovering Food-Derived Bioactive Peptides: Reviews and Perspectives. *TrAC Trends Anal. Chem.* **2023**, *162*, 117051. [CrossRef]
163. Geng, H.; Chen, F.; Ye, J.; Jiang, F. Applications of Molecular Dynamics Simulation in Structure Prediction of Peptides and Proteins. *Comput. Struct. Biotechnol. J.* **2019**, *17*, 1162–1170. [CrossRef]
164. Jalily Hasani, H.; Barakat, K. Homology Modeling: An Overview of Fundamentals and Tools. *Int. Rev. Model. Simul. (IREMOS)* **2017**, *10*, 129. [CrossRef]
165. Muhammed, M.T.; Aki-Yalcin, E. Homology Modeling in Drug Discovery: Overview, Current Applications, and Future Perspectives. *Chem Biol. Drug Des.* **2019**, *93*, 12–20. [CrossRef] [PubMed]
166. John, B.; Sali, A. Comparative Protein Structure Modeling by Iterative Alignment, Model Building and Model Assessment. *Nucleic Acids Res.* **2003**, *31*, 3982–3992. [CrossRef]
167. MacCallum, J.L.; Pérez, A.; Schnieders, M.J.; Hua, L.; Jacobson, M.P.; Dill, K.A. Assessment of Protein Structure Refinement in CASP9. *Proteins* **2011**, *79* (Suppl. S10), 74–90. [CrossRef]
168. Schmidt, T.; Bergner, A.; Schwede, T. Modelling Three-Dimensional Protein StructuRes. for Applications in Drug Design. *Drug Discov. Today* **2014**, *19*, 890–897. [CrossRef]
169. Majumder, P. Computational Methods Used in Prediction of Protein Structure. In *Statistical Modelling and Machine Learning Principles for Bioinformatics Techniques, Tools, and Applications*; Srinivasa, K.G., Siddesh, G.M., Manisekhar, S.R., Eds.; Springer: Singapore, 2020; pp. 119–133, ISBN 9789811524455.
170. Fiser, A. Template-Based Protein Structure Modeling. In *Computational Biology*; Fenyö, D., Ed.; Humana Press: Totowa, NJ, USA, 2010; pp. 73–94, ISBN 978-1-60761-842-3.
171. Xiang, Z. Advances in Homology Protein Structure Modeling. *Curr. Protein. Pept. Sci.* **2006**, *7*, 217–227. [PubMed]
172. Floudas, C.A.; Fung, H.K.; McAllister, S.R.; Mönnigmann, M.; Rajgaria, R. Advances in Protein Structure Prediction and de Novo Protein Design: A Review. *Chem. Eng. Sci.* **2006**, *61*, 966–988. [CrossRef]
173. Tiwari, A.; Chauhan, R.P.; Agarwal, A.; Ramteke, P.W. Molecular Modeling of Proteins: Methods, Recent Advances, and Future Prospects. In *Computer-Aided Drug Design*; Singh, D.B., Ed.; Springer: Singapore, 2020; pp. 23–47, ISBN 9789811568152.
174. Abdou, Y.T.; Saleeb, S.M.; Abdel-Raouf, K.M.A.; Allam, M.; Adel, M.; Amleh, A. Characterization of a Novel Peptide Mined from the Red Sea Brine Pools and Modified to Enhance Its Anticancer Activity. *BMC Cancer* **2023**, *23*, 699. [CrossRef]
175. Roy, A.; Kucukural, A.; Zhang, Y. I-TASSER: A Unified Platform for Automated Protein Structure and Function Prediction. *Nat. Protoc.* **2010**, *5*, 725–738. [CrossRef]

176. Durme, J.V.; Maurer-Stroh, S.; Gallardo, R.; Wilkinson, H.; Rousseau, F.; Schymkowitz, J. Accurate Prediction of DnaK-Peptide Binding via Homology Modelling and Experimental Data. *PLoS Comput. Biol.* **2009**, *5*, e1000475. [CrossRef]
177. Chakraborty, S.; Bhattacharya, S.; Ghosh, S.; Bera, A.K.; Haldar, U.; Pal, A.K.; Mukhopadhyay, B.P.; Banerjee, A. Structural and Interactional Homology of Clinically Potential Trypsin Inhibitors: Molecular Modelling of Cucurbitaceae Family Peptides Using the X-Ray Structure of MCTI-II. *Protein Eng. Des. Sel.* **2000**, *13*, 551–555. [CrossRef]
178. Ostberg, N.; Kaznessis, Y. Protegrin Structure–Activity Relationships: Using Homology Models of Synthetic Sequences to Determine Structural Characteristics Important for Activity. *Peptides* **2005**, *26*, 197–206. [CrossRef]
179. Rodrigues, J.P.G.L.M.; Melquiond, A.S.J.; Karaca, E.; Trellet, M.; van Dijk, M.; van Zundert, G.C.P.; Schmitz, C.; de Vries, S.J.; Bordogna, A.; Bonati, L.; et al. Defining the Limits of Homology Modeling in Information-Driven Protein Docking. *Proteins Struct. Funct. Bioinform.* **2013**, *81*, 2119–2128. [CrossRef]
180. Monzon, A.M.; Zea, D.J.; Marino-Buslje, C.; Parisi, G. Homology Modeling in a Dynamical World. *Protein Sci.* **2017**, *26*, 2195–2206. [CrossRef] [PubMed]
181. Haddad, Y.; Adam, V.; Heger, Z. Ten Quick Tips for Homology Modeling of High-Resolution Protein 3D Structures. *PLoS Comput. Biol.* **2020**, *16*, e1007449. [CrossRef]
182. Stanzione, F.; Giangreco, I.; Cole, J.C. Chapter Four—Use of Molecular Docking Computational Tools in Drug Discovery. In *Progress in Medicinal Chemistry*; Witty, D.R., Cox, B., Eds.; Elsevier: Amsterdam, The Netherlands, 2021; Volume 60, pp. 273–343.
183. Andreani, J.; Quignot, C.; Guerois, R. Structural Prediction of Protein Interactions and Docking Using Conservation and Coevolution. *WIRes. Comput. Mol. Sci.* **2020**, *10*, e1470. [CrossRef]
184. Hetényi, C.; van der Spoel, D. Efficient Docking of Peptides to Proteins without Prior Knowledge of the Binding Site. *Protein Sci.* **2002**, *11*, 1729–1737. [CrossRef]
185. Singh, S.; Bani Baker, Q.; Singh, D.B. Chapter 18—Molecular Docking and Molecular Dynamics Simulation. In *Bioinformatics*; Singh, D.B., Pathak, R.K., Eds.; Academic Press: Cambridge, MA, USA, 2022; pp. 291–304, ISBN 978-0-323-89775-4.
186. Morris, G.M.; Lim-Wilby, M. Molecular Docking. In *Molecular Modeling of Proteins*; Kukol, A., Ed.; Humana Press: Totowa, NJ, USA, 2008; pp. 365–382, ISBN 978-1-59745-177-2.
187. Pantsar, T.; Poso, A. Binding Affinity via Docking: Fact and Fiction. *Molecules* **2018**, *23*, 1899. [CrossRef]
188. Agu, P.C.; Afiukwa, C.A.; Orji, O.U.; Ezeh, E.M.; Ofoke, I.H.; Ogbu, C.O.; Ugwuja, E.I.; Aja, P.M. Molecular Docking as a Tool for the Discovery of Molecular Targets of Nutraceuticals in Diseases Management. *Sci. Rep.* **2023**, *13*, 13398. [CrossRef]
189. Meng, X.-Y.; Zhang, H.-X.; Mezei, M.; Cui, M. Molecular Docking: A Powerful Approach for Structure-Based Drug Discovery. *Curr. Comput. Aided Drug Des.* **2011**, *7*, 146–157.
190. Paggi, J.M.; Pandit, A.; Dror, R.O. The Art and Science of Molecular Docking. *Annu. Rev. Biochem.* **2024**, *93*, 389–410. [CrossRef]
191. Atkovska, K.; Samsonov, S.A.; Paszkowski-Rogacz, M.; Pisabarro, M.T. Multipose Binding in Molecular Docking. *Int. J. Mol. Sci.* **2014**, *15*, 2622–2645. [CrossRef] [PubMed]
192. Kundrotas, P.J.; Zhu, Z.; Janin, J.; Vakser, I.A. Templates Are Available to Model Nearly All Complexes of Structurally Characterized Proteins. *Proc. Natl. Acad. Sci. USA* **2012**, *109*, 9438–9441. [CrossRef]
193. Lensink, M.F.; Velankar, S.; Wodak, S.J. Modeling Protein-Protein and Protein-Peptide Complexes: CAPRI 6th Edition. *Proteins* **2017**, *85*, 359–377. [CrossRef]
194. Ciemny, M.; Kurcinski, M.; Kamel, K.; Kolinski, A.; Alam, N.; Schueler-Furman, O.; Kmiecik, S. Protein–Peptide Docking: Opportunities and Challenges. *Drug Discov. Today* **2018**, *23*, 1530–1537. [CrossRef]
195. Agrawal, P.; Singh, H.; Srivastava, H.K.; Singh, S.; Kishore, G.; Raghava, G.P.S. Benchmarking of Different Molecular Docking Methods for Protein-Peptide Docking. *BMC Bioinform.* **2019**, *19*, 426. [CrossRef]
196. Corbeil, C.R.; Williams, C.I.; Labute, P. Variability in Docking Success Rates Due to Dataset Preparation. *J. Comput. Aided Mol. Des.* **2012**, *26*, 775–786. [CrossRef]
197. Pagadala, N.S.; Syed, K.; Tuszynski, J. Software for Molecular Docking: A Review. *Biophys. Rev.* **2017**, *9*, 91–102. [CrossRef]
198. Trott, O.; Olson, A.J. AutoDock Vina: Improving the Speed and Accuracy of Docking with a New Scoring Function, Efficient Optimization, and Multithreading. *J. Comput. Chem.* **2010**, *31*, 455–461. [CrossRef] [PubMed]
199. Kellenberger, E.; Rodrigo, J.; Muller, P.; Rognan, D. Comparative Evaluation of Eight Docking Tools for Docking and Virtual Screening Accuracy. *Proteins Struct. Funct. Bioinform.* **2004**, *57*, 225–242. [CrossRef] [PubMed]
200. Weng, G.; Gao, J.; Wang, Z.; Wang, E.; Hu, X.; Yao, X.; Cao, D.; Hou, T. Comprehensive Evaluation of Fourteen Docking Programs on Protein–Peptide Complexes. *J. Chem. Theory Comput.* **2020**, *16*, 3959–3969. [CrossRef] [PubMed]
201. Zhao, H.; Jiang, D.; Shen, C.; Zhang, J.; Zhang, X.; Wang, X.; Nie, D.; Hou, T.; Kang, Y. Comprehensive Evaluation of 10 Docking Programs on a Diverse Set of Protein–Cyclic Peptide Complexes. *J. Chem. Inf. Model.* **2024**, *64*, 2112–2124. [CrossRef] [PubMed]
202. Gao, J.; Li, T.; Chen, D.; Gu, H.; Mao, X. Identification and Molecular Docking of Antioxidant Peptides from Hemp Seed Protein Hydrolysates. *LWT* **2021**, *147*, 111453. [CrossRef]

203. Wen-Tao, C.H.E.N.; Zhang, Y.-Y.; Qiang, Q.; Zou, P.; Xu, Y.; Sun, C.; Badar, I.H. Characterizations and Molecular Docking Mechanism of the Interactions between Peptide FDGDF (Phe-Asp-Gly-Asp-Phe) and SOD Enzyme. *Heliyon* **2024**, *10*, e24515. [CrossRef] [PubMed]
204. London, N.; Raveh, B.; Schueler-Furman, O. Peptide Docking and Structure-Based Characterization of Peptide Binding: From Knowledge to Know-How. *Curr. Opin. Struct. Biol.* **2013**, *23*, 894–902. [CrossRef]
205. Meyer, K.; Selbach, M. Peptide-Based Interaction Proteomics. *Mol. Cell Proteom.* **2020**, *19*, 1070–1075. [CrossRef] [PubMed]
206. Petsalaki, E.; Russell, R.B. Peptide-Mediated Interactions in Biological Systems: New Discoveries and Applications. *Curr. Opin Biotechnol.* **2008**, *19*, 344–350. [CrossRef]
207. Raghavender, U.S.; Rathore, R.S. Protein-Peptide Interactions in Regulatory Events. In *Encyclopedia of Bioinformatics and Computational Biology*; Ranganathan, S., Gribskov, M., Nakai, K., Schönbach, C., Eds.; Academic Press: Oxford, UK, 2019; pp. 688–696, ISBN 978-0-12-811432-2.
208. Mudgil, P.; Al Dhaheri, M.K.O.; Alsubousi, M.S.M.; Khan, H.; Redha, A.A.; Yap, P.-G.; Gan, C.-Y.; Maqsood, S. Molecular Docking Studies on α-Amylase Inhibitory Peptides from Milk of Different Farm Animals. *J. Dairy Sci.* **2024**, *107*, 2633–2652. [CrossRef]
209. Auwal, S.M.; Abidin, N.Z.; Zarei, M.; Tan, C.P.; Saari, N. Identification, Structure-Activity Relationship and in Silico Molecular Docking Analyses of Five Novel Angiotensin I-Converting Enzyme (ACE)-Inhibitory Peptides from Stone Fish (*Actinopyga* Lecanora) Hydrolysates. *PLoS ONE* **2019**, *14*, e0197644. [CrossRef]
210. Zhou, P.; Jin, B.; Li, H.; Huang, S.-Y. HPEPDOCK: A Web Server for Blind Peptide–Protein Docking Based on a Hierarchical Algorithm. *Nucleic Acids Res.* **2018**, *46*, W443–W450. [CrossRef]
211. Kurcinski, M.; Badaczewska-Dawid, A.; Kolinski, M.; Kolinski, A.; Kmiecik, S. Flexible Docking of Peptides to Proteins Using CABS-dock. *Protein Sci.* **2020**, *29*, 211–222. [CrossRef] [PubMed]
212. Zhang, Y.; Sanner, M.F. AutoDock CrankPep: Combining Folding and Docking to Predict Protein–Peptide Complexes. *Bioinformatics* **2019**, *35*, 5121–5127. [CrossRef]
213. Lee, H.; Heo, L.; Lee, M.S.; Seok, C. GalaxyPepDock: A Protein–Peptide Docking Tool Based on Interaction Similarity and Energy Optimization. *Nucleic Acids Res.* **2015**, *43*, W431–W435. [CrossRef] [PubMed]
214. London, N.; Raveh, B.; Cohen, E.; Fathi, G.; Schueler-Furman, O. Rosetta FlexPepDock Web Server—High Resolution Modeling of Peptide–Protein Interactions. *Nucleic Acids Res.* **2011**, *39*, W249–W253. [CrossRef]
215. Donsky, E.; Wolfson, H.J. PepCrawler: A Fast RRT-Based Algorithm for High-Resolution Refinement and Binding Affinity Estimation of Peptide Inhibitors. *Bioinformatics* **2011**, *27*, 2836–2842. [CrossRef]
216. Dominguez, C.; Boelens, R.; Bonvin, A.M.J.J. HADDOCK: A Protein-Protein Docking Approach Based on Biochemical or Biophysical Information. *J. Am. Chem. Soc.* **2003**, *125*, 1731–1737. [CrossRef] [PubMed]
217. de Vries, S.J.; Rey, J.; Schindler, C.E.M.; Zacharias, M.; Tuffery, P. The pepATTRACT Web Server for Blind, Large-Scale Peptide–Protein Docking. *Nucleic Acids Res.* **2017**, *45*, W361–W364. [CrossRef]
218. Alam, N.; Goldstein, O.; Xia, B.; Porter, K.A.; Kozakov, D.; Schueler-Furman, O. High-Resolution Global Peptide-Protein Docking Using Fragments-Based PIPER-FlexPepDock. *PLoS Comput. Biol.* **2017**, *13*, e1005905. [CrossRef]
219. Porter, K.A.; Xia, B.; Beglov, D.; Bohnuud, T.; Alam, N.; Schueler-Furman, O.; Kozakov, D. ClusPro PeptiDock: Efficient Global Docking of Peptide Recognition Motifs Using FFT. *Bioinformatics* **2017**, *33*, 3299–3301. [CrossRef]
220. Antes, I. DynaDock: A New Molecular Dynamics-Based Algorithm for Protein-Peptide Docking Including Receptor Flexibility. *Proteins* **2010**, *78*, 1084–1104. [CrossRef]
221. Ben-Shimon, A.; Niv, M.Y. AnchorDock: Blind and Flexible Anchor-Driven Peptide Docking. *Structure* **2015**, *23*, 929–940. [CrossRef]
222. Xu, X.; Yan, C.; Zou, X. MDockPeP: An Ab-Initio Protein-Peptide Docking Server. *J. Comput. Chem.* **2018**, *39*, 2409–2413. [CrossRef]
223. Lamiable, A.; Thévenet, P.; Rey, J.; Vavrusa, M.; Derreumaux, P.; Tufféry, P. PEP-FOLD3: Faster de Novo Structure Prediction for Linear Peptides in Solution and in Complex. *Nucleic Acids Res.* **2016**, *44*, W449–W454. [CrossRef] [PubMed]
224. Antunes, D.A.; Moll, M.; Devaurs, D.; Jackson, K.R.; Lizée, G.; Kavraki, L.E. DINC 2.0: A New Protein-Peptide Docking Webserver Using an Incremental Approach. *Cancer Res.* **2017**, *77*, e55–e57. [CrossRef] [PubMed]
225. Jones, G.; Willett, P.; Glen, R.C.; Leach, A.R.; Taylor, R. Development and Validation of a Genetic Algorithm for Flexible Docking1. *J. Mol. Biol.* **1997**, *267*, 727–748. [CrossRef] [PubMed]
226. Raveh, B.; London, N.; Zimmerman, L.; Schueler-Furman, O. Rosetta FlexPepDock Ab-Initio: Simultaneous Folding, Docking and Refinement of Peptides onto Their Receptors. *PLoS ONE* **2011**, *6*, e18934. [CrossRef]
227. Li, H.; Lu, L.; Chen, R.; Quan, L.; Xia, X.; Lü, Q. PaFlexPepDock: Parallel Ab-Initio Docking of Peptides onto Their Receptors with Full Flexibility Based on Rosetta. *PLoS ONE* **2014**, *9*, e94769. [CrossRef]
228. Obarska-Kosinska, A.; Iacoangeli, A.; Lepore, R.; Tramontano, A. PepComposer: Computational Design of Peptides Binding to a given Protein Surface. *Nucleic. Acids Res.* **2016**, *44*, W522. [CrossRef]
229. London, N.; Raveh, B.; Schueler-Furman, O. Modeling Peptide-Protein Interactions. *Methods Mol. Biol.* **2012**, *857*, 375–398. [CrossRef]

230. Khramushin, A.; Marcu, O.; Alam, N.; Shimony, O.; Padhorny, D.; Brini, E.; Dill, K.A.; Vajda, S.; Kozakov, D.; Schueler-Furman, O. Modeling Beta-Sheet Peptide-Protein Interactions: Rosetta FlexPepDock in CAPRI Rounds 38-45. *Proteins* **2020**, *88*, 1037–1049. [CrossRef]
231. Tuffery, P.; Derreumaux, P. Flexibility and Binding Affinity in Protein–Ligand, Protein–Protein and Multi-Component Protein Interactions: Limitations of Current Computational Approaches. *J. R Soc. Interface* **2012**, *9*, 20–33. [CrossRef] [PubMed]
232. Antunes, D.A.; Devaurs, D.; Kavraki, L.E. Understanding the Challenges of Protein Flexibility in Drug Design. *Expert Opin. Drug Discov.* **2015**, 10. [CrossRef]
233. Buonfiglio, R.; Recanatini, M.; Masetti, M. Protein Flexibility in Drug Discovery: From Theory to Computation. *Chem. Med. Chem* **2015**, *10*, 1141–1148. [CrossRef]
234. Maus, H.; Hinze, G.; Hammerschmidt, S.J.; Schirmeister, T.; Basché, T. Conformational Selection and Induced Fit: The Behavior of Two Homologous Proteases. *ChemistryEurope* **2023**, *1*, e202300060. [CrossRef]
235. Belorkar, S.A.; Jogaiah, S. Chapter 1—Enzymes—Past, Present, and Future. In *Protocols and Applications in Enzymology*; Belorkar, S.A., Jogaiah, S., Eds.; Academic Press: Cambridge, MA, USA, 2022; pp. 1–15, ISBN 978-0-323-91268-6.
236. Koshland, D.E.; Némethy, G.; Filmer, D. Comparison of Experimental Binding Data and Theoretical Models in Proteins Containing Subunits. *Biochemistry* **1966**, *5*, 365–385. [CrossRef] [PubMed]
237. Miller, E.B.; Murphy, R.B.; Sindhikara, D.; Borrelli, K.W.; Grisewood, M.J.; Ranalli, F.; Dixon, S.L.; Jerome, S.; Boyles, N.A.; Day, T.; et al. Reliable and Accurate Solution to the Induced Fit Docking Problem for Protein–Ligand Binding. *J. Chem. Theory Comput.* **2021**, *17*, 2630–2639. [CrossRef]
238. Kuzu, G.; Keskin, O.; Gursoy, A.; Nussinov, R. Expanding the Conformational Selection Paradigm in Protein-Ligand Docking. *Methods Mol. Biol.* **2012**, *819*, 59–74. [CrossRef] [PubMed]
239. Galburt, E.A.; Tomko, E.J. Conformational Selection and Induced Fit as a Useful Framework for Molecular Motor Mechanisms. *Biophys. Chem.* **2017**, *223*, 11. [CrossRef]
240. Morando, M.A.; Saladino, G.; D.'Amelio, N.; Pucheta-Martinez, E.; Lovera, S.; Lelli, M.; López-Méndez, B.; Marenchino, M.; Campos-Olivas, R.; Gervasio, F.L. Conformational Selection and Induced Fit Mechanisms in the Binding of an Anticancer Drug to the C-Src Kinase. *Sci. Rep.* **2016**, *6*, 24439. [CrossRef]
241. Trellet, M.; Melquiond, A.S.J.; Bonvin, A.M.J.J. A Unified Conformational Selection and Induced Fit Approach to Protein-Peptide Docking. *PLoS ONE* **2013**, *8*, e58769. [CrossRef]
242. Chaudhury, S.; Gray, J.J. Conformer Selection and Induced Fit in Flexible Backbone Protein-Protein Docking Using Computational and NMR Ensembles. *J. Mol. Biol.* **2008**, *381*, 1068–1087. [CrossRef]
243. Mamidi, A.S.; Surolia, A. Mixed Mechanism of Conformational Selection and Induced Fit as a Molecular Recognition Process in the Calreticulin Family of Proteins. *PLoS Comput. Biol.* **2022**, *18*, e1010661. [CrossRef] [PubMed]
244. Verdonk, M.L.; Cole, J.C.; Hartshorn, M.J.; Murray, C.W.; Taylor, R.D. Improved Protein-Ligand Docking Using GOLD. *Proteins* **2003**, *52*, 609–623. [CrossRef] [PubMed]
245. Kurcinski, M.; Oleniecki, T.; Ciemny, M.P.; Kuriata, A.; Kolinski, A.; Kmiecik, S. CABS-Flex Standalone: A Simulation Environment for Fast Modeling of Protein Flexibility. *Bioinformatics* **2019**, *35*, 694–695. [CrossRef] [PubMed]
246. Sapundzhi, F.; Prodanova, K.; Lazarova, M. Survey of the Scoring Functions for Protein-Ligand Docking. *AIP Conf. Proc.* **2019**, *2172*, 100008. [CrossRef]
247. Schneider, M.; Rosam, M.; Glaser, M.; Patronov, A.; Shah, H.; Back, K.C.; Daake, M.A.; Buchner, J.; Antes, I. BiPPred: Combined Sequence- and Structure-Based Prediction of Peptide Binding to the Hsp70 Chaperone BiP. *Proteins Struct. Funct. Bioinform.* **2016**, *84*, 1390–1407. [CrossRef]
248. Jiménez-García, B.; Pons, C.; Fernández-Recio, J. pyDockWEB: A Web Server for Rigid-Body Protein–Protein Docking Using Electrostatics and Desolvation Scoring. *Bioinformatics* **2013**, *29*, 1698–1699. [CrossRef]
249. Alam, N.; Schueler-Furman, O. Modeling Peptide-Protein Structure and Binding Using Monte Carlo Sampling Approaches: Rosetta FlexPepDock and FlexPepBind. In *Modeling Peptide-Protein Interactions: Methods and Protocols*; Schueler-Furman, O., London, N., Eds.; Springer: New York, NY, USA, 2017; pp. 139–169, ISBN 978-1-4939-6798-8.
250. Ciemny, M.P.; Debinski, A.; Paczkowska, M.; Kolinski, A.; Kurcinski, M.; Kmiecik, S. Protein-Peptide Molecular Docking with Large-Scale Conformational Changes: The P53-MDM2 Interaction. *Sci. Rep.* **2016**, *6*, 37532. [CrossRef]
251. Kozakov, D.; Hall, D.R.; Xia, B.; Porter, K.A.; Padhorny, D.; Yueh, C.; Beglov, D.; Vajda, S. The ClusPro Web Server for Protein–Protein Docking. *Nat. Protoc.* **2017**, *12*, 255–278. [CrossRef]
252. Yu, J.; Andreani, J.; Ochsenbein, F.; Guerois, R. Lessons from (Co-)Evolution in the Docking of Proteins and Peptides for CAPRI Rounds 28–35. *Proteins Struct. Funct. Bioinform.* **2017**, *85*, 378–390. [CrossRef]
253. Huang, S.-Y. Exploring the Potential of Global Protein–Protein Docking: An Overview and Critical Assessment of Current Programs for Automatic Ab Initio Docking. *Drug Discov. Today* **2015**, *20*, 969–977. [CrossRef] [PubMed]
254. Yan, C.; Xu, X.; Zou, X. Fully Blind Docking at the Atomic Level for Protein-Peptide Complex Structure Prediction. *Structure* **2016**, *24*, 1842–1853. [CrossRef] [PubMed]

255. Acuner Ozbabacan, S.E.; Engin, H.B.; Gursoy, A.; Keskin, O. Transient Protein–Protein Interactions. *Protein Eng. Des. Sel.* **2011**, *24*, 635–648. [CrossRef]
256. Lavi, A.; Ngan, C.H.; Movshovitz-Attias, D.; Bohnuud, T.; Yueh, C.; Beglov, D.; Schueler-Furman, O.; Kozakov, D. Detection of Peptide-Binding Sites on Protein Surfaces: The First Step towards the Modeling and Targeting of Peptide-Mediated Interactions. *Proteins* **2013**, *81*, 2096–2105. [CrossRef]
257. Perkins, J.R.; Diboun, I.; Dessailly, B.H.; Lees, J.G.; Orengo, C. Transient Protein-Protein Interactions: Structural, Functional, and Network Properties. *Structure* **2010**, *18*, 1233–1243. [CrossRef]
258. Haslam, N.J.; Shields, D.C. Peptide-Binding Domains: Are Limp Handshakes Safest? *Sci. Signal.* **2012**, *5*, pe40. [CrossRef]
259. Staneva, I.; Wallin, S. Binding Free Energy Landscape of Domain-Peptide Interactions. *PLoS Comput. Biol.* **2011**, *7*, e1002131. [CrossRef] [PubMed]
260. Bottegoni, G.; Rocchia, W.; Cavalli, A. Application of Conformational Clustering in Protein-Ligand Docking. *Methods Mol. Biol.* **2012**, *819*, 169–186. [CrossRef]
261. Kozakov, D.; Clodfelter, K.H.; Vajda, S.; Camacho, C.J. Optimal Clustering for Detecting Near-Native Conformations in Protein Docking. *Biophys. J.* **2005**, *89*, 867–875. [CrossRef]
262. Chermak, E.; Donato, R.D.; Lensink, M.F.; Petta, A.; Serra, L.; Scarano, V.; Cavallo, L.; Oliva, R. Introducing a Clustering Step in a Consensus Approach for the Scoring of Protein-Protein Docking Models. *PLoS ONE* **2016**, *11*, e0166460. [CrossRef]
263. Kurcinski, M.; Pawel Ciemny, M.; Oleniecki, T.; Kuriata, A.; Badaczewska-Dawid, A.E.; Kolinski, A.; Kmiecik, S. CABS-Dock Standalone: A Toolbox for Flexible Protein–Peptide Docking. *Bioinformatics* **2019**, *35*, 4170–4172. [CrossRef]
264. Osguthorpe, D.J.; Sherman, W.; Hagler, A.T. Generation of Receptor Structural Ensembles for Virtual Screening Using Binding Site Shape Analysis and Clustering. *Chem. Biol. Drug Des.* **2012**, *80*, 182–193. [CrossRef] [PubMed]
265. Monti, A.; Vitagliano, L.; Caporale, A.; Ruvo, M.; Doti, N. Targeting Protein–Protein Interfaces with Peptides: The Contribution of Chemical Combinatorial Peptide Library Approaches. *Int. J. Mol. Sci.* **2023**, *24*, 7842. [CrossRef]
266. Klebe, G. Virtual Ligand Screening: Strategies, Perspectives and Limitations. *Drug Discov. Today* **2006**, *11*, 580–594. [CrossRef] [PubMed]
267. Matter, H.; Sotriffer, C. Applications and Success Stories in Virtual Screening. In *Virtual Screening*; John Wiley & Sons, Ltd.: Hoboken, NJ, USA, 2011; pp. 319–358. ISBN 978-3-527-63332-6.
268. Tripathi, N.M.; Bandyopadhyay, A. High Throughput Virtual Screening (HTVS) of Peptide Library: Technological Advancement in Ligand Discovery. *Eur. J. Med. Chem.* **2022**, *243*, 114766. [CrossRef]
269. Bhunia, S.S.; Saxena, M.; Saxena, A.K. Ligand- and Structure-Based Virtual Screening in Drug Discovery. In Biophysical and Computational Tools in Drug Discovery. In *Biophysical and Computational Tools in Drug Discovery*; Saxena, A.K., Ed.; Springer International Publishing: Cham, Switzerland, 2021; pp. 281–339, ISBN 978-3-030-85281-8.
270. Amarasinghe, K.N.; De Maria, L.; Tyrchan, C.; Eriksson, L.A.; Sadowski, J.; Petrović, D. Virtual Screening Expands the Non-Natural Amino Acid Palette for Peptide Optimization. *J. Chem. Inf. Model.* **2022**, *62*, 2999–3007. [CrossRef]
271. Carlsson, J.; Luttens, A. Structure-Based Virtual Screening of Vast Chemical Space as a Starting Point for Drug Discovery. *Curr. Opin. Struct. Biol.* **2024**, *87*, 102829. [CrossRef] [PubMed]
272. Basciu, A.; Athar, M.; Kurt, H.; Neville, C.; Malloci, G.; Muredda, F.C.; Bosin, A.; Ruggerone, P.; Bonvin, A.M.J.J.; Vargiu, A.V. Toward the Prediction of Binding Events in Very Flexible, Allosteric, Multidomain Proteins. *J. Chem. Inf. Model.* **2025**, *65*, 2052–2065. [CrossRef]
273. Maia, E.H.B.; Assis, L.C.; de Oliveira, T.A.; da Silva, A.M.; Taranto, A.G. Structure-Based Virtual Screening: From Classical to Artificial Intelligence. *Front. Chem.* **2020**, *8*, 343. [CrossRef]
274. Zhao, W.; Su, L.; Huo, S.; Yu, Z.; Li, J.; Liu, J. Virtual Screening, Molecular Docking and Identification of Umami Peptides Derived from Oncorhynchus Mykiss. *Food Sci. Hum. Wellness* **2023**, *12*, 89–93. [CrossRef]
275. Xu, X.; Kao, W.-L.; Wang, A.; Lee, H.-J.; Duan, R.; Holmes, H.; Gallazzi, F.; Ji, J.; Sun, H.; Heng, X.; et al. In Silico Screening of Protein-Binding Peptides with an Application to Developing Peptide Inhibitors against Antibiotic Resistance. *PNAS Nexus* **2024**, *3*, pgae541. [CrossRef] [PubMed]
276. Gimeno, A.; Ojeda-Montes, M.J.; Tomás-Hernández, S.; Cereto-Massagué, A.; Beltrán-Debón, R.; Mulero, M.; Pujadas, G.; Garcia-Vallvé, S. The Light and Dark Sides of Virtual Screening: What Is There to Know? *Int. J. Mol. Sci.* **2019**, *20*, 1375. [CrossRef] [PubMed]
277. Acharya, C.; Coop, A.; Polli, J.E.; MacKerell, A.D. Recent Advances in Ligand-Based Drug Design: Relevance and Utility of the Conformationally Sampled Pharmacophore Approach. *Curr. Comput. Aided Drug Des.* **2011**, *7*, 10–22. [CrossRef] [PubMed]
278. Phillips, P.D.; Andersen, T.; McDougal, O.M.; Phillips, P.D.; Andersen, T.; McDougal, O.M. Assessing the Utility and Limitations of High Throughput Virtual Screening. *AIMSMOLES* **2016**, *3*, 238–245. [CrossRef]
279. Krüger, D.M.; Glas, A.; Bier, D.; Pospiech, N.; Wallraven, K.; Dietrich, L.; Ottmann, C.; Koch, O.; Hennig, S.; Grossmann, T.N. Structure-Based Design of Non-Natural Macrocyclic Peptides That Inhibit Protein–Protein Interactions. *J. Med. Chem.* **2017**, *60*, 8982–8988. [CrossRef]

280. Mollica, A.; Zengin, G.; Durdagi, S.; Ekhteiari Salmas, R.; Macedonio, G.; Stefanucci, A.; Dimmito, M.P.; Novellino, E. Combinatorial Peptide Library Screening for Discovery of Diverse α-Glucosidase Inhibitors Using Molecular Dynamics Simulations and Binary QSAR Models. *J. Biomol. Struct. Dyn.* **2019**, *37*, 726–740. [CrossRef]
281. Essegian, D.J.; Cunningham, T.A.; Zerio, C.J.; Chapman, E.; Schatz, J.; Schürer, S.C. Molecular Dynamics Simulations Identify Tractable Lead-like Phenyl-Piperazine Scaffolds as eIF4A1 ATP-Competitive Inhibitors. *ACS Omega* **2021**, *6*, 24432–24443. [CrossRef]
282. Hussain, A.; Shaw, P.E.; Hirst, J.D. Molecular Dynamics Simulations and in Silico Peptide Ligand Screening of the Elk-1 ETS Domain. *J. Cheminformatics* **2011**, *3*, 49. [CrossRef]
283. Stefanucci, A.; Dimmito, M.P.; Zengin, G.; Luisi, G.; Mirzaie, S.; Novellino, E.; Mollica, A. Discovery of Novel Amide Tripeptides as Pancreatic Lipase Inhibitors by Virtual Screening. *New J. Chem.* **2019**, *43*, 3208–3217. [CrossRef]
284. Li, Y.; Gao, X.; Pan, D.; Liu, Z.; Xiao, C.; Xiong, Y.; Du, L.; Cai, Z.; Lu, W.; Dang, Y.; et al. Identification and Virtual Screening of Novel Anti-Inflammatory Peptides from Broccoli Fermented by Lactobacillus Strains. *Front. Nutr.* **2023**, *9*, 1118900. [CrossRef]
285. Duffy, F.J.; Devocelle, M.; Shields, D.C. Computational Approaches to Developing Short Cyclic Peptide Modulators of Protein-Protein Interactions. *Methods Mol. Biol.* **2015**, *1268*, 241–271. [CrossRef] [PubMed]
286. Shtatland, T.; Guettler, D.; Kossodo, M.; Pivovarov, M.; Weissleder, R. PepBank—A Database of Peptides Based on Sequence Text Mining and Public Peptide Data Sources. *BMC Bioinform.* **2007**, *8*, 280. [CrossRef]
287. Das, D.; Jaiswal, M.; Khan, F.N.; Ahamad, S.; Kumar, S. PlantPepDB: A Manually Curated Plant Peptide Database. *Sci. Rep.* **2020**, *10*, 2194. [CrossRef]
288. Quiroz, C.; Saavedra, Y.B.; Armijo-Galdames, B.; Amado-Hinojosa, J.; Olivera-Nappa, Á.; Sanchez-Daza, A.; Medina-Ortiz, D. Peptipedia: A User-Friendly Web Application and a Comprehensive Database for Peptide Research Supported by Machine Learning Approach. *Database* **2021**, *2021*, baab055. [CrossRef]
289. Minkiewicz, P.; Iwaniak, A.; Darewicz, M. BIOPEP-UWM Database of Bioactive Peptides: Current Opportunities. *Int. J. Mol. Sci.* **2019**, *20*, 5978. [CrossRef]
290. Wang, G. The Antimicrobial Peptide Database Is 20 Years Old: Recent Developments and Future Directions. *Protein Sci.* **2023**, *32*, e4778. [CrossRef]
291. Wang, J.; Yin, T.; Xiao, X.; He, D.; Xue, Z.; Jiang, X.; Wang, Y. StraPep: A Structure Database of Bioactive Peptides. *Database* **2018**, *2018*, bay038. [CrossRef] [PubMed]
292. Liu, L.; Yang, L.; Cao, S.; Gao, Z.; Yang, B.; Zhang, G.; Zhu, R.; Wu, D. CyclicPepedia: A Knowledge Base of Natural and Synthetic Cyclic Peptides. *Brief Bioinform.* **2024**, *25*, bbae190. [CrossRef]
293. Alder, B.J.; Wainwright, T.E. Studies in Molecular Dynamics. *I. Gen. Method. J. Chem. Phys.* **1959**, *31*, 459–466. [CrossRef]
294. De Vivo, M.; Masetti, M.; Bottegoni, G.; Cavalli, A. Role of Molecular Dynamics and Related Methods in Drug Discovery. *J. Med. Chem.* **2016**, *59*, 4035–4061. [CrossRef]
295. Karplus, M.; McCammon, J.A. Molecular Dynamics Simulations of Biomolecules. *Nat. Struct. Biol.* **2002**, *9*, 646–652. [CrossRef]
296. Wieczorek, G.; Niedzialek, D. Molecular Dynamics. In *eLS*; John Wiley & Sons, Ltd: West Sussex, UK, 2020; pp. 1–18, ISBN 978-0-470-01590-2.
297. Gioia, D.; Bertazzo, M.; Recanatini, M.; Masetti, M.; Cavalli, A. Dynamic Docking: A Paradigm Shift in Computational Drug Discovery. *Molecules* **2017**, *22*, 2029. [CrossRef] [PubMed]
298. Salmaso, V.; Moro, S. Bridging Molecular Docking to Molecular Dynamics in Exploring Ligand-Protein Recognition Process: An Overview. *Front. Pharmacol.* **2018**, *9*, 923. [CrossRef]
299. De Vivo, M.; Cavalli, A. Recent Advances in Dynamic Docking for Drug Discovery. *WIRes. Comput. Mol. Sci.* **2017**, *7*, e1320. [CrossRef]
300. Salo-Ahen, O.M.H.; Alanko, I.; Bhadane, R.; Bonvin, A.M.J.J.; Honorato, R.V.; Hossain, S.; Juffer, A.H.; Kabedev, A.; Lahtela-Kakkonen, M.; Larsen, A.S.; et al. Molecular Dynamics Simulations in Drug Discovery and Pharmaceutical Development. *Processes* **2021**, *9*, 71. [CrossRef]
301. González, M.A. Force Fields and Molecular Dynamics Simulations. *Jdn* **2011**, *12*, 169–200. [CrossRef]
302. Abraham, M.J.; Murtola, T.; Schulz, R.; Páll, S.; Smith, J.C.; Hess, B.; Lindahl, E. GROMACS: High Performance Molecular Simulations through Multi-Level Parallelism from Laptops to Supercomputers. *SoftwareX* **2015**, *1–2*, 19–25. [CrossRef]
303. Case, D.A.; Cheatham, T.E.; Darden, T.; Gohlke, H.; Luo, R.; Merz, K.M.; Onufriev, A.; Simmerling, C.; Wang, B.; Woods, R.J. The Amber Biomolecular Simulation Programs. *J. Comput. Chem.* **2005**, *26*, 1668–1688. [CrossRef]
304. Brooks, B.R.; Brooks, C.L.; MacKerell, A.D.; Nilsson, L.; Petrella, R.J.; Roux, B.; Won, Y.; Archontis, G.; Bartels, C.; Boresch, S.; et al. CHARMM: The Biomolecular Simulation Program. *J. Comput. Chem.* **2009**, *30*, 1545–1614. [CrossRef]
305. Phillips, J.C.; Braun, R.; Wang, W.; Gumbart, J.; Tajkhorshid, E.; Villa, E.; Chipot, C.; Skeel, R.D.; Kalé, L.; Schulten, K. Scalable Molecular Dynamics with NAMD. *J. Comput. Chem.* **2005**, *26*, 1781–1802. [CrossRef]

306. Eastman, P.; Swails, J.; Chodera, J.D.; McGibbon, R.T.; Zhao, Y.; Beauchamp, K.A.; Wang, L.-P.; Simmonett, A.C.; Harrigan, M.P.; Stern, C.D.; et al. OpenMM 7: Rapid Development of High Performance Algorithms for Molecular Dynamics. *PLoS Comput. Biol.* **2017**, *13*, e1005659. [CrossRef]
307. Chiu, M.; Herbordt, M.C. Molecular Dynamics Simulations on High-Performance Reconfigurable Computing Systems. *ACM Trans. Reconfigurable Technol. Syst.* **2010**, *3*, 1–37. [CrossRef]
308. Liu, W.; Schmidt, B.; Voss, G.; Müller-Wittig, W. Accelerating Molecular Dynamics Simulations Using Graphics Processing Units with CUDA. *Comput. Phys. Commun.* **2008**, *179*, 634–641. [CrossRef]
309. Nobile, M.S.; Cazzaniga, P.; Tangherloni, A.; Besozzi, D. Graphics Processing Units in Bioinformatics, Computational Biology and Systems Biology. *Brief Bioinform.* **2017**, *18*, 870–885. [CrossRef] [PubMed]
310. Hollingsworth, S.A.; Dror, R.O. Molecular Dynamics Simulation for All. *Neuron* **2018**, *99*, 1129–1143. [CrossRef]
311. Sinha, S.; Tam, B.; Wang, S.M. Applications of Molecular Dynamics Simulation in Protein Study. *Membranes* **2022**, *12*, 844. [CrossRef] [PubMed]
312. Elghobashi-Meinhardt, N. Exploring Peptide–Solvent Interactions: A Computational Study. *Molecules* **2018**, *23*, 2355. [CrossRef]
313. Wałęsa, R.; Broda, M.A. The Influence of Solvent on Conformational Properties of Peptides with Aib Residue—A DFT Study. *J. Mol. Model* **2017**, *23*, 349. [CrossRef]
314. Nguyen, H.D.; Marchut, A.J.; Hall, C.K. Solvent Effects on the Conformational Transition of a Model Polyalanine Peptide. *Protein Sci.* **2004**, *13*, 2909–2924. [CrossRef]
315. Efremov, R.G.; Nolde, D.E.; Vergoten, G.; Arseniev, A.S. A Solvent Model for Simulations of Peptides in Bilayers. I. Membrane-Promoting α-Helix Formation. *Biophys. J.* **1999**, *76*, 2448–2459. [CrossRef]
316. Florová, P.; Sklenovský, P.; Banáš, P.; Otyepka, M. Explicit Water Models Affect the Specific Solvation and Dynamics of Unfolded Peptides While the Conformational Behavior and Flexibility of Folded Peptides Remain Intact. *J. Chem. Theory Comput.* **2010**, *6*, 3569–3579. [CrossRef] [PubMed]
317. Jalali, S.; Yang, Y.; Mahmoudinobar, F.; Singh, S.M.; Nilsson, B.L.; Dias, C. Using All-Atom Simulations in Explicit Solvent to Study Aggregation of Amphipathic Peptides into Amyloid-like Fibrils. *J. Mol. Liq.* **2022**, *347*, 118283. [CrossRef]
318. Chen, J.; Brooks, C.L.; Khandogin, J. Recent Advances in Implicit Solvent Based Methods for Biomolecular Simulations. *Curr. Opin. Struct. Biol.* **2008**, *18*, 140–148. [CrossRef] [PubMed]
319. Juneja, A.; Ito, M.; Nilsson, L. Implicit Solvent Models and Stabilizing Effects of Mutations and Ligands on the Unfolding of the Amyloid β-Peptide Central Helix. *J. Chem. Theory Comput.* **2013**, *9*, 834–846. [CrossRef]
320. Kleinjung, J.; Fraternali, F. Design and Application of Implicit Solvent Models in Biomolecular Simulations. *Curr Opin Struct Biol.* **2014**, *25*, 126–134. [CrossRef]
321. Zhang, J.; Zhang, H.; Wu, T.; Wang, Q.; van der Spoel, D. Comparison of Implicit and Explicit Solvent Models for the Calculation of Solvation Free Energy in Organic Solvents. *J. Chem. Theory Comput* **2017**, *13*, 1034–1043. [CrossRef]
322. Zhang, H.; Juraskova, V.; Duarte, F. Modelling Chemical Processes in Explicit Solvents with Machine Learning Potentials. *Nat. Commun.* **2024**, *15*, 6114. [CrossRef]
323. Xiong, Y.; Izadi, S.; Onufriev, A.V. A Fast Polarizable Water Model for Atomistic Simulations. *J. Chem. Theory Comput.* **2022**, *18*, 6324–6333. [CrossRef]
324. Huang, K.; Duan, L.; Zhang, J.Z.H. From Implicit to Explicit: An Interaction-Reorganization Approach to Molecular Solvation Energy. *J. Chem. Theory Comput.* **2024**, *20*, 10961–10971. [CrossRef]
325. Alibakhshi, A.; Hartke, B. Improved Prediction of Solvation Free Energies by Machine-Learning Polarizable Continuum Solvation Model. *Nat. Commun.* **2021**, *12*, 3584. [CrossRef]
326. Lindorff-Larsen, K.; Piana, S.; Dror, R.O.; Shaw, D.E. How Fast-Folding Proteins Fold. *Science* **2011**, *334*, 517–520. [CrossRef] [PubMed]
327. Melo, M.C.R.; Bernardi, R.C. Fostering Discoveries in the Era of Exascale Computing: How the next Generation of Supercomputers Empowers Computational and Experimental Biophysics Alike. *Biophys. J.* **2023**, *122*, 2833–2840. [CrossRef] [PubMed]
328. Chen, C.H.; Melo, M.C.; Berglund, N.; Khan, A.; de la Fuente-Nunez, C.; Ulmschneider, J.P.; Ulmschneider, M.B. Understanding and Modelling the Interactions of Peptides with Membranes: From Partitioning to Self-Assembly. *Curr. Opin. Struct. Biol.* **2020**, *61*, 160–166. [CrossRef] [PubMed]
329. Rathee, P.; Moorkkannur, S.N.; Prabhakar, R. Chapter Seven—Structural Studies of Catalytic Peptides Using Molecular Dynamics Simulations. In *Methods in Enzymology*; Korendovych, I.V., Ed.; Peptide Catalysts, Including Catalytic Amyloids; Academic Press: Cambridge, MA, USA, 2024; Volume 697, pp. 151–180.
330. Zhang, S.; Lv, S.; Fu, X.; Han, L.; Han, W.; Li, W. Molecular Dynamics Simulations Study of the Interactions between Human Dipeptidyl-Peptidase III and Two Substrates. *Molecules* **2021**, *26*, 6492. [CrossRef]
331. Aslan, A.; Ari Yuka, S. Therapeutic Peptides for Coronary Artery Diseases: In Silico Methods and Current Perspectives. *Amino Acids* **2024**, *56*, 37. [CrossRef]

332. Allsopp, R.; Pavlova, A.; Cline, T.; Salyapongse, A.M.; Gillilan, R.E.; Di, Y.P.; Deslouches, B.; Klauda, J.B.; Gumbart, J.C.; Tristram-Nagle, S. Antimicrobial Peptide Mechanism Studied by Scattering-Guided Molecular Dynamics Simulation. *J. Phys. Chem. B* **2022**, *126*, 6922–6935. [CrossRef]
333. Maleš, M.; Zoranić, L. Simulation Study of the Effect of Antimicrobial Peptide Associations on the Mechanism of Action with Bacterial and Eukaryotic Membranes. *Membranes* **2022**, *12*, 891. [CrossRef]
334. Palmer, N.; Maasch, J.R.M.A.; Torres, M.D.T.; de la Fuente-Nunez, C. Molecular Dynamics for Antimicrobial Peptide Discovery. *Infect. Immun.* **2021**, *89*, 10–1128. [CrossRef]
335. Stephani, J.C.; Gerhards, L.; Khairalla, B.; Solov'yov, I.A.; Brand, I. How Do Antimicrobial Peptides Interact with the Outer Membrane of Gram-Negative Bacteria? Role of Lipopolysaccharides in Peptide Binding, Anchoring, and Penetration. *ACS Infect. Dis.* **2024**, *10*, 763–778. [CrossRef]
336. Aghazadeh, H.; Ganjali Koli, M.; Ranjbar, R.; Pooshang Bagheri, K. Interactions of GF-17 Derived from LL-37 Antimicrobial Peptide with Bacterial Membranes: A Molecular Dynamics Simulation Study. *J. Comput. Aided Mol. Des.* **2020**, *34*, 1261–1273. [CrossRef]
337. Benfield, A.H.; Henriques, S.T. Mode-of-Action of Antimicrobial Peptides: Membrane Disruption vs. Intracellular Mechanisms. *Front. Med. Technol* **2020**, *2*, 610997. [CrossRef]
338. Bennett, W.F.D.; Hong, C.K.; Wang, Y.; Tieleman, D.P. Antimicrobial Peptide Simulations and the Influence of Force Field on the Free Energy for Pore Formation in Lipid Bilayers. *J. Chem. Theory Comput.* **2016**, *12*, 4524–4533. [CrossRef]
339. Cruz, V.L.; Ramos, J.; Melo, M.N.; Martinez-Salazar, J. Bacteriocin AS-48 Binding to Model Membranes and Pore Formation as Revealed by Coarse-Grained Simulations. *Biochim. Biophys. Acta* **2013**, *1828*, 2524–2531. [CrossRef]
340. Su, J.; Marrink, S.J.; Melo, M.N. Localization Preference of Antimicrobial Peptides on Liquid-Disordered Membrane Domains. *Front. Cell Dev. Biol.* **2020**, *8*, 350. [CrossRef]
341. Hazam, P.K.; Jerath, G.; Kumar, A.; Chaudhary, N.; Ramakrishnan, V. Effect of Tacticity-Derived Topological Constraints in Bactericidal Peptides. *Biochim. Biophys. Acta (BBA) Biomembr.* **2017**, *1859*, 1388–1395. [CrossRef]
342. Chen, J.-N.; Jiang, F.; Wu, Y.-D. Accurate Prediction for Protein–Peptide Binding Based on High-Temperature Molecular Dynamics Simulations. *J. Chem. Theory Comput.* **2022**, *18*, 6386–6395. [CrossRef] [PubMed]
343. Ouyang, J.; Sheng, Y.; Wang, W. Recent Advances of Studies on Cell-Penetrating Peptides Based on Molecular Dynamics Simulations. *Cells* **2022**, *11*, 4016. [CrossRef]
344. Reid, L.M.; Verma, C.S.; Essex, J.W. The Role of Molecular Simulations in Understanding the Mechanisms of Cell-Penetrating Peptides. *Drug Discov. Today* **2019**, *24*, 1821–1835. [CrossRef] [PubMed]
345. Tran, D.P.; Tada, S.; Yumoto, A.; Kitao, A.; Ito, Y.; Uzawa, T.; Tsuda, K. Using Molecular Dynamics Simulations to Prioritize and Understand AI-Generated Cell Penetrating Peptides. *Sci. Rep.* **2021**, *11*, 10630. [CrossRef]
346. Damjanovic, J.; Miao, J.; Huang, H.; Lin, Y.-S. Elucidating Solution StructuRes. of Cyclic Peptides Using Molecular Dynamics Simulations. *Chem. Rev.* **2021**, *121*, 2292–2324. [CrossRef] [PubMed]
347. Chen, S.; Lin, T.; Basu, R.; Ritchey, J.; Wang, S.; Luo, Y.; Li, X.; Pei, D.; Kara, L.B.; Cheng, X. Design of Target Specific Peptide Inhibitors Using Generative Deep Learning and Molecular Dynamics Simulations. *Nat. Commun.* **2024**, *15*, 1611. [CrossRef]
348. Noé, F.; Tkatchenko, A.; Müller, K.-R.; Clementi, C. Machine Learning for Molecular Simulation. *Annu. Rev. Phys. Chem.* **2020**, *71*, 361–390. [CrossRef]
349. Prašnikar, E.; Ljubič, M.; Perdih, A.; Borišek, J. Machine Learning Heralding a New Development Phase in Molecular Dynamics Simulations. *Artif. Intell. Rev.* **2024**, *57*, 102. [CrossRef]
350. Best, R.B.; Hummer, G. Reaction Coordinates and Rates from Transition Paths. *Proc. Natl. Acad. Sci. USA* **2005**, *102*, 6732–6737. [CrossRef]
351. Chmiela, S.; Sauceda, H.E.; Müller, K.-R.; Tkatchenko, A. Towards Exact Molecular Dynamics Simulations with Machine-Learned Force Fields. *Nat. Commun.* **2018**, *9*, 3887. [CrossRef] [PubMed]
352. Zhang, J.; Chen, D.; Xia, Y.; Huang, Y.-P.; Lin, X.; Han, X.; Ni, N.; Wang, Z.; Yu, F.; Yang, L.; et al. Artificial Intelligence Enhanced Molecular Simulations. *J. Chem. Theory Comput.* **2023**, *19*, 4338–4350. [CrossRef] [PubMed]
353. Weber, J.K.; Morrone, J.A.; Kang, S.; Zhang, L.; Lang, L.; Chowell, D.; Krishna, C.; Huynh, T.; Parthasarathy, P.; Luan, B.; et al. Unsupervised and Supervised AI on Molecular Dynamics Simulations Reveals Complex Characteristics of HLA-A2-Peptide Immunogenicity. *Brief. Bioinform.* **2024**, *25*, bbad504. [CrossRef]
354. Behler, J.; Parrinello, M. Generalized Neural-Network Representation of High-Dimensional Potential-Energy Surfaces. *Phys. Rev. Lett.* **2007**, *98*, 146401. [CrossRef]
355. Konovalov, K.A.; Unarta, I.C.; Cao, S.; Goonetilleke, E.C.; Huang, X. Markov State Models to Study the Functional Dynamics of Proteins in the Wake of Machine Learning. *JACS Au* **2021**, *1*, 1330–1341. [CrossRef]
356. Jephthah, S.; Pesce, F.; Lindorff-Larsen, K.; Skepö, M. Force Field Effects in Simulations of Flexible Peptides with Varying Polyproline II Propensity. *J. Chem. Theory Comput.* **2021**, *17*, 6634–6646. [CrossRef]

357. Lopes, P.E.M.; Guvench, O.; MacKerell, A.D. Current Status of Protein Force Fields for Molecular Dynamics. *Methods Mol. Biol.* **2015**, *1215*, 47–71. [CrossRef] [PubMed]
358. Ponder, J.W.; Case, D.A. Force Fields for Protein Simulations. In *Advances in Protein Chemistry*; Elsevier: Amsterdam, The Netherlands, 2003; Volume 66, pp. 27–85, ISBN 978-0-12-034266-2.
359. Wang, J.; Wolf, R.M.; Caldwell, J.W.; Kollman, P.A.; Case, D.A. Development and Testing of a General Amber Force Field. *J. Comput. Chem.* **2004**, *25*, 1157–1174. [CrossRef] [PubMed]
360. Best, R.B.; Zhu, X.; Shim, J.; Lopes, P.E.M.; Mittal, J.; Feig, M.; MacKerell, A.D., Jr. Optimization of the Additive CHARMM All-Atom Protein Force Field Targeting Improved Sampling of the Backbone ϕ, ψ and Side-Chain X1 and X2 Dihedral Angles. *J. Chem. Theory Comput.* **2012**, *8*, 3257–3273. [CrossRef]
361. Oostenbrink, C.; Soares, T.A.; van der Vegt, N.F.A.; van Gunsteren, W.F. Validation of the 53A6 GROMOS Force Field. *Eur. Biophys. J.* **2005**, *34*, 273–284. [CrossRef]
362. Jorgensen, W.L.; Maxwell, D.S.; Tirado-Rives, J. Development and Testing of the OPLS All-Atom Force Field on Conformational Energetics and Properties of Organic Liquids. *J. Am. Chem. Soc.* **1996**, *118*, 11225–11236. [CrossRef]
363. Huang, J.; Lopes, P.E.M.; Roux, B.; MacKerell, A.D., Jr. Recent Advances in Polarizable Force Fields for Macromolecules: Microsecond Simulations of Proteins Using the Classical Drude Oscillator Model. *J. Phys. Chem. Lett.* **2014**, *5*, 3144–3150. [CrossRef] [PubMed]
364. Jing, Z.; Liu, C.; Cheng, S.Y.; Qi, R.; Walker, B.D.; Piquemal, J.-P.; Ren, P. Polarizable Force Fields for Biomolecular Simulations: Recent Advances and Applications. *Annu. Rev. Biophys.* **2019**, *48*, 371–394. [CrossRef]
365. Ponder, J.W.; Wu, C.; Ren, P.; Pande, V.S.; Chodera, J.D.; Schnieders, M.J.; Haque, I.; Mobley, D.L.; Lambrecht, D.S.; DiStasio, R.A., Jr.; et al. Current Status of the AMOEBA Polarizable Force Field. *J. Phys. Chem. B* **2010**, *114*, 2549–2564. [CrossRef]
366. Robustelli, P.; Piana, S.; Shaw, D.E. Developing a Molecular Dynamics Force Field for Both Folded and Disordered Protein States. *Proc. Natl. Acad. Sci. USA* **2018**, *115*, E4758. [CrossRef]
367. Lindorff-Larsen, K.; Piana, S.; Palmo, K.; Maragakis, P.; Klepeis, J.L.; Dror, R.O.; Shaw, D.E. Improved Side-Chain Torsion Potentials for the Amber ff99SB Protein Force Field. *Proteins Struct. Funct. Bioinform.* **2010**, *78*, 1950–1958. [CrossRef]
368. Liu, H.; Song, D.; Zhang, Y.; Yang, S.; Luo, R.; Chen, H.-F. Extensive Tests and Evaluation of the CHARMM36IDPSFF Force Field for Intrinsically Disordered Proteins and Folded Proteins. *Phys. Chem. Chem. Phys.* **2019**, *21*, 21918–21931. [CrossRef] [PubMed]
369. Huang, J.; Rauscher, S.; Nawrocki, G.; Ran, T.; Feig, M.; de Groot, B.L.; Grubmüller, H.; MacKerell, A.D. CHARMM36m: An Improved Force Field for Folded and Intrinsically Disordered Proteins. *Nat. Methods* **2017**, *14*, 71–73. [CrossRef]
370. Arnautova, Y.A.; Jagielska, A.; Scheraga, H.A. A New Force Field (ECEPP-05) for Peptides, Proteins, and Organic Molecules. *J. Phys. Chem. B* **2006**, *110*, 5025–5044. [CrossRef] [PubMed]
371. Li, S.; Elcock, A.H. Residue-Specific Force Field (RSFF2) Improves the Modeling of Conformational Behavior of Peptides and Proteins. *J. Phys. Chem. Lett.* **2015**, *6*, 2127–2133. [CrossRef]
372. Geng, H.; Jiang, F.; Wu, Y.-D. Accurate Structure Prediction and Conformational Analysis of Cyclic Peptides with Residue-Specific Force Fields. *J. Phys. Chem. Lett.* **2016**, *7*, 1805–1810. [CrossRef] [PubMed]
373. Miao, J.; Ghosh, A.P.; Ho, M.N.; Li, C.; Huang, X.; Pentelute, B.L.; Baleja, J.D.; Lin, Y.-S. Assessing the Performance of Peptide Force Fields for Modeling the Solution Structural Ensembles of Cyclic Peptides. *J. Phys. Chem. B* **2024**, *128*, 5281–5292. [CrossRef]
374. Hickey, J.L.; Sindhikara, D.; Zultanski, S.L.; Schultz, D.M. Beyond 20 in the 21st Century: Prospects and Challenges of Non-Canonical Amino Acids in Peptide Drug Discovery. *ACS Med. Chem. Lett.* **2023**, *14*, 557–565. [CrossRef]
375. Chen, M.; Jiang, X.; Zhang, L.; Chen, X.; Wen, Y.; Gu, Z.; Li, X.; Zheng, M. The Emergence of Machine Learning Force Fields in Drug Design. *Med. Res. Rev.* **2024**, *44*, 1147–1182. [CrossRef]
376. Lazim, R.; Suh, D.; Choi, S. Advances in Molecular Dynamics Simulations and Enhanced Sampling Methods for the Study of Protein Systems. *Int. J. Mol. Sci.* **2020**, *21*, 6339. [CrossRef]
377. Shen, W.; Zhou, T.; Shi, X. Enhanced Sampling in Molecular Dynamics Simulations and Their Latest Applications—A Review. *Nano Res.* **2023**, *16*, 13474–13497. [CrossRef]
378. Bernardi, R.C.; Melo, M.C.R.; Schulten, K. Enhanced Sampling Techniques in Molecular Dynamics Simulations of Biological Systems. *Biochim. Biophys. Acta* **2015**, *1850*, 872–877. [CrossRef] [PubMed]
379. Hukushima, K. Extended Ensemble Monte Carlo Approach to Hardly Relaxing Problems. *Comput. Phys. Commun.* **2002**, *147*, 77–82. [CrossRef]
380. Kästner, J. Umbrella Sampling. *WIRes. Comput. Mol. Sci.* **2011**, *1*, 932–942. [CrossRef]
381. Mori, T.; Miyashita, N.; Im, W.; Feig, M.; Sugita, Y. Molecular Dynamics Simulations of Biological Membranes and Membrane Proteins Using Enhanced Conformational Sampling Algorithms. *Biochim. Biophys. Acta* **2016**, *1858*, 1635–1651. [CrossRef] [PubMed]
382. Laio, A.; Gervasio, F.L. Metadynamics: A Method to Simulate Rare Events and Reconstruct the Free Energy in Biophysics, Chemistry and Material Science. *Rep. Prog. Phys.* **2008**, *71*, 126601. [CrossRef]
383. Pfaendtner, J. Metadynamics to Enhance Sampling in Biomolecular Simulations. *Methods Mol. Biol.* **2019**, *2022*, 179–200. [CrossRef]

384. Hamelberg, D.; Mongan, J.; McCammon, J.A. Accelerated Molecular Dynamics: A Promising and Efficient Simulation Method for Biomolecules. *J. Chem. Phys.* **2004**, *120*, 11919–11929. [CrossRef]
385. Tóth, J.; Tomán, H.; Hajdu, A. Efficient Sampling-Based Energy Function Evaluation for Ensemble Optimization Using Simulated Annealing. *Pattern Recognit.* **2020**, *107*, 107510. [CrossRef]
386. Wang, J.; Arantes, P.R.; Bhattarai, A.; Hsu, R.V.; Pawnikar, S.; Huang, Y.M.; Palermo, G.; Miao, Y. Gaussian Accelerated Molecular Dynamics: Principles and Applications. *WIRes. Comput. Mol. Sci.* **2021**, *11*, e1521. [CrossRef]
387. Barducci, A.; Pfaendtner, J.; Bonomi, M. Tackling Sampling Challenges in Biomolecular Simulations. *Methods Mol. Biol.* **2015**, *1215*, 151–171. [CrossRef]
388. Turner, M.; Mutter, S.T.; Kennedy-Britten, O.D.; Platts, J.A. Replica Exchange Molecular Dynamics Simulation of the Coordination of Pt(II)-Phenanthroline to Amyloid-β. *RSC Adv.* **2019**, *9*, 35089–35097. [CrossRef] [PubMed]
389. Leahy, C.T.; Kells, A.; Hummer, G.; Buchete, N.-V.; Rosta, E. Peptide Dimerization-Dissociation Rates from Replica Exchange Molecular Dynamics. *J. Chem. Phys.* **2017**, *147*, 152725. [CrossRef]
390. Clayton, J.; Baweja, L.; Wereszczynski, J. Peptide Dynamics and Metadynamics: Leveraging Enhanced Sampling Molecular Dynamics to Robustly Model Long-Timescale Transitions. *Methods Mol. Biol.* **2022**, *2405*, 151. [CrossRef]
391. Hao, G.-F.; Xu, W.-F.; Yang, S.-G.; Yang, G.-F. Multiple Simulated Annealing-Molecular Dynamics (MSA-MD) for Conformational Space Search of Peptide and Miniprotein. *Sci. Rep.* **2015**, *5*, 15568. [CrossRef]
392. Wang, J.; Miao, Y. Peptide Gaussian Accelerated Molecular Dynamics (Pep-GaMD): Enhanced Sampling and Free Energy and Kinetics Calculations of Peptide Binding. *J. Chem. Phys* **2020**, *153*, 154109. [CrossRef]
393. Frazee, N.; Billlings, K.R.; Mertz, B. Gaussian Accelerated Molecular Dynamics Simulations Facilitate Prediction of the Permeability of Cyclic Peptides. *PLoS ONE* **2024**, *19*, e0300688. [CrossRef]
394. Barducci, A.; Bonomi, M.; Parrinello, M. Metadynamics. *WIRes. Comput. Mol. Sci.* **2011**, *1*, 826–843. [CrossRef]
395. Sugita, Y.; Okamoto, Y. Replica-Exchange Molecular Dynamics Method for Protein Folding. *Chem. Phys. Lett.* **1999**, *314*, 141–151. [CrossRef]
396. Faller, C.E.; Reilly, K.A.; Hills, R.D., Jr.; Guvench, O. Peptide Backbone Sampling Convergence with the Adaptive Biasing Force Algorithm. *J. Phys. Chem. B* **2013**, *117*, 518–526. [CrossRef]
397. Do, P.-C.; Lee, E.H.; Le, L. Steered Molecular Dynamics Simulation in Rational Drug Design. *J. Chem. Inf. Model.* **2018**, *58*, 1473–1482. [CrossRef]
398. Izrailev, S.; Stepaniants, S.; Isralewitz, B.; Kosztin, D.; Lu, H.; Molnar, F.; Wriggers, W.; Schulten, K. Steered Molecular Dynamics. In *Proceedings of the Computational Molecular Dynamics: Challenges, Methods, Ideas*; Deuflhard, P., Hermans, J., Leimkuhler, B., Mark, A.E., Reich, S., Skeel, R.D., Eds.; Springer: Berlin/Heidelberg, Germany, 1999; pp. 39–65.
399. Miao, Y.; Feixas, F.; Eun, C.; McCammon, J.A. Accelerated Molecular Dynamics Simulations of Protein Folding. *J. Comput. Chem.* **2015**, *36*, 1536–1549. [CrossRef]
400. Perez, D.; Uberuaga, B.P.; Shim, Y.; Amar, J.G.; Voter, A.F. Chapter 4 Accelerated Molecular Dynamics Methods: Introduction and Recent Developments. In *Annual Reports in Computational Chemistry*; Wheeler, R.A., Ed.; Elsevier: Amsterdam, The Netherlands, 2009; Volume 5, pp. 79–98.
401. Chong, L.T.; Saglam, A.S.; Zuckerman, D.M. Path-Sampling Strategies for Simulating Rare Events in Biomolecular Systems. *Curr. Opin. Struct. Biol.* **2017**, *43*, 88–94. [CrossRef]
402. Valsson, O.; Parrinello, M. Variationally Enhanced Sampling. In *Handbook of Materials Modeling: Methods: Theory and Modeling*; Andreoni, W., Yip, S., Eds.; Springer International Publishing: Cham, Switzerland, 2020; pp. 621–634, ISBN 978-3-319-44677-6.
403. Terziyska, M.; Deseva, I.; Terziyski, Z. Food-Derived Bioactive Peptides And Artificial Intelligence Techniques For Their Prediction: A Brief Review. *Int. J. Sci. Technol. Res.* **2021**, *10*, 141–147.
404. Chen, Z.; Wang, R.; Guo, J.; Wang, X. The Role and Future Prospects of Artificial Intelligence Algorithms in Peptide Drug Development. *Biomed. Pharmacother.* **2024**, *175*, 116709. [CrossRef]
405. Goles, M.; Daza, A.; Cabas-Mora, G.; Sarmiento-Varón, L.; Sepúlveda-Yañez, J.; Anvari-Kazemabad, H.; Davari, M.D.; Uribe-Paredes, R.; Olivera-Nappa, Á.; Navarrete, M.A.; et al. Peptide-Based Drug Discovery through Artificial Intelligence: Towards an Autonomous Design of Therapeutic Peptides. *Brief. Bioinform.* **2024**, *25*, bbae275. [CrossRef]
406. Gao, W.; Mahajan, S.P.; Sulam, J.; Gray, J.J. Deep Learning in Protein Structural Modeling and Design. *Patterns* **2020**, *1*, 100142. [CrossRef]
407. Kussmann, M. Prediction, Discovery, and Characterization of Plant- and Food-Derived Health-Beneficial Bioactive Peptides. *Nutrients* **2022**, *14*, 4810. [CrossRef]
408. Raschka, S.; Kaufman, B. Machine Learning and AI-Based Approaches for Bioactive Ligand Discovery and GPCR-Ligand Recognition. *Methods* **2020**, *180*, 89–110. [CrossRef]
409. Ye, J.; Li, A.; Zheng, H.; Yang, B.; Lu, Y. Machine Learning Advances in Predicting Peptide/Protein-Protein Interactions Based on Sequence Information for Lead Peptides Discovery. *Adv. Biol.* **2023**, *7*, 2200232. [CrossRef]

410. Yin, S.; Mi, X.; Shukla, D. Leveraging Machine Learning Models for Peptide-Protein Interaction Prediction. *arXiv* **2024**, arXiv:2310.18249v2.
411. Watson, J.L.; Juergens, D.; Bennett, N.R.; Trippe, B.L.; Yim, J.; Eisenach, H.E.; Ahern, W.; Borst, A.J.; Ragotte, R.J.; Milles, L.F.; et al. De Novo Design of Protein Structure and Function with RFdiffusion. *Nature* **2023**, *620*, 1089–1100. [CrossRef] [PubMed]
412. Zhao, L.; He, Q.; Song, H.; Zhou, T.; Luo, A.; Wen, Z.; Wang, T.; Lin, X. Protein A-like Peptide Design Based on Diffusion and ESM2 Models. *Molecules* **2024**, *29*, 4965. [CrossRef] [PubMed]
413. Aithani, L.; Alcaide, E.; Bartunov, S.; Cooper, C.D.O.; Doré, A.S.; Lane, T.J.; Maclean, F.; Rucktooa, P.; Shaw, R.A.; Skerratt, S.E. Advancing Structural Biology through Breakthroughs in AI. *Curr. Opin. Struct. Biol.* **2023**, *80*, 102601. [CrossRef]
414. Schauperl, M.; Denny, R.A. AI-Based Protein Structure Prediction in Drug Discovery: Impacts and Challenges. *J. Chem. Inf. Model.* **2022**, *62*, 3142–3156. [CrossRef]
415. Wang, S.; Li, W.; Liu, S.; Xu, J. RaptorX-Property: A Web Server for Protein Structure Property Prediction. *Nucleic Acids Res.* **2016**, *44*, W430–W435. [CrossRef]
416. Zhou, X.; Zheng, W.; Li, Y.; Pearce, R.; Zhang, C.; Bell, E.W.; Zhang, G.; Zhang, Y. I-TASSER-MTD: A Deep-Learning-Based Platform for Multi-Domain Protein Structure and Function Prediction. *Nat. Protoc.* **2022**, *17*, 2326–2353. [CrossRef] [PubMed]
417. Du, Z.; Su, H.; Wang, W.; Ye, L.; Wei, H.; Peng, Z.; Anishchenko, I.; Baker, D.; Yang, J. The trRosetta Server for Fast and Accurate Protein Structure Prediction. *Nat. Protoc.* **2021**, *16*, 5634–5651. [CrossRef]
418. Timmons, P.B.; Hewage, C.M. APPTEST Is a Novel Protocol for the Automatic Prediction of Peptide Tertiary Structures. *Brief. Bioinform.* **2021**, *22*, bbab308. [CrossRef]
419. Baek, M.; DiMaio, F.; Anishchenko, I.; Dauparas, J.; Ovchinnikov, S.; Lee, G.R.; Wang, J.; Cong, Q.; Kinch, L.N.; Schaeffer, R.D.; et al. Accurate Prediction of Protein StructuRes. and Interactions Using a Three-Track Neural Network. *Science* **2021**, *373*, 871–876. [CrossRef]
420. Lin, Z.; Akin, H.; Rao, R.; Hie, B.; Zhu, Z.; Lu, W.; Smetanin, N.; Verkuil, R.; Kabeli, O.; Shmueli, Y.; et al. Evolutionary-Scale Prediction of Atomic-Level Protein Structure with a Language Model. *Science* **2023**, *379*, 1123–1130. [CrossRef]
421. Ko, J.; Lee, J. Can AlphaFold2 Predict Protein-Peptide Complex StructuRes. Accurately? *bioRxiv* **2021**. [CrossRef]
422. Tsaban, T.; Varga, J.K.; Avraham, O.; Ben-Aharon, Z.; Khramushin, A.; Schueler-Furman, O. Harnessing Protein Folding Neural Networks for Peptide–Protein Docking. *Nat. Commun.* **2022**, *13*, 176. [CrossRef]
423. Mirdita, M.; Schütze, K.; Moriwaki, Y.; Heo, L.; Ovchinnikov, S.; Steinegger, M. ColabFold: Making Protein Folding Accessible to All. *Nat. Methods* **2022**, *19*, 679–682. [CrossRef]
424. Zhang, C.; Zhang, C.; Shang, T.; Zhu, N.; Wu, X.; Duan, H. HighFold: Accurately Predicting StructuRes. of Cyclic Peptides and Complexes with Head-to-Tail and Disulfide Bridge Constraints. *Brief. Bioinform.* **2024**, *25*, bbae215. [CrossRef]
425. Rettie, S.A.; Campbell, K.V.; Bera, A.K.; Kang, A.; Kozlov, S.; De La Cruz, J.; Adebomi, V.; Zhou, G.; DiMaio, F.; Ovchinnikov, S.; et al. Cyclic Peptide Structure Prediction and Design Using AlphaFold. *bioRxiv* **2023**. bioRxiv:2023.02.25.529956. [CrossRef]
426. Wu, R.; Ding, F.; Wang, R.; Shen, R.; Zhang, X.; Luo, S.; Su, C.; Wu, Z.; Xie, Q.; Berger, B.; et al. High-Resolution de Novo Structure Prediction from Primary Sequence. *bioRxiv* **2022**. bioRxiv:2022.07.21.500999.
427. Cordoves-Delgado, G.; García-Jacas, C.R. Predicting Antimicrobial Peptides Using ESMFold-Predicted StructuRes. and ESM-2-Based Amino Acid FeatuRes. with Graph Deep Learning. *J. Chem. Inf. Model.* **2024**, *64*, 4310–4321. [CrossRef] [PubMed]
428. Abdin, O.; Kim, P.M. Direct Conformational Sampling from Peptide Energy Landscapes through Hypernetwork-Conditioned Diffusion. *Nat. Mach. Intell.* **2024**, *6*, 775–786. [CrossRef]
429. Yang, Z.; Zeng, X.; Zhao, Y.; Chen, R. AlphaFold2 and Its Applications in the Fields of Biology and Medicine. *Sig. Transduct. Target Ther.* **2023**, *8*, 1–14. [CrossRef]
430. Chang, L.; Mondal, A.; Singh, B.; Martínez-Noa, Y.; Perez, A. Revolutionizing Peptide-Based Drug Discovery: Advances in the Post-AlphaFold Era. *WIRes. Comput. Mol. Sci.* **2024**, *14*, e1693. [CrossRef]
431. Desai, D.; Kantliwala, S.V.; Vybhavi, J.; Ravi, R.; Patel, H.; Patel, J. Review of AlphaFold 3: Transformative Advances in Drug Design and Therapeutics. *Cureus* **2024**, *16*, e63646. [CrossRef] [PubMed]
432. Mikhaylov, V.; Brambley, C.A.; Keller, G.L.J.; Arbuiso, A.G.; Weiss, L.I.; Baker, B.M.; Levine, A.J. Accurate Modeling of Peptide-MHC StructuRes. with AlphaFold. *Structure* **2024**, *32*, 228–241.e4. [CrossRef]
433. McDonald, E.F.; Jones, T.; Plate, L.; Meiler, J.; Gulsevin, A. Benchmarking AlphaFold2 on Peptide Structure Prediction. *Structure* **2023**, *31*, 111–119.e2. [CrossRef] [PubMed]
434. Johansson-Åkhe, I.; Wallner, B. Improving Peptide-Protein Docking with AlphaFold-Multimer Using Forced Sampling. *Front. Bioinform.* **2022**, *2*, 959160. [CrossRef]
435. Kosugi, T.; Ohue, M. Design of Cyclic Peptides Targeting Protein–Protein Interactions Using AlphaFold. *Int. J. Mol. Sci.* **2023**, *24*, 13257. [CrossRef] [PubMed]
436. Mao, J.; Guan, S.; Chen, Y.; Zeb, A.; Sun, Q.; Lu, R.; Dong, J.; Wang, J.; Cao, D. Application of a Deep Generative Model Produces Novel and Diverse Functional Peptides against Microbial Resistance. *Comput. Struct. Biotechnol. J.* **2022**, *21*, 463–471. [CrossRef] [PubMed]

437. Wan, F.; Kontogiorgos-Heintz, D.; de la Fuente-Nunez, C. Deep Generative Models for Peptide Design. *Digit. Discov.* **2022**, *1*, 195–208. [CrossRef]
438. Alhumaid, N.K.; Tawfik, E.A. Reliability of AlphaFold2 Models in Virtual Drug Screening: A Focus on Selected Class A GPCRs. *Int. J. Mol. Sci.* **2024**, *25*, 10139. [CrossRef]
439. Borkakoti, N.; Thornton, J.M. AlphaFold2 Protein Structure Prediction: Implications for Drug Discovery. *Curr. Opin. Struct. Biol.* **2023**, *78*, 102526. [CrossRef]
440. Singh, S.; Gupta, H.; Sharma, P.; Sahi, S. Advances in Artificial Intelligence (AI)-Assisted Approaches in Drug Screening. *Artif. Intell. Chem.* **2024**, *2*, 100039. [CrossRef]
441. Ding, X.; Chen, X.; Sullivan, E.E.; Shay, T.F.; Gradinaru, V. Fast, Accurate Ranking of Engineered Proteins by Target-Binding Propensity Using Structure Modeling. *Mol. Ther.* **2024**, *32*, 1687–1700. [CrossRef] [PubMed]
442. Chang, L.; Perez, A. Ranking Peptide Binders by Affinity with AlphaFold. *Angew. Chem. Int. Ed.* **2023**, *62*, e202213362. [CrossRef]
443. Scardino, V.; Filippo, J.I.D.; Cavasotto, C.N. How Good Are AlphaFold Models for Docking-Based Virtual Screening? *iScience* **2023**, *26*, 105920. [CrossRef]
444. Abramson, J.; Adler, J.; Dunger, J.; Evans, R.; Green, T.; Pritzel, A.; Ronneberger, O.; Willmore, L.; Ballard, A.J.; Bambrick, J.; et al. Accurate Structure Prediction of Biomolecular Interactions with AlphaFold 3. *Nature* **2024**, *630*, 493–500. [CrossRef] [PubMed]
445. Simpkin, A.J.; Mesdaghi, S.; Sánchez Rodríguez, F.; Elliott, L.; Murphy, D.L.; Kryshtafovych, A.; Keegan, R.M.; Rigden, D.J. Tertiary Structure Assessment at CASP15. *Proteins* **2023**, *91*, 1616–1635. [CrossRef]
446. Miller, R.D.; Iinishi, A.; Modaresi, S.M.; Yoo, B.-K.; Curtis, T.D.; Lariviere, P.J.; Liang, L.; Son, S.; Nicolau, S.; Bargabos, R.; et al. Computational Identification of a Systemic Antibiotic for Gram-Negative Bacteria. *Nat. Microbiol.* **2022**, *7*, 1661–1672. [CrossRef]
447. Zhong, G.; Wang, Z.-J.; Yan, F.; Zhang, Y.; Huo, L. Recent Advances in Discovery, Bioengineering, and Bioactivity-Evaluation of Ribosomally Synthesized and Post-Translationally Modified Peptides. *ACS Biomed. Chem. Au* **2022**, *3*, 1–31. [CrossRef]
448. Süssmuth, R.D.; Kulike-Koczula, M.; Gao, P.; Kosol, S. Fighting Antimicrobial Resistance: Innovative Drugs in Antibacterial Research. *Angew. Chem. Int. Ed.* **2025**, *64*, e202414325. [CrossRef]

biomolecules

MDPI

Article

Exploring the Allosteric Pathways of Asciminib in the Dual Inhibition of BCR-ABL1

Jie Ming, Hongwei Gao * and Jiuyu Zhan

School of Life Science, Ludong University, Yantai 264025, China; 20192504430@m.ldu.edu.cn (J.M.); zhanjy@ldu.edu.cn (J.Z.)
* Correspondence: gaohongw369@ldu.edu.cn

Abstract: The BCR-ABL1 fusion protein is a critical therapeutic target in Chronic Myeloid Leukemia (CML). Current monotherapy approaches involve types of inhibitors that can be categorized into ATP competitive inhibitors and allosteric inhibitors. However, resistance mutations in the tyrosine kinase domain of BCR-ABL1 have limited the effectiveness of these drugs. Research indicates that dual inhibition of BCR-ABL1 by combining these two types of inhibitors effectively addresses the issue of drug resistance as there are no overlapping resistance mechanisms. However, the underlying reasons for the observed synergistic effects have not yet been thoroughly elucidated. In this study, we employed molecular dynamics simulation to observe the synergistic interactions of BCR-ABL1 by the allosteric inhibitor asciminib and ATP competitive inhibitors nilotinib and ponatinib. Our study reveals that when asciminib binds to BCR-ABL1, nilotinib and ponatinib exhibit more substantial binding stability compared to monotherapy. At the atomic level, we have elucidated the reasons for the enhanced binding affinity of nilotinib and ponatinib when using a co-inhibition therapy. Our study reveals the allosteric communication pathway between asciminib and ponatinib, providing more detailed insights into the effectiveness of combination therapy. These findings provide valuable insights into combination therapies, aiding in the rational use of medications and guiding the design of novel inhibitors.

Keywords: BCR-ABL1; CML; tyrosine kinase inhibitors; allosteric inhibition; combination therapy

1. Introduction

Chronic myeloid leukemia (CML) is a clonal myeloproliferative neoplasm that accounts for approximately 15% of newly diagnosed leukemia cases in adults [1]. It is characterized by the presence of the Philadelphia chromosome, a genetic abnormality resulting from a reciprocal translocation between chromosomes 9 and 22, which leads to the fusion of the Abelson tyrosine kinase (ABL1) gene with the breakpoint cluster region (BCR) gene, encoding the constitutively active tyrosine kinase (BCR-ABL1) and driving the uncontrolled proliferation of myeloid cells [2,3]. In the treatment of CML there has been a concentrated effort to target the BCR-ABL1 fusion protein, which serves as a key therapeutic target [3,4]. The initial therapeutic approach centered on the use of inhibitors designed to obstruct ATP binding to the tyrosine kinase active site specifically, consequently suppressing the activity of BCR-ABL1 [5].

In 2001, imatinib was approved by the FDA for front-line treatment of CML, marking a significant milestone as the first-generation BCR-ABL1 inhibitor [6]. The success of imatinib validated the feasibility of targeted therapy for CML through kinase inhibitors. However, the emergence of resistance due to point mutations with the kinase domain has hindered imatinib's ability to completely cure CML [7].

To address this problem, the second-generation tyrosine kinase inhibitors (TKIs) nilotinib and dasatinib were developed using distinct design approaches [8,9]. Nilotinib, based on the structure of imatinib, underwent structural optimization to enhance its interactions with residues at the ATP site of inactive ABL, thereby increasing its binding affinity to ABL1 [10–12]. Like imatinib, nilotinib is classified as a "DFG-out" inhibitor [10]. Dasatinib, on the other hand, targets a conserved region distant from the ATP site to achieve binding with the active conformation of ABL1, making it a "DFG-in" inhibitor [13,14]. Although second-generation inhibitors possess higher potency than imatinib and are effective against most imatinib-resistant mutants, they are ineffective against the "gatekeeper" T334I mutation (T315I in Abl 1a numbering) [15–18]. The third-generation inhibitors such as ponatinib have been developed to target the T334I mutation, binding at the ATP site of the inactive conformation of ABL1 as a "DFG-out" inhibitor [11]. However, ponatinib has been associated with severe adverse reactions, and resistance can emerge due to the emergence of T334M mutation or compound mutations, rendering ponatinib ineffective [19–22]. The fourth-generation inhibitor asciminib (Figure 1) was approved by the FAD in 2021 for the clinical treatment of CML [23,24]. Asciminib mimics the myristoyl group, binding to the myristoyl pocket (allosteric site), thereby restoring the ABL1 protein's autoinhibitory mechanism and shifting the protein conformation to an inactive state, which inhibits the activity of BCR-ABL1 [23,25,26]. The allosteric site is distinct from the ATP site, which lays the foundation for the combination therapy of ATP competitive inhibitors with allosteric inhibitors (Figure 2) [27].

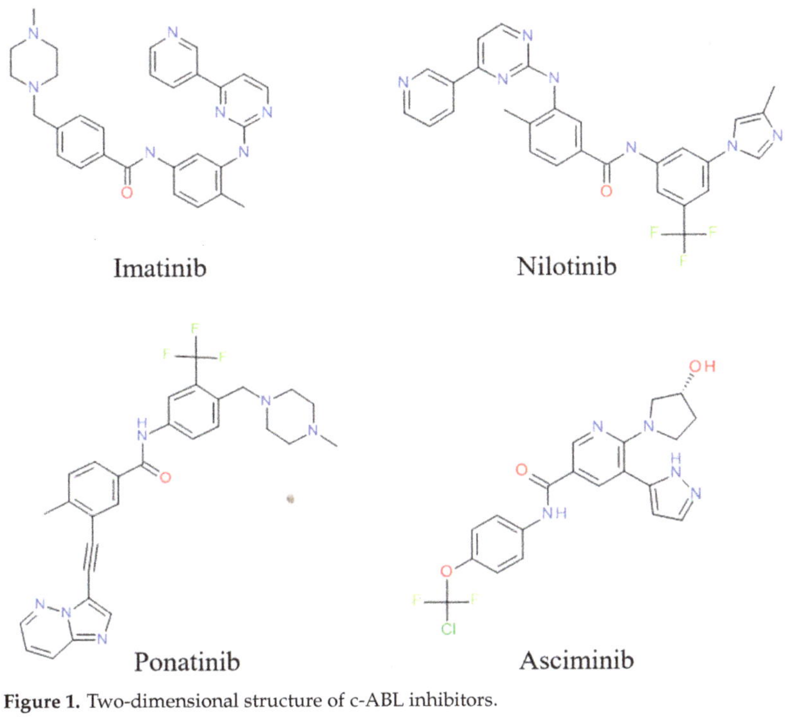

Figure 1. Two-dimensional structure of c-ABL inhibitors.

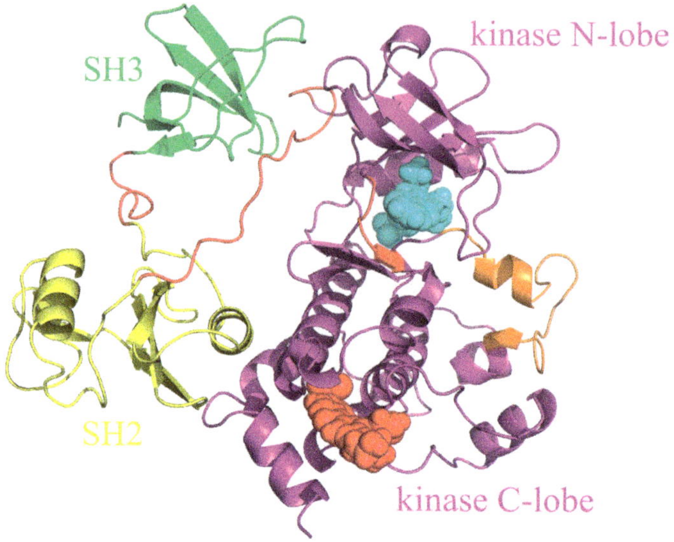

Figure 2. The three-dimensional structure of the ABL1 protein (PDB ID code 5MO4). The inhibitor nilotinib (cyan) is bound to the ATP site, whereas asciminib (red) is bound to the allosteric site in the kinase domain. Color coding is as follows: SH3 domain (green), SH2 domain (yellow), kinase domain (deep purple), interdomain linkers (tv-red), and activation loop (orange).

Resistance and drug toxicity have limited the application of TKIs in the clinical treatment of CML. However, the combination therapy of ATP competitive inhibitors and allosteric inhibitors holds promise in overcoming these issues [28]. Jahnke et al. provides evidence that asciminib can co-bind with ATP site inhibitors, including imatinib, nilotinib, and dasatinib, to the ABL protein. Importantly, this co-binding does not compromise the individual inhibitory mechanisms of these agents [29]. The study indicates the feasibility of combining allosteric inhibitors with ATP competitive inhibitors for the treatment of ABL protein. Combination therapy with drugs that lack overlapping resistance mechanisms can effectively reduce the emergence of drug resistance [28,30]. In 2024, Okamoto et al. showed that the combined use of the ATP competitive inhibitor imatinib and the allosteric asciminib prevented the development of resistant mutations [31]. Another potential advantage of combination therapy is that it leverages synergistic effects between drugs, effectively reducing the required dosages. Wylie et al. and Gleixner et al. indicate that the co-administration of asciminib with nilotinib or ponatinib enhances the inhibition against BCR-ABL1 and suppresses the emergence of resistance mutations [32–34]. The addition of asciminib significantly reduces the concentration of ponatinib required to achieve inhibition [34]. As the field of combination therapy advances, it is essential to gain a deeper understanding of the structural basis for the synergistic effects of drug molecules under dual inhibition to further develop TKIs with enhanced efficacy [33,35].

In this study, we aim to explore the inhibitory mechanism of allosteric inhibitor asciminib in conjunction with ATP competitive inhibitors to induce conformational inactivation of ABL1. Through molecular dynamics simulations, we investigate how the binding of asciminib affects the inhibitory activity of ATP competitive inhibitors nilotinib and ponatinib against wild-type (WT) ABL1 at the molecular level. In this study, we aim to decipher the inhibitor-residue interactions under co-inhibition, enhancing our understanding of the communication between the ATP binding site and the allosteric site of ABL1.

2. Materials and Methods

2.1. System Preparation

For this study, five different systems were considered: ABL1–asciminib, ABL1–nilotinib, ABL1–nilotinib–asciminib, ABL1–ponatinib, and ABL1–ponatinib–asciminib. The initial structure of the ABL1 protein was retrieved from Protein Data Bank (PDB code:5MO4), which corresponds to the ABL1 kinase domain with the T334I and D382N mutations in complex with asciminib and nilotinib [32]. Two reverse mutations were carried out, and the missing residues were restored using Discovery Studio 2020 (DS 2020) to obtain the complete WT ABL1 structure. Subsequently, DS 2020 was utilized for molecular docking to position ponatinib into the ATP binding site of the WT ABL1 protein. The ionizable residues within the ABL1 structure were assigned their protonation states using the H++ server [36].

2.2. Molecular Dynamics (MD) Simulations

MD simulations were carried out on the systems with AMBER 18 [37]. The partial charges and missing parameter fields for the inhibitors were determined using the Antechamber module in AMBER 18 software. The inhibitors were parameterized with the General AMBER Force Field (GAFF) [38], while ABL1 was parameterized with the ff14SB force [39]. The t-leap module was employed to ensure the electrical neutrality of the system, with the protein complex positioned at the center of the TIP3P water model [40].

All of the MD simulations were executed by the AMBER 18 software [41]. A two-stage minimization process was employed to prepare a reasonable initial model. Initially, the complex was minimized under constraints (500 kcal mol^{-1} $Å^{-2}$) for 10,000 steps. Subsequently, the minimization was repeated for another 10,000 steps without any restrictions. The system was gradually heated from 0 K to 310 K over a period of 300 ps, followed by an equilibration simulation at a constant temperature of 310 K for 1ns. Under NPT conditions, we conducted 200 ns MD simulations for four complex systems to obtain trajectories. Subsequently, the trajectories were analyzed using the cpptraj module.

2.3. MM-GB/SA Calculations

Determining the binding free energy for protein–ligand interactions is a significant approach within the field of computational biophysics. To calculate the free binding energy between inhibitors and the ABL1 protein, we employed the MM-GBSA (Molecular Mechanics Generalized Born Surface Area) method in AMBER 18 [42–45]. The binding free energy (ΔG_{bind}) was derived from an average of 10,000 snapshots extracted from each simulation trajectory ranging from 100 ns to 200 ns. The calculation is based on the following formula:

$$\Delta G_{bind} = G_{complex} - \left(G_{receptor} + G_{ligand}\right) \tag{1}$$

$$\Delta G_{bind} = \Delta E_{MM} + \Delta G_{sol} - T\Delta S \tag{2}$$

$$\Delta E_{MM} = \Delta E_{int} + \Delta E_{ele} + \Delta E_{vdW} \tag{3}$$

$$\Delta G_{sol} = \Delta G_{GB} + \Delta G_{SA} \tag{4}$$

$$\Delta G_{SA} = \gamma SASA \tag{5}$$

where the terms $G_{complex}$, $G_{receptor}$, and G_{ligand} symbolize the free energy of the complex, the receptor, and the ligand. The terms ΔE_{MM}, ΔG_{sol}, and $T\Delta S$ symbolize the gas-phase energy, solvation free energy, and a vibrational entropy term. ΔE_{MM} is divided into internal interaction (ΔE_{int}), electrostatic interaction (ΔE_{ele}), and van der Waals energy (ΔE_{vdW}). Solvation free energy (ΔG_{sol}) is the summation of polar solvation energy (ΔG_{GB}) and nonpolar solvation energy (ΔG_{SA}). The ΔG_{GB} contribution was assessed by solving the GB equation, and the ΔG_{SA} was determined from the solvent accessible surface area (SASA). We performed free energy calculations utilizing the Generalized Born (GB) model with the igb = 5 parameter set. For the calculation of the nonpolar solvation free energy term, γ is set to 0.0072 kcal mol^{-1} $Å^{-2}$ and SASA is the solvent accessible surface area determined by a probe radius of 1.4 Å.

When analyzing a ternary complex, we treat the inhibitor of interest as the ligand and consider the protein bound to the second small molecule as the receptor. Thus, to compute the binding free energy of nilotinib in the ABL1–nilotinib–asciminib system, nilotinib is designated as the ligand, while the ABL1–asciminib complex is regarded as the receptor.

To further elucidate the interactions between the protein and inhibitors, we dissected the total binding free energy to the residue level using the MM-GB/SA method in AMBER 18, based on the stable trajectory obtained from molecular dynamics simulations.

2.4. Residue Interaction Network

In this study, the webPSN web server was employed to analyze the residue interaction networks of the systems, enhancing our comprehension of allosteric communication within biomolecular systems [46]. Based on the final 100 ns of the MD trajectories, PSNTools was utilized to generate networks detailing the interactions between protein residues [47]. In our study, a residue network was constructed by treating each amino acid residue and ligand as a node, with non-covalent interactions between residues serving as edges to connect these nodes. The webPSN server was also used to compute differences in residue networks between single- and dual-inhibition systems, thereby identifying allosteric communication paths caused by asciminib binding to the ABL1 protein.

3. Results and Discussion

3.1. Correlation of ATP Binding Site with Allosteric Site

The stability of the complex system was assessed by measuring the root mean square deviation (RMSD) of the complex structure. The average RMSD for all C-α atoms of the trajectories for all the systems demonstrate that equilibration was achieved after 100 ns and can be used for the following analyses (Figure 3).

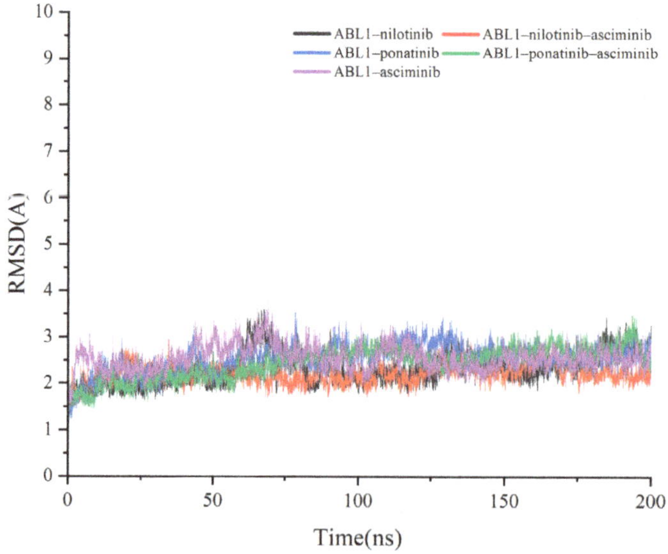

Figure 3. RMSD of Cα atoms of the protein backbone atoms.

Dynamic cross correlation (DCC) analyses are a popular method for analyzing communications among separate parts of a molecular system [48,49]. The method can decipher the hidden dynamics of residues in the ATP binding site and allosteric site of BCR-ABL1. To investigate the effect of double-drugging for BCR-ABL1, we focus on the positive correlation of atoms within 5 Å from inhibitors (Figure 4). An occurrence of positive concerted motion between residues indicates the presence of strong interactions among them, which means coordinated movement in the same direction. As shown in Figure 4, the results show a stabilizing pathway between the ATP binding site and the allosteric site of BCR-ABL1, suggesting that the co-binding of asciminib and nilotinib enhances the communication of ATP binding site and allosteric site of BCR-ABL1. When asciminib binds to ABL1 protein, it promotes cooperative movements between nilotinib and a larger number of residues at the ATP binding site. This suggests that more residues are engaged in allosteric communication. However, the dynamic correlation coefficient (DCC) only indicates the dynamic coupling between the ATP site and the allosteric site. To further analyze how asciminib binding to ABL1 affects the binding affinity of ATP inhibitors at the ATP site, a more in-depth analysis through the calculation of binding free energy is necessary.

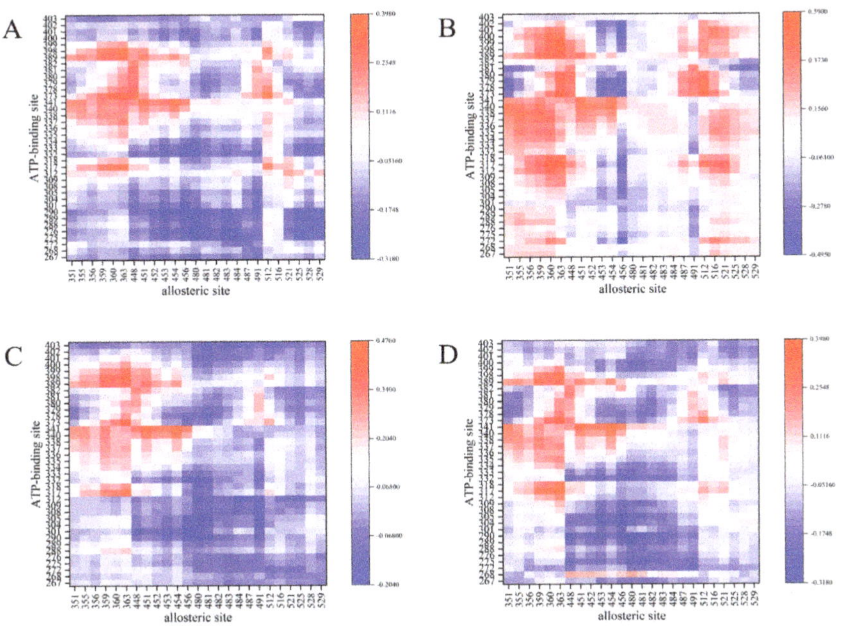

Figure 4. Dynamic cross correlation of ATP binding site and allosteric site of the BCR-ABL1. (**A**) ABL1–nilotinib, (**B**) ABL1–nilotinib–asciminib, (**C**) ABL1–ponatinib, (**D**) ABL1–ponatinib–asciminib. The color relates to the degree of correlation, the higher the correlation, the darker the color. Red is a highly positive correlation; blue is a highly negative correlation.

3.2. Binding Free Energy Calculations

To evaluate the binding energy of the ATP competitive inhibitors and allosteric inhibitors when they bind to BCR-ABL1 alone, as well as during co-inhibition, the MM-GB/SA method in AMBER18 was used in calculating the four systems and is shown in Table 1. The binding free energy (ΔG_{bind}) indicates that the value of binding free energy of nilotinib or ponatinib to ABL1 was higher when bound alone. The results demonstrate that the co-inhibition of BCR-ABL1 by nilotinib or ponatinib and asciminib results in a more favorable binding than when each inhibitor is bound alone.

Table 1. Binding energy of four complex at the ABL1 ATP binding site (kcal mol^{-1}).

System	ΔE_{vdw}	ΔE_{ele}	ΔG_{GB}	ΔG_{SA}	$-T\Delta S$	ΔG_{bind} ^{a}
ABL1–nilotinib	−71.75	−34.5284	45.075	−9.1536	31.5174	−38.8396
ABL1–nilotinib–asciminib	−72.3503	−37.3234	47.6184	−9.4476	30.2918	−41.2111
ABL1–ponatinib	−72.3503	−23.5641	29.3745	−9.0356	29.3277	−44.5496
ABL1–ponatinib–asciminib	−73.4241	−30.6382	34.2425	−9.3242	27.285	−51.859

^{a}: $\Delta G_{bind} = \Delta G_{MM-GB/SA} - T\Delta S$. $\Delta G_{MM-GB/SA} = \Delta E_{vdw} + \Delta E_{ele} + \Delta G_{GB} + \Delta G_{SA}$.

From the comparison between the single-inhibition system and the co-inhibition system, it was found that the enhancement in the binding affinity of ponatinib with ABL1 under co-inhibition is primarily attributed to the strengthened electrostatic interactions between ponatinib and the residues of the ATP binding site.

In the co-inhibition system, the binding affinity of asciminib to ABL1 is slightly reduced compared with that in the ABL1–asciminib complex alone, a result that aligns with experimental observations (Table 2). When asciminib is co-bound with an ATP competitive inhibitor to the ABL1 protein, a weak antagonism is observed; however, this does not prevent the simultaneous engagement of both inhibitors at ABL1, and their combination still yields synergistic therapeutic efficacy [29].

Table 2. Binding energy of three complex at the ABL1 allosteric site (kcal mol^{-1}).

System	ΔE_{vdw}	ΔE_{ele}	ΔG_{GB}	ΔG_{SA}	$-T\Delta S$	ΔG_{bind} ^{a}
ABL1–ascminib	−43.7724	−24.8216	33.4048	−5.7043	22.6313	−18.2622
ABL1–nilotinib–asciminib	−43.9577	−22.5531	35.7391	−5.6030	22.0603	−14.3145
ABL1–ponatinib–asciminib	−44.9838	−19.3029	33.4972	−5.6402	21.3559	−15.0738

^{a}: $\Delta G_{bind} = \Delta G_{MM-GB/SA} - T\Delta S$. $\Delta G_{MM-GB/SA} = \Delta E_{vdw} + \Delta E_{ele} + \Delta G_{GB} + \Delta G_{SA}$.

In order to further provide a detailed understanding of each residue that contributes significantly to the binding of ATP competitive inhibitors to BCR-ABL1, both when they bind alone and during co-inhibition with asciminib, we decomposed the total binding free energy into the binding free energies of individual residues. The energy of key residues, which includes van der Waals energy, electrostatic interaction, polar solvation energy, and nonpolar solvation energy, is characterized by nonpolar solvation and van der Waals interaction energies as positive contributors to the binding of inhibitors. In contrast, polar solvation energy acts as a negative factor affecting it. The key residues with binding free energy ($\Delta G_{bind} \leq -1$ kcal/mol) are listed in Tables S1–S4. Residues showing a free energy change of 20% or more are considered to have a significant impact on the allosteric communication pathway (Table 3). The co-binding of nilotinib with ABL1 and asciminib results in decreased binding affinity for key residues PHE378, ASH400, and PHE401. Interestingly, when asciminib binds to ABL1 it diminishes the binding affinity of nilotinib for key residues that are crucial in allosteric communication, as seen in the ABL1–nilotinib–asciminib system compared to the ABL1–nilotinib system alone. In other words, the enhanced binding affinity of nilotinib for ABL1 in the presence of asciminib is mainly attributed to non-key residues rather than key residues. In light of the DCC analysis, the binding of asciminib to ABL1 protein has been shown to engage a larger number of residues at the ATP binding site in cooperative movements with residues at the allosteric site. When compared to the nilotinib–ABL1 system, this suggests that interactions between a greater number of residues and nilotinib have been enhanced. However, there is no significant qualitative improvement observed. The increased binding affinity between nilotinib and ABL1 protein represents a quantitative advancement rather than a qualitative one. In contrast, co-binding of ponatinib with ABL1 and asciminib enhances the binding affinity for key residues LEU267, VAL275, THR334, and GLU335, but weakens it for the key residue PHE336.The improved binding affinity of ponatinib with residues LEU267, VAL275, and THR334 is attributed to enhanced van der Waals interaction energies, indicating a closer approach of these residues to ponatinib. In contrast, the increased binding affinity between GLU335 and ponatinib stems from strengthened electrostatic interactions. The result indicates that the presence of asciminib enhances the binding affinity between key residues and ponatinib, thereby strengthening the overall binding affinity of ponatinib to ABL1.

Table 3. Summary of binding free energy on key residues affected by allosteric modulation (kcal mol^{-1}).

Inhibitor	Residues	$\Delta G_{bind}{}^{1}$	$\Delta G_{bind}{}^{2}$	Percentage Change
nilotinib	PHE378	−1.014	−0.695	−31%
	ASP400	−2.865	−2.305	−20%
	PHE401	−1.926	−1.55	−20%
ponatinib	LEU267	−0.844	−1.575	87%
	VAL275	−0.847	−1.573	86%
	THR334	−1.256	−1.76	40%
	GLU335	−0.574	−1.014	77%
	PHE336	−1.669	−0.87	−48%

$\Delta G_{bind}{}^{1}$: the binding free energy of the residues within the ABL1–nilotinib system or the ABL1–ponatinib system. $\Delta G_{bind}{}^{2}$: the binding free energy of the residues within the ABL1–nilotinib–asciminib system or ABL1–ponatinib–asciminib system.

3.3. The H-Bonding Ability of Inhibitors at the ATP Binding Site

To evaluate the impact of the binding of the allosteric inhibitor asciminib on the H-bonding ability of ATP competitive inhibitors nilotinib and ponatinib at the ATP site, hydrogen bond occupancies were used for four systems, and the results are shown in Figure 5. In the ABL1–nilotinib system, the hydrogen bond occupancy value for GLU305 in their interaction with nilotinib was 58.44%. In the ABL1–nilotinib–asciminib system, the value was 74.45% for GLU305. The hydrogen bond interaction between nilotinib and GLU305 at the ATP site was enhanced when asciminib was also bound. In the ABL1–ponatinib system, the hydrogen bond occupancy value for MET337 in the interaction with ponatinib was 82.84%. In the ABL1–ponatinib–asciminib system, the value changed to 64.14%. The results revealed that the presence of asciminib impeded the formation of hydrogen bonds between ponatinib and MET337. From the hydrogen bond occupancy results, it can be inferred that the binding of asciminib to ABL1 affects the hydrogen occupancies between inhibitors and residues at the ATP site, thereby influencing the binding affinity between inhibitors and individual residues. However, when considering the inhibitors as a whole, these changes in hydrogen bond occupancies do not significantly impact the binding affinity between the inhibitors and the ABL1 protein.

Further analysis from the perspective of the entire inhibitor molecule was conducted to observe the allosteric inhibitor asciminib on the hydrogen bonding ability of ATP competitive inhibitors at the ATP site. The average number of hydrogen bonds formed by ATP competitive inhibitors in the systems was calculated as follows: nilotinib in ABL1–nilotinib (3.32), nilotinib in ABL1–nilotinib–asciminib (3.40), ponatinib in ABL1–ponatinib (2.03), ponatinib in ABL1–ponatinib–asciminib (1.92). From the perspective of the inhibitors' overall hydrogen bonding ability, the changes in hydrogen bonding ability between the inhibitors and the ABL1 protein, induced by asciminib binding to ABL1, do not significantly affect the binding affinity of ATP competitive inhibitors to ABL1.

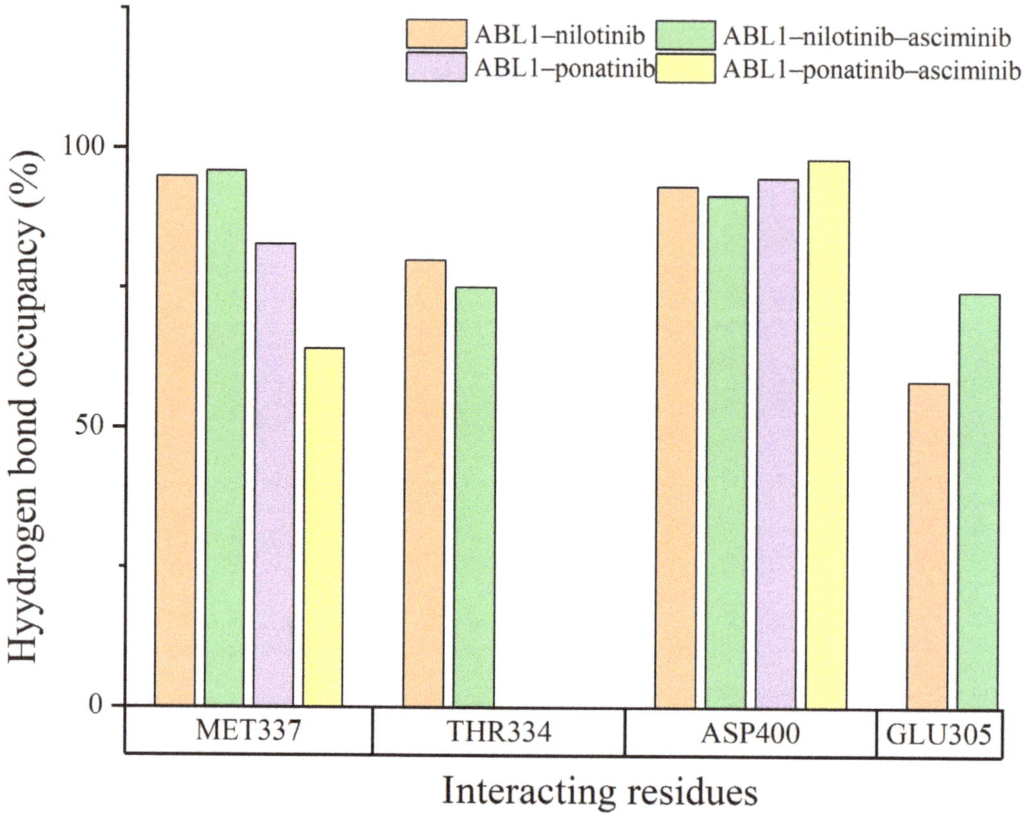

Figure 5. The hydrogen bond occupancies for key residues at the ABL1 binding interface with ATP competitive inhibitors. Only residues with hydrogen bond occupancies exceeding 40% are depicted.

3.4. The Hydrophobic Interaction of Inhibitors at the ATP Binding Site

To further analyze the impact of the interaction between ATP competitive inhibitors and residues at the ATP site when co-inhibited with allosteric inhibitor asciminib in ABL1, we focused on the hydrophobic interactions formed by inhibitors and residues, which are a significant factor in evaluating the binding of ATP competitive inhibitors to the ATP site of ABL1, and the results are shown in Figure 6. In the ABL1–nilotinib–asciminib system, nilotinib forms hydrophobic interactions with residues ASP400, TYR272, ALA288, LYS304, VAL308, VAL275, LEU267, and MET337 at the ATP site, leading to a more stable binding of nilotinib in the ATP site. Additionally, the hydrophobic interactions between nilotinib and LEU267, TYR272, VAL275, and ASP400 resulted in a displacement of nilotinib position compared to the ABL1–nilotinib system. In the ABL1–ponatinib–asciminib system, the hydrophobic interactions formed between ponatinib and residues LEU267, VAL275, MET309, and VAL318 at the ATP site enhanced the binding stability of ponatinib with ABL1. Based on the analysis of binding free energies of the residues, it was found that LEU267 and VAL318 play a crucial role in the binding of ponatinib to ABL1 in the ABL1–ponatinib–asciminib system. The results suggest that these hydrophobic interactions are crucial for maintaining the binding affinity of inhibitors, as they help stabilize the interaction between inhibitors and residues in the ATP site.

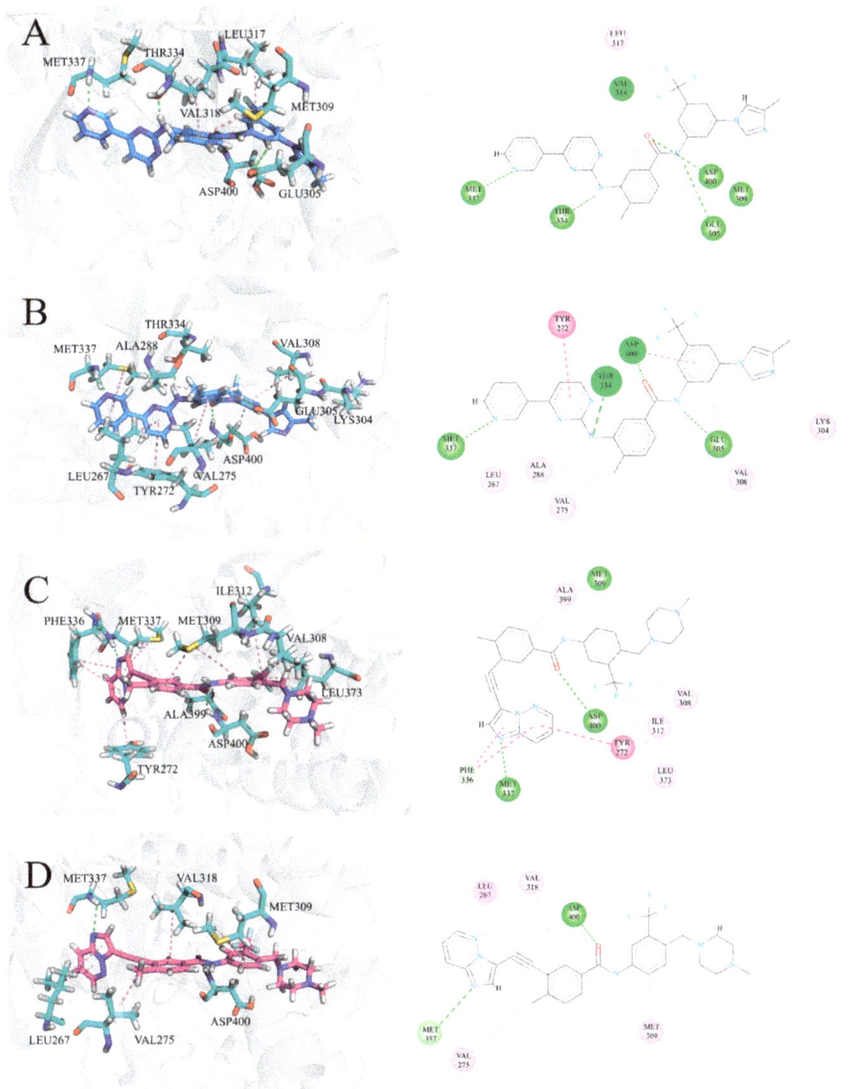

Figure 6. Binding modes of inhibitors to ABL1 are illustrated, with nilotinib and ponatinib shown in marine and light magenta. Residues with interactive relationships are displayed in stick mode, and H-bonds and hydrophobic interactions are marked with green and pink dotted lines. Nitrogen atoms are marked in blue, sulfur atoms in yellow, and oxygen atoms in red. The left side of the figure shows the three-dimensional structures, and the right side shows the two-dimensional representations. In the right-side figure: Inhibitors are shown as lines, residues as balls, hydrogen bonds as green dashed lines, hydrophobic interactions as pink dashed lines, and residues are colored by interaction type. (**A**) Nilotinib binding to residues in the ABL1–nilotinib system. (**B**) Nilotinib binding to residues in the ABL1–nilotinib–asciminib system. (**C**) Ponatinib binding to residues in the ABL1–ponatinib system. (**D**) Ponatinib binding to residues in the ABL1–ponatinib–asciminib system.

3.5. The Electrostatic Interaction of Inhibitors at the ATP Binding Site

The enhancement of electrostatic interactions between ATP competitive inhibitors (nilotinib and ponatinib) and ABL1 is a significant factor contributing to the increased binding affinity of these inhibitors with ABL1 when they co-inhibit ABL1 with asciminib. To further elucidate the impact of these interactions, this study focused on the electrostatic interactions between ATP competitive inhibitors and each residue at the ATP site during the co-inhibition of ABL1 by ATP competitive inhibitors and asciminib. Summarizing the electrostatic interactions between ATP competitive inhibitors and ABL1 with an occupancy rate greater than 40% in both the single-inhibition system and the co-inhibition system, we identified residues that demonstrate enhanced interactions with ATP competitive inhibitors compared to the single-inhibition system, as shown in Figure 7.

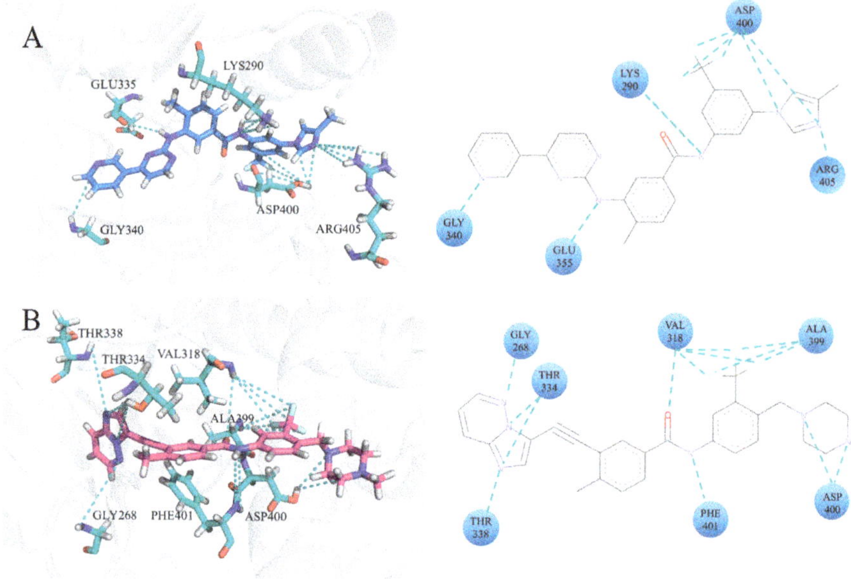

Figure 7. Electrostatic interactions between nilotinib and ponatinib with the ATP binding site residues are depicted. Nilotinib is shown in marine blue, while ponatinib is represented in light magenta. Residues with interactive relationships are displayed in stick mode, and electrostatic interactions are marked with cyan dotted lines. Nitrogen atoms are marked in blue and oxygen atoms in red. The left panel presents the three-dimensional view, whereas the right panel provides a two-dimensional representation. In the right-side figure, inhibitors are shown as lines, residues are shown as blue balls, and electrostatic interactions are shown as cyan dashed lines. (**A**) Nilotinib binding to residues in the ABL1–nilotinib–asciminib system. (**B**) Ponatinib binding to residues in the ABL1–ponatinib–asciminib system.

In the ABL1–nilotinib–asciminib system, the residues that exhibit enhanced electrostatic interactions with nilotinib are rather sparsely distributed. This scattered distribution pattern suggests that the electrostatic contributions to binding are not localized to a single region but rather spread across various parts of the protein, which have implications for the overall binding affinity and specificity of the complex. In the ABL1–ponatinib–asciminib system, residues VAL318, ALA399, ASP400, and PHE401 encircled ponatinib, forming stronger electrostatic with the retroamide group and the trifluoromethyl group. The retroamide group is known to enhance the potency of ponatinib and anchor the inhibitor in the ATP site [11]. In contrast, the trifluoromethyl group contributes to exploring new hydrophobic contacts in the deep region of the ATP site [11]. The results suggest that the enhanced electrostatic interactions between the retroamide group and residues play a crucial role in increasing the affinity of ponatinib for ABL1 during the co-inhibition of ABL1 by ponatinib and asciminib.

In this study, we employed molecular docking and molecular dynamics simulations to establish four systems that model the inhibition of ABL tyrosine kinase by ATP competitive inhibitors—nilotinib and ponatinib—either alone or in combination with asciminib within the human body. Through dynamic cross correlation (DCC) analyses, we determined that the ATP site and allosteric site are not entirely independent. Instead, they exhibit positive concerted motion, indicating the presence of a stable signaling pathway between the two. This suggests that the binding of asciminib to ABL1 can influence the binding of ATP competitive inhibitors to ABL1 through signal transduction mechanisms.

Nilotinib, developed through the structural optimization of imatinib, retains the same hydrogen bonding capabilities as imatinib while favorably expanding hydrophobic interactions at the ATP site, thereby enhancing the therapeutic effect against imatinib-resistant mutations [12]. Interestingly, when co-bound with asciminib to ABL1, nilotinib and imatinib exhibit distinctly different binding affinities with ABL1 [30]. The binding free energy of key residues decreases, indicating a weakened binding capacity between nilotinib and these residues. This suggests that the presence of asciminib affects the binding of nilotinib with key residues at the ATP site, leading to a reduction in their binding affinity. The enhancement of nilotinib's binding affinity to ABL1 in the presence of asciminib is indeed an intriguing phenomenon. This increase is not attributed to key residues but rather to the improvement in binding free energy of non-key residues, which is a noteworthy observation. Similarly, the presence of asciminib has a minimal impact on the overall hydrogen bonding ability of nilotinib but significantly affects its hydrophobic interactions. Nilotinib forms hydrophobic interactions with a more significant number of residues at the ATP site, thereby enhancing its binding affinity to ABL1. In DCC analyses, when nilotinib co-inhibits ABL1 with asciminib there is a significant positive concerted motion between more residues at the ATP site and residues at the allosteric site compared to when nilotinib inhibits ABL1 alone.

Synthesizing our analysis, we venture a hypothesis: given that asciminib binds to the allosteric site of ABL1, resulting in reduced binding energy between nilotinib and key residues at the ATP site, we would have expected asciminib to hinder the binding of nilotinib to the ATP site of ABL1. However, the structural advantages of nilotinib allow it to maintain its binding affinity with ABL1. The residues at the ATP site do not effectively "push" nilotinib out; instead, they enhance interactions with nilotinib, forming additional hydrophobic contacts and strengthening nilotinib's hydrophobic capabilities at the ATP site. This enhancement is sufficient to explain the increased binding affinity between nilotinib and ABL1, which relies on the binding free energy of non-key residues. The strengthened hydrophobic interactions between the ATP site residues and nilotinib bind nilotinib with more residues, leading to concerted motion. This also accounts for the observed increase in residues exhibiting positive concerted motion at the ATP site under co-inhibition.

Ponatinib is based on AP23464, a trisubstituted purine analog, and has undergone structural optimization to enhance its drug efficacy [11]. Based on the results, it is evident that the binding of asciminib to ABL1 enhances the binding affinity of ponatinib with ABL1, with the improvement in binding affinity of ponatinib relying on strengthening interactions with key residues. The combination of asciminib with ABL1 intensifies the electrostatic interactions between partial groups of ponatinib and residues at the ATP site. The combined therapy of asciminib and ponatinib in the treatment of ABL1 initiates communication between residues at the allosteric site and those at the ATP site. The binding of asciminib to ABL1 enhances the strong interactions between partial groups of ponatinib and key residues through signal transduction, thereby influencing the binding of ponatinib to ABL1.

3.6. Allosteric Communications Between Asciminib and Ponatinib

Based on the ABL1–ponatinib system and the ABL1–ponatinib–asciminib system, we used PSNTools to calculate the interactions between residues over the time range of 100 ns to 200 ns, exploring the potential communication mechanisms between asciminib and ponatinib.

To explore the allosteric communication pathways by which asciminib binding to ABL1 enhances ponatinib's binding affinity to ABL1, we screened the global metapath using asciminib and ponatinib as start and end points, creating new metapaths. The relevant information of the filtered paths is summarized in Table 4. The increased number of nodes and links in the metapath of the ABL1–ponatinib–asciminib system, compared to the ABL1–ponatinib system, indicate that asciminib binding to ABL1 can broaden communication paths between residue pairs and activate allosteric communication pathways. The decrease in the number of shortest paths indicates that the binding of asciminib to ABL1 optimizes the pathways for signal transmission, thereby enhancing the efficiency of signal propagation. The average path length, which refers to the average number of nodes in the filtered paths pool, indicates that a higher value suggests that signal transmission in the residue network involves more amino acids, highlighting the intricate nature of ABL1′s allosteric communication function. The changes in these values indicate that the ABL1 protein exhibits greater stability and efficiency under the combined inhibition of ponatinib and asciminib.

Table 4. Summary of filtered paths differences between the ABL1–ponatinib system and the ABL1–ponatinib–asciminib system.

Value	ABL1–Ponatinib–Asciminib System	ABL1–Ponatinib System
Number Of Nodes in MetaPath	73	45
Specific Nodes in MetaPath	48 (65.75%)	20 (44.44%)
Shared Nodes in MetaPath	25 (34.25%)	25 (55.56%)
Number Of Links MetaPath	73	45
Specific Links in MetaPath	53 (72.60%)	25 (55.56%)
Shared Links in MetaPath	20 (27.40%)	20 (44.44%)
Number of Shortest Paths	2039	2359
Length Of Smallest Path	3	4
Average Path Length	18.78	16.16
Length of Longest Path	38	31
Minimum Path Force	2.75	3.80
Average Path Force	7.34	8.17
Maximum Path Force	11.17	13.62
Minimum Path Correlation	0.80	0.80
Average Path Correlation	0.89	0.88
Maximum Path Correlation	0.94	0.93
Minimum % Of Corr. Nodes	2.78	3.45
Average % Of Corr. Nodes	9.27	10.51
Maximum % Of Corr. Nodes	100	75
Minimum Path Hubs %	10	0
Average Path Hubs %	42.66	38.64
Maximum Path Hubs %	80	75

Our study focuses on the allosteric communication pathway between the allosteric site and the ATP binding site (Figure 8). This pathway is initiated by residue SER522 at the allosteric site, traverses the SH2 domain, and upon reaching residue TRP118 in the SH3 domain diverges into two distinct signaling routes. One signaling pathway passes through residues ASN113 → ASN115 → GLU117 → TYR245 → GLN319 → GLU335 → VAL318

→ ponatinib. Another pathway passes through residues THR98 → GLY95 → TYR283 → LYS282 → ASP260 → GLU257 → ARG326 → GLU327 → ARG258 → TYR331 → MET263 → TYR276 → LYS266 → GLY270 → GLY269 → GLY268 → GLN271 → TYR272 → PHE401 → ponatinib.

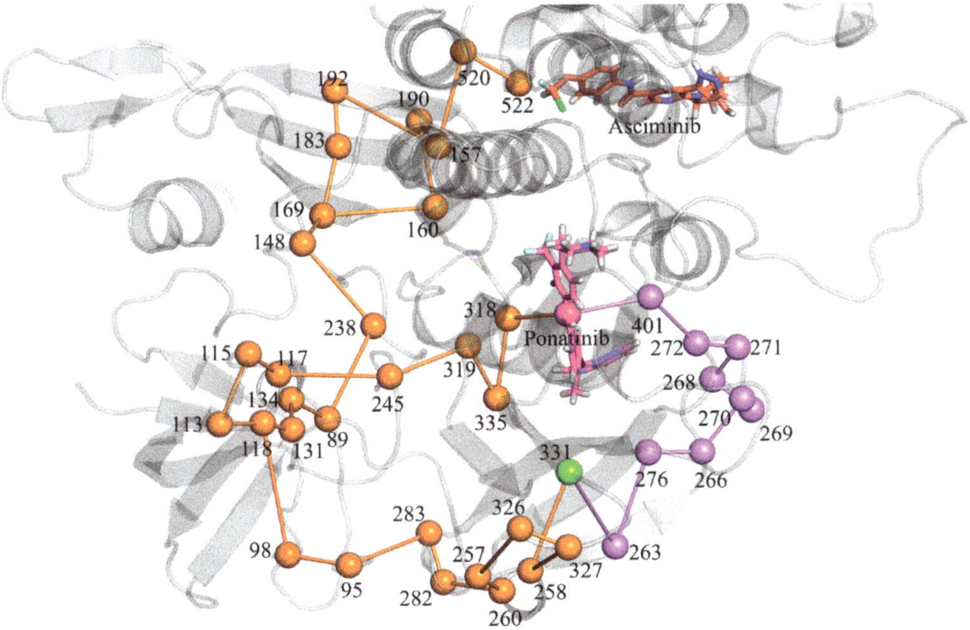

Figure 8. Three-dimensional diagram of the communication pathway between asciminib and ponatinib. Nodes and links specific to the ABL1–ponatinib–asciminib system network are colored orange, those specific to the ABL1–ponatinib system network are colored purple, and nodes and links shared by both networks are colored green. Asciminib is shown as a red stick, ponatinib is displayed as a light magenta stick and node, where nitrogen atoms are blue, chlorine atoms are green, and fluorine atoms are cyan.

The first pathway is associated with the enhanced binding affinity between GLU335 and ponatinib (Figure 9). Specifically, the carboxyl group of GLU335 forms a hydrogen bond with the nitrogen atom of GLN319. This interaction draws the negatively charged carboxyl group of GLU335 closer to ponatinib, thereby facilitating electrostatic interactions. This explains why electrostatic interactions play a crucial role in the enhanced binding affinity between GLU335 and ponatinib.

The second pathway is associated with enhanced binding affinity between residues LEU267, VAL275, and ponatinib (Figure 10). The TRP118 → TYR331 segment is unique to the ABL1–ponatinib–asciminib system, while TYR331 → PHE401 is specific to the ABL1-ponatinib system. This indicates that asciminib binding to ABL1 disrupts the TYR331 → PHE401 signaling pathway. The allosteric communication in the ABL1–ponatinib–asciminib system causes the protein conformation MET263 → TYR276 to be more open than in the ABL1–ponatinib system. This allows residues LEU267 and VAL275 to move closer to the hydrophobic groups of ponatinib, strengthening the hydrophobic interactions between them.

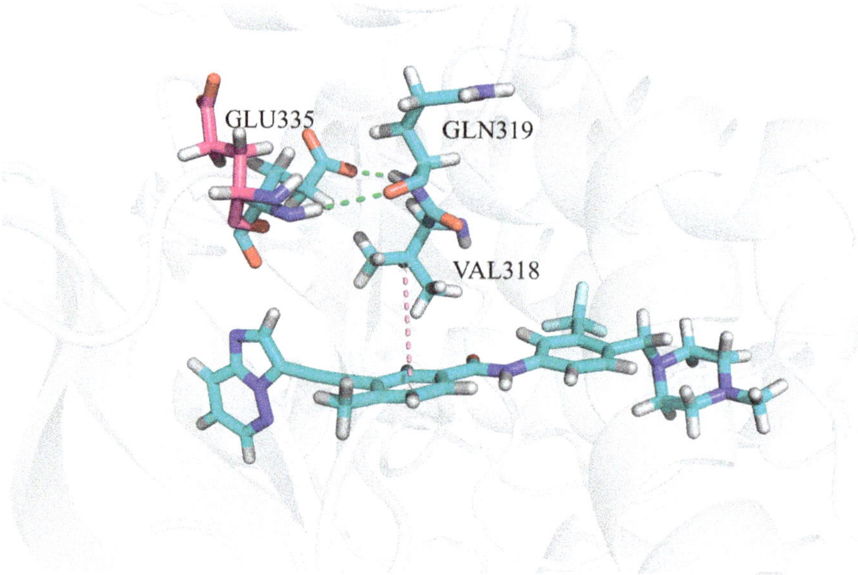

Figure 9. Interactions of residues GLU335, GLN319, and VAL318 with ponatinib. Residues in the ABL1–ponatinib system are shown in light magenta, while residues in the ABL1–ponatinib–asciminib system are shown in cyan. Residues are displayed in stick mode. H-bonds and hydrophobic interactions are marked with green and pink dotted lines, respectively. Nitrogen atoms are displayed in blue, and oxygen atoms are in red.

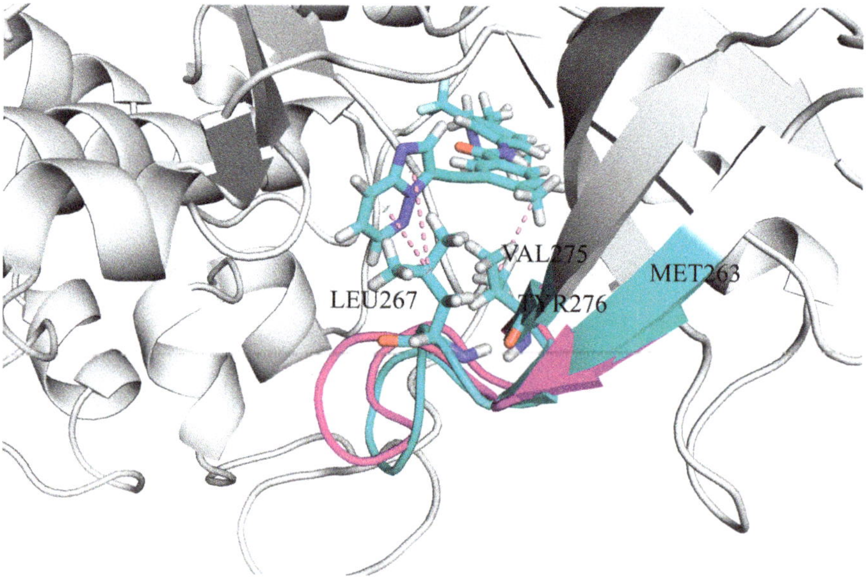

Figure 10. Interactions of ponatinib with residues LEU267 and VAL275. Residues in the ABL1–ponatinib system (MET263 → TYR276) are shown in light magenta, while residues in the ABL1–ponatinib–asciminib system (MET263 → TYR276) are shown in cyan. Residues LEU267 and VAL275 are displayed in stick mode, and hydrophobic interactions are marked with pink dotted lines. Nitrogen atoms are displayed in blue, and oxygen atoms are in red.

4. Conclusions

Mutations in the tyrosine kinase domain of the BCR-ABL1 fusion protein pose a significant challenge to the treatment of chronic myeloid leukemia (CML). Combination therapy is emerging as a promising new strategy in addressing this issue. The combination therapy of asciminib with ATP competitive inhibitors has been shown to effectively control and eliminate CML xenograft tumors, with no recurrence observed after treatment cessation. Interestingly, in the combined treatment of asciminib with nilotinib or ponatinib, the required doses of nilotinib or ponatinib were significantly reduced [50]. This suggests that we can achieve therapeutic effects with lower drug doses, thereby reducing the potential side effects and toxicity associated with the drugs. This study employed molecular dynamics (MD) simulation and molecular mechanics generalized born surface area (MM-GB/SA) to investigate the interactions between inhibitors and residues at the ATP site when asciminib binds to ABL1. This study elucidates the molecular mechanisms underlying the impact of asciminib binding to ABL1 on the binding affinity of ATP competitive inhibitors, including nilotinib and ponatinib, with ABL1. The study also clarifies that the difference in binding affinity between imatinib and nilotinib with ABL1, when asciminib is bound to ABL1, is attributed to the inherent binding affinity of the inhibitors themselves with ABL1. The successful combination therapy of asciminib with ponatinib has effectively addressed the issue of ponatinib's limited use due to its severe side effects. This study indicates that the enhancement of ponatinib's binding affinity to ABL1 relies on the strengthened interactions between certain groups of ponatinib and residues at the ATP site. The research findings contribute to the potential for further structural optimization of ponatinib, with the expectation of developing inhibitors with lower toxic side effects when used in combination with asciminib for the treatment of ABL1.

The study provides a molecular understanding of the combined treatment of ABL1, and our results can aid in the further design and development of these inhibitors to enhance their inhibitory effects in the treatment of CML. This understanding is crucial for optimizing therapeutic strategies and potentially leads to the discovery of inhibitors with lower toxic side effects, which is particularly important in the context of combination therapies involving asciminib and other ATP competitive inhibitors like nilotinib and ponatinib.

Supplementary Materials: The following supporting information can be downloaded at: https://www.mdpi.com/article/10.3390/biom15091214/s1, Figure S1: Dynamic Cross-Correlation of Residues in the Allosteric Communication Path; Table S1: Binding free energy decomposition on individual residues in the ABL1- nilotinib system (kcal mol^{-1}); Table S2: Binding free energy decomposition on individual residues in the ABL1-nilotinib-asciminib system (kcal mol^{-1}); Table S3: Binding free energy decomposition on individual residues in the ABL1-ponatinib system (kcal mol^{-1}); Table S4: Binding free energy decomposition on individual residues in the ABL1-ponatinib-asciminib system (kcal mol^{-1}).

Author Contributions: Conceptualization, Methodology, Validation, Formal analysis, Investigation, Writing—Original Draft, Writing—Review & Editing, Visualization, J.M.; Software, Data Curation, Project administration, J.Z.; formal analysis, Resources, Supervision, Funding acquisition, H.G. All authors have read and agreed to the published version of the manuscript.

Funding: This research was funded by the Natural Science Foundation of Shandong, China [Grant No. ZR2023MC059], and the High-end Talent Introduction "Double Hundred Plan" Special Foundation of Yantai City [Grant No. 612211012002]." And The APC was funded by the Natural Science Foundation of Shandong, China [Grant No. ZR2023MC059], and the High-end Talent Introduction "Double Hundred Plan" Special Foundation of Yantai City [Grant No. 612211012002].

Institutional Review Board Statement: Not applicable.

Informed Consent Statement: Not applicable.

Data Availability Statement: Data and software used in this work are all openly available. The initial structure of the ABL1 protein (PDB code:5MO4) was obtained from the RCSB Protein Data Bank (https://www.rcsb.org) (accessed on 11 March 2024). The structures of nilotinib, ponatinib, and asciminib were obtained from PubChem (https://pubchem.ncbi.nlm.nih.gov) (accessed on 13 March 2024). All of software utilized in this work, Discovery Studio 2020, H++ server, Amber18, and WebPSN are available since we have the license for each one.

Conflicts of Interest: The authors declare no conflicts of interest.

References

1. Jabbour, E.; Kantarjian, H. Chronic myeloid leukemia: 2020 update on diagnosis, therapy and monitoring. *Am. J. Hematol.* **2020**, *95*, 691–709. [CrossRef] [PubMed]
2. Rowley, J.D. A New Consistent Chromosomal Abnormality in Chronic Myelogenous Leukaemia identified by Quinacrine Fluorescence and Giemsa Staining. *Nature* **1973**, *243*, 290–293. [CrossRef] [PubMed]
3. Braun, T.P.; Eide, C.A.; Druker, B.J. Response and Resistance to BCR-ABL1-Targeted Therapies. *Cancer Cell* **2020**, *37*, 530–542. [CrossRef]
4. El-Tanani, M.; Nsairat, H.; Matalka, I.I.; Lee, Y.F.; Rizzo, M.; Aljabali, A.A.; Mishra, V.; Mishra, Y.; Hromić-Jahjefendić, A.; Tambuwala, M.M. The impact of the BCR-ABL oncogene in the pathology and treatment of chronic myeloid leukemia. *Pathol.-Res. Pract.* **2024**, *254*, 155161. [CrossRef]
5. Han, H.-J.; Kim, J.J.; Pyne, D.; Travas, A.; Ambalavanan, A.; Kimura, S.; Deininger, M.W.; Kim, J.-W.; Kim, D.D.H. In vitro evidence of synergistic efficacy with asciminib combined with reduced dose of ATP-binding pocket tyrosine kinase inhibitors according to the ABL1 kinase domain mutation profile. *Leukemia* **2023**, *38*, 412–415. [CrossRef]
6. Hunter, T. Treatment for chronic myelogenous leukemia: The long road to imatinib. *J. Clin. Investig.* **2007**, *117*, 2036–2043. [CrossRef]
7. Chandrasekhar, C.; Kumar, P.S.; Sarma, P.V.G.K. Novel mutations in the kinase domain of BCR-ABL gene causing imatinib resistance in chronic myeloid leukemia patients. *Sci. Rep.* **2019**, *9*, 1–17. [CrossRef]
8. Saglio, G. Second-generation TKIs: Which and when? *Leuk. Suppl.* **2012**, *1*, S40–S42. [CrossRef]
9. Abruzzese, E.; Breccia, M.; Latagliata, R. Second-Generation Tyrosine Kinase Inhibitors in First-Line Treatment of Chronic Myeloid Leukaemia (CML). *BioDrugs* **2013**, *28*, 17–26. [CrossRef]
10. Blay, J.-Y.; von Mehren, M. Nilotinib: A Novel, Selective Tyrosine Kinase Inhibitor. *Semin. Oncol.* **2011**, *38*, S3–S9. [CrossRef] [PubMed]
11. Nascimento, M.; Moura, S.; Parra, L.; Vasconcellos, V.; Costa, G.; Leite, D.; Dias, M.; Fernandes, T.V.A.; Hoelz, L.; Pimentel, L.; et al. Ponatinib: A Review of the History of Medicinal Chemistry behind Its Development. *Pharmaceuticals* **2024**, *17*, 1361. [CrossRef]
12. Weisberg, E.; Manley, P.; Mestan, J.; Cowan-Jacob, S.; Ray, A.; Griffin, J.D. AMN107 (nilotinib): A novel and selective inhibitor of BCR-ABL. *Br. J. Cancer* **2006**, *94*, 1765–1769. [CrossRef] [PubMed]
13. Lambert, G.K.; Duhme-Klair, A.-K.; Morgan, T.; Ramjee, M.K. The background, discovery and clinical development of BCR-ABL inhibitors. *Drug. Discov. Today* **2013**, *18*, 992–1000. [CrossRef] [PubMed]
14. Kwarcinski, F.E.; Brandvold, K.R.; Phadke, S.; Beleh, O.M.; Johnson, T.K.; Meagher, J.L.; Seeliger, M.A.; Stuckey, J.A.; Soellner, M.B. Conformation-Selective Analogues of Dasatinib Reveal Insight into Kinase Inhibitor Binding and Selectivity. *ACS. Chem. Biol.* **2016**, *11*, 1296–1304. [CrossRef]
15. Gorre, M.E.; Mohammed, M.; Ellwood, K.; Hsu, N.; Paquette, R.; Rao, P.N.; Sawyers, C.L. Clinical Resistance to STI-571 Cancer Therapy Caused by BCR-ABL Gene Mutation or Amplification. *Science* **2001**, *293*, 876–880. [CrossRef]
16. Manley, P.W.; Barys, L.; Cowan-Jacob, S.W. The specificity of asciminib, a potential treatment for chronic myeloid leukemia, as a myristate-pocket binding ABL inhibitor and analysis of its interactions with mutant forms of BCR-ABL1 kinase. *Leuk. Res.* **2020**, *98*, 106458. [CrossRef] [PubMed]
17. Tamai, M.; Inukai, T.; Kojika, S.; Abe, M.; Kagami, K.; Harama, D.; Shinohara, T.; Watanabe, A.; Oshiro, H.; Akahane, K.; et al. T315I mutation of BCR-ABL1 into human Philadelphia chromosome-positive leukemia cell lines by homologous recombination using the CRISPR/Cas9 system. *Sci. Rep.* **2018**, *8*, 9966. [CrossRef]
18. Mian, A.; Schüll, M.; Zhao, Z.; Oancea, C.; Hundertmark, A.; Beissert, T.; Ottmann, O.G.; Ruthardt, M. The gatekeeper mutation T315I confers resistance against small molecules by increasing or restoring the ABL-kinase activity accompanied by aberrant transphosphorylation of endogenous BCR, even in loss-of-function mutants of BCR/ABL. *Leukemia* **2009**, *23*, 1614–1621. [CrossRef]
19. Zabriskie, M.S.; Eide, C.A.; Tantravahi, S.K.; Vellore, N.A.; Estrada, J.; Nicolini, F.E.; Khoury, H.J.; Larson, R.A.; Konopleva, M.; Cortes, J.E.; et al. BCR-ABL1 Compound Mutations Combining Key Kinase Domain Positions Confer Clinical Resistance to Ponatinib in Ph Chromosome-Positive Leukemia. *Cancer Cell* **2014**, *26*, 428–442. [CrossRef]

20. Georgoulia, P.S.; Todde, G.; Bjelic, S.; Friedman, R. The catalytic activity of Abl1 single and compound mutations: Implications for the mechanism of drug resistance mutations in chronic myeloid leukaemia. *Biochim. Biophys. Acta (BBA)-Gen. Subj.* **2019**, *1863*, 732–741. [CrossRef]
21. Chen, J.; Liu, H.; Wang, F.; Zhang, Y.; Chen, X.; Nie, D.; Li, Y.; Tan, Y.; Xu, Y.; Ma, X. Dynamic Evolution of Ponatinib Resistant BCR-ABL1 T315 and Compound Mutations. *Blood* **2019**, *134*, 3796. [CrossRef]
22. Heiblig, M.; Rea, D.; Chrétien, M.-L.; Charbonnier, A.; Rousselot, P.; Coiteux, V.; Escoffre-Barbe, M.; Dubruille, V.; Huguet, F.; Cayssials, E.; et al. Ponatinib evaluation and safety in real-life chronic myelogenous leukemia patients failing more than two tyrosine kinase inhibitors: The PEARL observational study. *Exp. Hematol.* **2018**, *67*, 41–48. [CrossRef]
23. Teng, M.; Luskin, M.R.; Cowan-Jacob, S.W.; Ding, Q.; Fabbro, D.; Gray, N.S. The Dawn of Allosteric BCR-ABL1 Drugs: From a Phenotypic Screening Hit to an Approved Drug. *J. Med. Chem.* **2022**, *65*, 7581–7594. [CrossRef] [PubMed]
24. Assanto, G.M.; Scalzulli, E.; Breccia, M. Asciminib in chronic myeloid leukemia. *Drugs Today* **2022**, *58*, 479–489. [CrossRef] [PubMed]
25. Schuld, P.; Grzesiek, S.; Schlotte, J.; Habazettl, J.M.; Jahnke, W.; Barys, L.; Cowan-Jacob, S.W.; Loo, A.; Wiget, A.; Manley, P.W. Structural and Biochemical Studies Confirming the Mechanism of Action of Asciminib, an Agent Specifically Targeting the ABL Myristoyl Pocket (STAMP). *Blood* **2020**, *136*, 34–35. [CrossRef]
26. Schoepfer, J.; Jahnke, W.; Berellini, G.; Buonamici, S.; Cotesta, S.; Cowan-Jacob, S.W.; Dodd, S.; Drueckes, P.; Fabbro, D.; Gabriel, T.; et al. Discovery of Asciminib (ABL001), an Allosteric Inhibitor of the Tyrosine Kinase Activity of BCR-ABL1. *J. Med. Chem.* **2018**, *61*, 8120–8135. [CrossRef] [PubMed]
27. Beyett, T.S.; To, C.; Heppner, D.E.; Rana, J.K.; Schmoker, A.M.; Jang, J.; De Clercq, D.J.H.; Gomez, G.; Scott, D.A.; Gray, N.S.; et al. Molecular basis for cooperative binding and synergy of ATP-site and allosteric EGFR inhibitors. *Nat. Commun.* **2022**, *13*, 1–11. [CrossRef] [PubMed]
28. Lindström, H.J.G.; Friedman, R. The effects of combination treatments on drug resistance in chronic myeloid leukaemia: An evaluation of the tyrosine kinase inhibitors axitinib and asciminib. *BMC. Cancer* **2020**, *20*, 1–14. [CrossRef]
29. Jahnke, W.; Paladini, J.; Habazettl, J.M.; Wiget, A.; Loo, A.; Jacob, S.W.C.; Grzesiek, S.; Manley, P.W. Correspondence on "Synergy and Antagonism between Allosteric and Active-Site Inhibitors of Abl Tyrosine Kinase". *Angew. Chem. Int. Ed.* **2022**, *61*, e202117276. [CrossRef]
30. Kim, C.; Ludewig, H.; Hadzipasic, A.; Kutter, S.; Nguyen, V.; Kern, D. A biophysical framework for double-drugging kinases. *Proc. Natl. Acad. Sci. USA* **2023**, *120*, e2304611120. [CrossRef]
31. Okamoto, N.; Yagi, K.; Imawaka, S.; Takaoka, M.; Aizawa, F.; Niimura, T.; Goda, M.; Miyata, K.; Kawada, K.; Izawa-Ishizawa, Y.; et al. Asciminib, a novel allosteric inhibitor of BCR-ABL1, shows synergistic effects when used in combination with imatinib with or without drug resistance. *Pharmacol. Res. Perspect.* **2024**, *12*, e1214. [CrossRef]
32. Wylie, A.A.; Schoepfer, J.; Jahnke, W.; Cowan-Jacob, S.W.; Loo, A.; Furet, P.; Marzinzik, A.L.; Pelle, X.; Donovan, J.; Zhu, W.; et al. The allosteric inhibitor ABL001 enables dual targeting of BCR–ABL1. *Nature* **2017**, *543*, 733–737. [CrossRef]
33. El Rashedy, A.A.; Appiah-Kubi, P.; Soliman, M.E.S. A Synergistic Combination Against Chronic Myeloid Leukemia: An Intra-molecular Mechanism of Communication in BCR–ABL1 Resistance. *Protein J.* **2019**, *38*, 142–150. [CrossRef] [PubMed]
34. Gleixner, K.V.; Filik, Y.; Berger, D.; Schewzik, C.; Stefanzl, G.; Sadovnik, I.; Degenfeld-Schonburg, L.; Eisenwort, G.; Schneeweiss-Gleixner, M.; Byrgazov, K.; et al. Asciminib and ponatinib exert synergistic anti-neoplastic effects on CML cells expressing BCR-ABL1 T315I-compound mutations. *Am. J. Cancer Res.* **2021**, *11*, 4470–4484. [PubMed]
35. El Rashedy, A.A.; Olotu, F.A.; Soliman, M.E.S. Dual Drug Targeting of Mutant Bcr-Abl Induces Inactive Conformation: New Strategy for the Treatment of Chronic Myeloid Leukemia and Overcoming Monotherapy Resistance. *Chem. Biodivers.* **2018**, *15*, e1700533. [CrossRef]
36. Gordon, J.C.; Myers, J.B.; Folta, T.; Shoja, V.; Heath, L.S.; Onufriev, A. H++: A server for estimating pKas and adding missing hydrogens to macromolecules. *Nucleic Acids Res.* **2005**, *33*, W368–W371. [CrossRef]
37. Case, D.A.; Cheatham, T.E., III; D.arden, T.; Gohlke, H.; Luo, R.; Merz, K.M., Jr.; Onufriev, A.; Simmerling, C.; Wang, B.; Woods, R.J. The Amber biomolecular simulation programs. *J. Comput. Chem.* **2005**, *26*, 1668–1688. [CrossRef]
38. Wang, J.; Wolf, R.M.; Caldwell, J.W.; Kollman, P.A.; Case, D.A. Development and testing of a general amber force field. *J. Comput. Chem.* **2004**, *25*, 1157–1174. [CrossRef]
39. Maier, J.A.; Martinez, C.; Kasavajhala, K.; Wickstrom, L.; Hauser, K.E.; Simmerling, C. ff14SB: Improving the accuracy of protein side chain and backbone parameters from ff99SB. *J. Chem. Theory Comput.* **2015**, *11*, 3696–3713. [CrossRef] [PubMed]
40. Florová, P.; Sklenovský, P.; Banáš, P.; Otyepka, M. Explicit Water Models Affect the Specific Solvation and Dynamics of Unfolded Peptides While the Conformational Behavior and Flexibility of Folded Peptides Remain Intact. *J. Chem. Theory Comput.* **2010**, *6*, 3569–3579. [CrossRef]
41. Amber 2024. Available online: https://ambermd.org/CiteAmber.php (accessed on 13 July 2024).

42. Kollman, P.A.; Massova, I.; Reyes, C.; Kuhn, B.; Huo, S.; Chong, L.; Lee, M.; Lee, T.; Duan, Y.; Wang, W.; et al. Calculating Structures and Free Energies of Complex Molecules: Combining Molecular Mechanics and Continuum Models. *Accounts Chem. Res.* **2000**, *33*, 889–897. [CrossRef] [PubMed]
43. Kong, X.; Sun, H.; Pan, P.; Zhu, F.; Chang, S.; Xu, L.; Li, Y.; Hou, T. Importance of protein flexibility in molecular recognition: A case study on Type-I1/2 inhibitors of ALK. *Phys. Chem. Chem. Phys.* **2018**, *20*, 4851–4863. [CrossRef]
44. Sun, H.; Li, Y.; Shen, M.; Tian, S.; Xu, L.; Pan, P.; Guan, Y.; Hou, T. Assessing the performance of MM/PBSA and MM/GBSA methods. Improved docking performance using high solute dielectric constant MM/GBSA and MM/PBSA rescoring. *Phys. Chem. Chem. Phys.* **2014**, *16*, 22035–22045. [CrossRef]
45. Federico, L.B.; Silva, G.M.; Gomes, S.Q.; Francischini, I.A.G.; Barcelos, M.P.; dos Santos, C.B.R.; Costa, L.T.; Rosa, J.M.C.; Silva, C.H.T.d.P.d. Potential colchicine binding site inhibitors unraveled by virtual screening, molecular dynamics and MM/PBSA. *Comput. Biol. Med.* **2021**, *137*, 104817. [CrossRef]
46. Seeber, M.; Felline, A.; Raimondi, F.; Mariani, S.; Fanelli, F. WebPSN: A web server for high-throughput investigation of structural communication in biomacromolecules. *Bioinformatics* **2014**, *31*, 779–781. [CrossRef]
47. Felline, A.; Seeber, M.; Fanelli, F. PSNtools for standalone and web-based structure network analyses of conformational ensembles. *Comput. Struct. Biotechnol. J.* **2022**, *20*, 640–649. [CrossRef]
48. Sun, D.; Wang, Z.; Zheng, Q.; Zhang, H. Exploring the inhibition mechanism on HIF-2 by inhibitor PT2399 and 0X3 using molecular dynamics simulations. *J. Mol. Recognit.* **2018**, *31*, e2730. [CrossRef] [PubMed]
49. Kasahara, K.; Fukuda, I.; Nakamura, H.; Agarwal, P.K. A Novel Approach of Dynamic Cross Correlation Analysis on Molecular Dynamics Simulations and Its Application to Ets1 Dimer–DNA Complex. *PLoS ONE* **2014**, *9*, e112419. [CrossRef] [PubMed]
50. Curik, N.; Laznicka, A.; Polivkova, V.; Krizkova, J.; Pokorna, E.; Semerak, P.; Suchankova, P.; Burda, P.; Hochhaus, A.; Polakova, K.M. Combination therapies with ponatinib and asciminib in a preclinical model of chronic myeloid leukemia blast crisis with compound mutations. *Leukemia* **2024**, *38*, 1415–1418. [CrossRef]

biomolecules

MDPI

Article

Ensemble-Based Binding Free Energy Profiling and Network Analysis of the KRAS Interactions with DARPin Proteins Targeting Distinct Binding Sites: Revealing Molecular Determinants and Universal Architecture of Regulatory Hotspots and Allosteric Binding

Mohammed Alshahrani [1], Vedant Parikh [1], Brandon Foley [1] and Gennady Verkhivker [1,2,*]

[1] Keck Center for Science and Engineering, Graduate Program in Computational and Data Sciences, Schmid College of Science and Technology, Chapman University, Orange, CA 92866, USA; alshahrani@chapman.edu (M.A.); vedpar31@gmail.com (V.P.); brfoley@chapman.edu (B.F.)

[2] Department of Biomedical and Pharmaceutical Sciences, Chapman University School of Pharmacy, Irvine, CA 92618, USA

* Correspondence: verkhivk@chapman.edu; Tel.: +1-714-516-4586

Abstract: KRAS is a pivotal oncoprotein that regulates cell proliferation and survival through interactions with downstream effectors such as RAF1. Despite significant advances in understanding KRAS biology, the structural and dynamic mechanisms of KRAS allostery remain poorly understood. In this study, we employ microsecond molecular dynamics simulations, mutational scanning, and binding free energy calculations together with dynamic network modeling to dissect how engineered DARPin proteins K27, K55, K13, and K19 engage KRAS through diverse molecular mechanisms ranging from effector mimicry to conformational restriction and allosteric modulation. Mutational scanning across all four DARPin systems identifies a core set of evolutionarily constrained residues that function as universal hotspots in KRAS recognition. KRAS residues I36, Y40, M67, and H95 consistently emerge as critical contributors to binding stability. Binding free energy computations show that, despite similar binding modes, K27 relies heavily on electrostatic contributions from major binding hotspots while K55 exploits a dense hydrophobic cluster enhancing its effector-mimetic signature. The allosteric binders K13 and K19, by contrast, stabilize a KRAS-specific pocket in the α3–loop–α4 motif, introducing new hinges and bottlenecks that rewire the communication architecture of KRAS without full immobilization. Network-based analysis reveals a strikingly consistent theme: despite their distinct mechanisms of recognition, all systems engage a unifying allosteric architecture that spans multiple functional motifs. This architecture is not only preserved across complexes but also mirrors the intrinsic communication framework of KRAS itself, where specific residues function as central hubs transmitting conformational changes across the protein. By integrating dynamic profiling, energetic mapping, and network modeling, our study provides a multi-scale mechanistic roadmap for targeting KRAS, revealing how engineered proteins can exploit both conserved motifs and isoform-specific features to enable precision modulation of KRAS signaling in oncogenic contexts.

Keywords: molecular dynamics; protein stability; mutational scanning; binding energetics; evolutionary mechanisms

1. Introduction

The GTPase KRAS (Kirsten rat sarcoma viral oncogene homolog) is a critical oncogene that is somatically mutated in approximately 10% of all human cancers, with particularly high prevalence in certain malignancies. Specifically, KRAS mutations are found in ~90% of pancreatic adenocarcinomas, ~40% of colorectal adenocarcinomas, ~35% of lung adenocarcinomas, and ~20% of multiple myeloma cases [1–6]. Among the RAS oncogene family, which includes KRAS, HRAS, and NRAS, KRAS is the most frequently mutated, accounting for ~30% of all RAS-driven cancers, while HRAS and NRAS mutations are less common, occurring in ~8% and ~3% of cases, respectively [7,8]. These mutations drive tumorigenesis by disrupting normal cellular signaling pathways, leading to uncontrolled cell proliferation and survival. RAS proteins function as binary molecular switches, cycling between an active GTP-bound state and an inactive GDP-bound state. This cycling is tightly regulated by two classes of proteins: Guanine nucleotide exchange factors (GEFs), which promote the exchange of GDP for GTP, activating RAS, and GTPase-activating proteins (GAPs), which enhance the intrinsic GTPase activity of RAS, facilitating GTP hydrolysis and returning RAS to its inactive state [9].

A complete GTPase reaction requires well-ordered conformations of the KRAS active site, which includes the phosphate-binding loop, P-loop (residues 10–17), switch I (residues 25–40), and switch II (residues 60–76) regions [10–12] (Figure 1). KRAS can adopt both open and closed states in its active (GTP-bound) form, depending on the conformation of its switch regions (Switch I and Switch II). These states are dynamic and play a critical role in regulating KRAS signaling and interactions with effector proteins. In the GTP-bound state, Switch I and Switch II undergo conformational changes that are critical for effector protein binding. Historically, it has been assumed that the GTP-bound state predominantly adopts a "closed" conformation (state II), which is competent for effector binding and signaling [13]. However, recent studies challenged this oversimplified view by demonstrating that even in the GTP-bound state, KRAS can sample multiple conformations, including an "open" state (state I) that is not competent for effector binding. Solution-state nuclear magnetic resonance spectroscopy (NMR) studies of RAS bound to the non-hydrolyzable GTP analog, GMPPNP, revealed two dominant conformational states in the switch I defined as state I (open conformation) and state II (closed conformation) that exist in the dynamic equilibrium (Supporting Information, Figure S1) [14–17]. In the state I, switch I region adopts an open conformation that is incompatible with effector binding. This conformation was described as "inactive" because it does not support downstream signaling, despite being in the GTP-bound state. In state II, switch I adopts a closed conformation that facilitates high-affinity binding to effector proteins like RAF1. These NMR studies demonstrated that the GTP-bound state of KRAS is not a single, static conformation but rather a dynamic ensemble of states, some of which are signaling-competent (closed) and others signaling-incompetent (open) (Supporting Information, Figure S1). Most oncogenic mutations in RAS proteins, particularly at residues G12, G13, and Q61, impair GTP hydrolysis and lock RAS in its active, GTP-bound conformation. These mutations introduce steric hindrance or alter the local structure of RAS, preventing GAPs from binding effectively. For example, mutations such as G12C, G12D, and Q61H alter the conformational dynamics of KRAS, leading to increased signaling activity and resistance to GAPs [18–21]. As a result, RAS proteins remain constitutively active, leading to the overactivation of downstream signaling pathways such as the MAP kinase pathway (RAS-RAF-MEK-ERK) and the PI3K pathway (PI3K-AKT-mTOR). These pathways drive cell proliferation, survival, and metabolism, contributing to tumorigenesis [22].

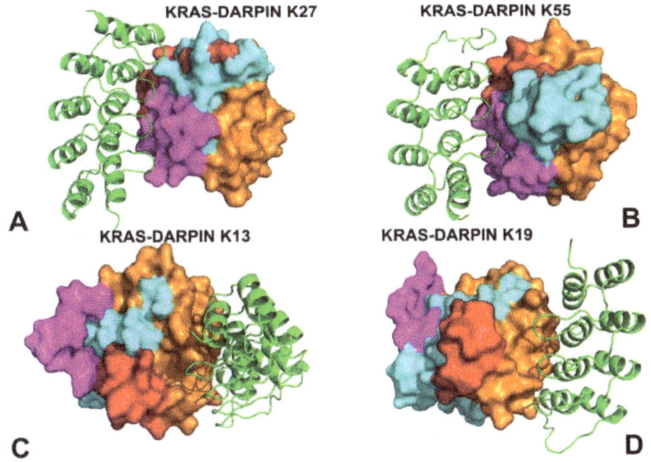

Figure 1. Structural overview and organization of the KRAS protein complexes with the DARPin proteins K27 (**A**), K55 (**B**), K13 (**C**), and K19 (**D**). The crystal structures of KRAS in complexes with DARPin K27 (pdb id 5O2S), DARPin K55 (pdb id 5O2T), DARPin K13 (pdb id 6H46), and DARPin K19 (pdb id 6H47) are used for the analysis and visualization. The KRAS protein is shown in surface representation, and the DARPin protein binding partners are shown in green ribbons. The functional KRAS regions are highlighted: switch I (residues 24–40 in magenta surface), switch II (residues 60–76 in red surface), and allosteric KRAS lobe (residues 87–166 in orange surface). The remaining KRAS residues are shown in cyan-colored surface representation.

The interaction between KRAS and RAF1 (CRAF) is a critical step in the activation of the MAPK pathway, which regulates cell growth, differentiation, and survival. RAF1 is a serine/threonine kinase that binds to KRAS through two domains: the RAS-binding domain (RBD) and the cysteine-rich domain (CRD). Structural studies have provided detailed insights into the molecular mechanisms underlying this interaction. High-resolution crystal structures of KRAS bound to the RBD of RAF1 have revealed that the RBD interacts primarily with the switch I (residues 25–40) and switch II (residues 60–76) regions of KRAS [23,24]. These regions undergo significant conformational changes during the GTPase cycle, transitioning between disordered (inactive) and ordered (active) states. The N-terminal region of BRAF, including the CR1 and CR2 domains, remains unresolved in the absence of RAS[20,] which suggests that RAS binding is essential for stabilizing these domains and enabling proper RAF activation. Recent cryo-electron microscopy (cryo-EM) studies of full-length BRAF in complex with 14-3-3 and MEK have provided unprecedented views of RAF's structural biology [25–29]. Oncogenic mutations in KRAS, such as G12V, G13D, and Q61R, do not disrupt the overall interaction with RAF1 but induce local rearrangements in the switch regions. X-ray crystallography of $KRAS^{Q61H}$:GTP revealed that a hyperdynamic switch II allows for a more stable interaction with switch I, suggesting that enhanced RAF activity arises from a combination of absent intrinsic GTP hydrolysis activity and increased affinity for RAF [30]. The nanodisc platforms and paramagnetic relaxation enhancement (PRE) analyses were employed to determine the structure of a hetero-tetrameric complex comprising KRAS and the RBD and CRD of activated RAF1 [5]. By leveraging these advanced techniques, this study investigated how the binding of the RBD or the RBD–CRD differentially modulates the dimerization modes of KRAS on both anionic and neutral membranes. A key discovery of this work is that RBD binding allosterically generates two distinct KRAS dimer interfaces, which exist in a dynamic equilibrium. One interface is favored when KRAS is free, while the other is stabilized when KRAS is in complex with the RBD–CRD [5]. This allosteric regulation highlights the plasticity of KRAS dimerization and its responsiveness to effector binding. Biochemical and structural analyses of variants identified in a KRAS-G12D revealed that attenuation of oncogenic KRAS can be mediated by protein instability and conformational rigidity, resulting in reduced binding affinity to effector proteins or reduced SOS-mediated nucleotide exchange activity. These studies define the landscape of single amino acid alterations that modulate the function of KRAS [31].

Designed Ankyrin Repeat Proteins (DARPins), such as K27 and K55, represent a class of synthetic binding proteins that engage the switch regions of KRAS with high affinity and specificity. These engineered proteins offer promising alternatives to traditional antibodies due to their small size, high stability, and modular recognition capabilities. Engineered protein binders such as DARPins K27 and K55 offer alternative strategies for modulating KRAS function: K27 stabilizes the GDP-bound inactive state, interfering with SOS1-mediated activation, while K55 mimics natural effectors, engaging the active-state conformation and disrupting downstream signaling [32]. These systems—natural regulators and synthetic binders alike—highlight the importance of both conserved structural motifs and flexible interaction networks in shaping KRAS functionality. Another study reported the design of allosteric DARPin binders K13 and K19 that specifically inhibit the KRAS isoform by binding to an allosteric site encompassing the region around KRAS-specific residue histidine 95 at the helix $\alpha3$/loop 7/helix $\alpha4$ interface [33]. Binding by the DARPin proteins in the allosteric lobe of KRAS influences KRAS-effector interactions, modulating KRAS nucleotide exchange and inhibiting KRAS dimerization at the plasma membrane. These results highlighted the importance of targeting the $\alpha3$/-$\alpha4$ interface, a previously untargeted site in RAS, for specifically inhibiting KRAS function [33]. A comprehensive analysis of KRAS energetics and allostery was undertaken using a combination of structural biology and biophysical techniques to map the energetic and allosteric landscape of KRAS [34]. This seminal study characterized allosteric energy landscapes that highlighted the thermodynamic consequences of mutations across KRAS, providing a quantitative framework to determine allosteric binding sites and their effect on KRAS binding. The study revealed the existence of extensive allosteric networks within KRAS that connect distant regions of the protein. These networks facilitate communication between the nucleotide-binding site, effector-binding site, and other regulatory regions. The authors also identified several cryptic pockets on the surface of KRAS that are transiently formed during conformational changes [34] and can be targeted by small molecules to modulate KRAS activity.

While crystallography and cryo-EM have provided valuable structural insights, computational studies and molecular modeling have played an equally important role in elucidating the dynamics, energetics, and molecular mechanisms of KRAS-RAF1 binding. MD simulations, free energy calculations, and docking studies have complemented experimental techniques, revealing the role of conformational flexibility, membrane interactions, and allosteric regulation. MD simulations have revealed how mutations affect the mobility of the switch I and switch II regions of KRAS, adopting multiple conformations that influence binding to RAF1 [35,36]. Multiscale coarse-grained and all-atom MD simulations of KRAS4b bound to the RBD and CRD domains of RAF-1, both in solution and anchored to a model plasma membrane, explored how RAS membrane orientation relates to the protein dynamics within the RAS-RBDCRD complex [37]. Solution MD simulations describe dynamic KRAS4b-CRD conformations, suggesting that the CRD has sufficient flexibility in this environment to substantially change its binding interface with KRAS4b. Quantitative spatial imaging and atomistic molecular dynamics simulations are used to examine molecular details of K-Ras plasma membrane binding [38] and the effects of amino acid variations on the structure and dynamics of KRAS-WT and the oncogenic mutants

G12D, G12V, and G13D of the HRAS and KRAS proteins. Based on data from μs-scale molecular dynamics simulations, we show that the overall structure of the proteins remains similar, but there are important differences in dynamics and interaction networks [39]. The insightful study employed all-atom molecular dynamics simulations of the C-RAF RBD and CRD regions when bound to oncogenic KRAS4B, suggesting that the membrane plays an integral role in regulating the configurational ensemble of the complex that samples a few states dynamically, reflecting a competition between CRAF CRD- and KRAS4B- membrane interactions [40]. Microsecond MD simulations are used to unveil the binding mechanisms of the FDA-approved MEK inhibitor trametinib with KRASG12D, providing insights for potential drug repurposing. The binding of trametinib was compared with clinical trial drug MRTX1133, which demonstrates exceptional activity against KRASG12D, for a better understanding of the interaction mechanism of trametinib with KRASG12D [41].

Multiple microsecond MD simulations and Markov State Model (MSM) analysis probed the kinetics of MRTX1133 binding to KRAS G12D, revealing the kinetically metastable states and potential pathways of MRTX1133 binding, while MM/GBSA analysis identified eight critical residues for MRTX1133 recognition and binding [42]. KRAS binding pockets and ligand interactions, identifying three distinct sites: the conserved nucleotide-binding site, the shallow switch-I/II pocket, and the allosteric Switch-II/α3 pocket [43]. In addition, this study identified KRAS-inhibitor interaction fingerprints aided by MD simulations that characterized the flexibility of these sites to accommodate diverse ligands [43]. MD simulations of the KRASG12C-AMG 510 complex investigated the impact of ligand binding on KRASG12C conformational changes [44]. Mapping simulation trajectories onto the Principal Component Analysis (PCA) model revealed that KRASG12C-AMG 510 predominantly adopted the inactive conformation, where AMG 510 reduced the flexibility of the switch regions, stabilizing the KRASG12C-AMG 510 complex in the inactive state [44]. MD simulations explored the structural dynamics and stability of wild-type KRAS and its oncogenic variants (G12C, G12D, G12V, G13D), followed by the free energy landscape analysis revealing unique conformational dynamics and altered thermodynamic stability in mutated KRAS variants [45]. Multiple MD simulations were performed on wild-type and mutant KRAS structures to investigate how G12C and G12D mutations stabilize the active state and how AMG-510 and MRTX1133 inhibitors force these mutants into the inactive state [46]. This study revealed that binding of AMG-510 (Sotorasib) can induce significant stabilization of the Switch-II region of KRAS-G12C, surpassing that of MRTX1133 bound with the KRAS-G12D mutant. The mutation probabilities of KRAS-G12 missense mutants and their long-timescale dynamics were assessed by atomistic MD simulations (170 μs) and MSM analysis, revealing allosteric hydrophobic signaling network in KRAS and allosteric modulation of protein dynamics among the G12X mutants, which is manifested in the switch regions that are responsible for the effector protein binding [47]. GaMD simulations were performed on the GDP-bound wild-type, G12A, G12D, and G12R KRAS to probe mutation-mediated impacts on conformational alterations of KRAS, showing that all three G12 mutations can alter the structural flexibility of the switch domains [48]. The analyses of the free energy landscapes indicated that the examined G12 mutations can induce more conformational states of KRAS and particularly result in more disordered switch domains [48]. GaMD simulations followed by deep learning (DL) were conducted to probe the effect of G12C mutation and binding of three proteins NF1, RAF1, and SOS1 on the conformational dynamics and free energy landscapes of KRAS4B, showing that mutation and binding alter contacts in key structural domains and enhance switch I and switch II mobility [49]. These findings highlighted the roles of partner binding and G12C in KRAS4B activity and allosteric regulation, providing theoretical insights into KRAS4B function. GaMD simulations and DL also explored the phosphorylation-mediated effect on conformational dynamics of the GTP-bound KRAS, revealing that the phosphorylation of pY32, pY64, and pY137 sites leads to more disordered states of the switch domains and can induce conformational transformations between the closed and open states [50]. Overall, structural and computational studies showed that in its active state, KRAS predominantly adopts a stable closed conformation that allows for productive interactions with effector proteins and signaling. KRAS can also transiently adopt open and intermediate states, especially during conformational fluctuations or when interacting with inhibitors. The balance between these states is influenced by nucleotide binding, oncogenic mutations, post-translational modifications, and interactions with partner proteins.

Despite the significant body of structural, biochemical, and computational studies of KRAS dynamics and binding, there are a number of open questions, from understanding the fundamental biophysics of KRAS dynamics and allosteric networks to addressing practical challenges in allosteric drug design and resistance. The important open questions include the following issues: (a) How do long-timescale conformational changes in KRAS, particularly in the switch regions, influence its binding to effectors (e.g., RAF1, PI3K, RALGDS) and regulators (e.g., GAPs, GEFs)? (b) What are allosteric networks that connect distal mutation sites to the switch regions, and how do these networks modulate KRAS activity? (c) How do KRAS mutations alter the free energy landscape of conformational states, and what are the thermodynamic drivers of mutation-induced oncogenic activation? (e) What are the energetic contributions of specific residues and interactions to KRAS stability and allosteric effector binding?

In this study, we attempted to address some of these issues by employing an integrative computational simulation strategy that employed multiple microsecond MD simulations of KRAS-DARPIN complexes, systematic mutational scanning of KRAS residues for binding and stability, rigorous MM-GBSA binding free energy analysis, and network modeling of allosteric networks to determine binding affinity hotspots and allosteric binding centers for KRAS complexes with a diverse panel of DARPIN proteins K27, K55, K13, and K19 (Supporting Information, Table S1). These simulations provide a detailed understanding of conformational flexibility and allosteric communication within the KRAS-DARPIN interactions and the interplay between allostery, binding, and dynamics. The ensemble-based mutational profiling of KRAS residues using a knowledge-based energy model enabled accurate identification of binding affinity and protein stability hotspots. The detailed MM-GBSA analysis of binding energetics reveals the thermodynamic drivers of binding and the energetic contributions of the binding affinity hotspots to KRAS stability and effector binding. Using MD-inferred conformational ensembles and dynamics-based network modeling, this study mapped potential allosteric hotspots and allosteric communication pathways in KRAS-DARPIN complexes. Consistent with experimental analysis, our results confirmed that the central β-sheet of KRAS acts as a hub for transmitting allosteric signals between distant functional sites, facilitating allosteric communication between the switch regions and the binding interface. Network-based dynamic modeling of allostery predicted key allosteric hotspots in which mutations or structural perturbations can dramatically affect the fidelity of allosteric communication. The results of this study provide important quantitative insights into principles of allosteric communication and allosteric binding in KRAS that are consistent with and can explain the experimental data. In particular, we find that allosteric communication hotspots are enriched close to the functional switch regions and leverage conformational plasticity in these regions for transmitting allosteric signals, also suggesting that local energetic propagation is the main allosteric mechanism.

2. Materials and Methods

2.1. Molecular Dynamics Simulations

The crystal and cryo-EM structures of the KRAS and KRAS complexes are obtained from the Protein Data Bank [51]. Multiple independent microsecond MD simulations are performed for the crystal structure of wild-type KRAS in complexes with DARPIN K27 (pdb id 5O2S), DARPIN K55 (pdb id 5O2T), DARPIN K13 (pdb id 6H46), and DARPIN K19 (pdb id 6H47) (Supporting Information, Table S2). For simulated structures, hydrogen atoms and missing residues were initially modeled using the MODELLER program [52] and subsequently reconstructed and optimized using a template-based loop prediction approach, ArchPRED [53]. The side chain rotamers were additionally optimized by the SCWRL4 tool [54]. Protonation states of ionizable residues were assigned using PROPKA, ensuring histidine residues were in the appropriate tautomeric state [55,56]. The tleap module in AMBER was used to generate the topology (prmtop) and coordinate (inpcrd) files. The system was solvated in a TIP3P water box with a 12 Å buffer and neutralized using Na+ or Cl- ions. The final system was saved for further simulations. Molecular mechanics parameters of proteins were assigned according to the ff14SB force field [57,58]. The protein structures were energy-minimized to remove steric clashes and optimize the structure.

All simulation systems were processed through several stages of structure preparation. First, two stages of minimization were performed: (a) with restraints on the protein backbone to relax solvent and ions, and (b) without restraints to minimize the entire system. The minimization was conducted using the pmemd.cuda module with the following parameters: 1000 cycles of steepest descent followed by 500 cycles of conjugate gradient minimization, a cutoff of 10 Å for non-bonded interactions, and the ff14SB force field for proteins [58,59]. After minimization, the system is heated from 100 K to 300 K, over 1 nanosecond of simulation time at a constant volume, with an integration time of 1 fs. Then, we relax the system at a constant pressure with protein restraints over 1 ns of simulation time at a constant pressure with restraints set to 1 kcal/mol $Å^2$. Finally, we relax the system with no restraints for 1 ns of simulation time at a constant pressure. This is followed by 2 μs of production period with 20,000 frames saved for analysis. We performed multiple independent 2 μs MD simulations under NPT conditions (300 K, 1 bar). The temperature was maintained using the Langevin thermostat, and pressure was controlled with the Monte Carlo barostat. A time step of 2 fs was used, and bonds involving hydrogen atoms were constrained using the SHAKE algorithm. MD trajectories were saved every 100 ps for analysis. The ff14SB force field was used for proteins, and the TIP3P model was used for water.

The CPPTRAJ software v6.4.4 in AMBER 18 was used to calculate the root mean squared deviation (RMSD) and root mean squared fluctuation (RMSF) of MD simulation trajectories in which the initial structure was used as the reference [59]. The structures were visualized using Visual Molecular Dynamics (VMD 1.9.3) [60] and PyMOL (Schrodinger, LLC. 2010. The PyMOL Molecular Graphics System, Version X X.).

To identify the dynamical coupling of the motions between protein segments, the cross-correlation coefficient (C_{ij}) was proposed for measuring the motion correlation between the C_α atom pair in residues i and j, which is defined as follows:

$$C_{ij} = \langle \Delta r_i \cdot \Delta r_j \rangle / \left(\langle \Delta r_i^2 \rangle \langle \Delta r_j^2 \rangle \right)^{1/2} \quad (1)$$

where Δr_i and Δr_j are the displacements from the mean position of the C_α atom pair in residues i and j, respectively, that are evaluated over the sampled period. Positive C_{ij} values are associated with correlated motion in the residue pair, whereas negative C_{ij} values stand for negatively correlated motion. The cross-correlation analysis in this work was implemented using the Bio3D package [61].

2.2. Mutational Scanning and Sensitivity Analysis of the KRAS Residues: Quantifying Effects of Mutations on KRAS Binding and Protein Stability

We conducted a systematic mutational scanning analysis of the KRAS residues in the KRAS complexes using conformational ensembles of KRAS-DARPIN protein complexes and averaging of mutation-induced energy changes. Every KRAS residue was systematically mutated using all substitutions, and corresponding protein stability and binding free energy changes were computed with the knowledge-based BeAtMuSiC approach [62–64]. This approach is based on statistical potential describing the pairwise inter-residue distances, backbone torsion angles, and solvent accessibilities, and considers the effect of the mutation on the strength of the interactions at the interface and on the overall stability of the complex. The binding free energy of a protein–protein complex can be expressed as the difference in the free energy of the complex and the free energy of the two protein binding partners:

$$\Delta G_{bind} = G^{com} - G^{A} - G^{B} \quad (2)$$

The change of the binding energy due to a mutation was calculated then as the following:

$$\Delta\Delta G_{bind} = \Delta G_{bind}^{mut} - \Delta G_{bind}^{wt} \quad (3)$$

We leveraged rapid calculations based on statistical potentials to compute the ensemble-averaged binding free energy changes using equilibrium samples from simulation trajectories. The binding free energy changes were obtained by averaging the results over 1000 and 10,000 equilibrium samples for each of the systems studied.

2.3. MM-GBSA Binding Free Energy Computations of KRAS-DARPIN Complexes

We calculated the ensemble-averaged changes in binding free energy using 1000 equilibrium samples obtained from simulation trajectories for each system under study. The binding free energies of the KRAS-DARPIN protein complexes were assessed using the MM-GBSA approach [65,66]. The energy decomposition analysis evaluates the contribution of each amino acid to the binding of KRAS to DARPIN proteins [67,68]. The binding free energy for the KRAS-DARPIN complexes was obtained using the following:

$$\Delta G_{bind} = G_{KRAS-DARPIN} - G_{KRAS} - G_{DARPIN} \quad (4)$$

$$\Delta G_{bind} = G_{KRAS-DARPIN} - G_{KRAS} - G_{DARPIN} \quad (5)$$

where ΔE_{MM} is total gas phase energy (sum of $\Delta E_{internal}$, $\Delta E_{electrostatic}$, and $\Delta Evdw$); $\Delta Gsol$ is sum of polar (ΔG_{GB}) and non-polar (ΔG_{SA}) contributions to solvation. Here, $G_{KRAS\text{–}DARPIN}$ represents the average over the snapshots of a single trajectory of the complex, G_{KRAS} and G_{DARPIN} correspond to the free energy of the KRAS and DARPIN protein, respectively. The polar and non-polar contributions to the solvation free energy are calculated using a Generalized Born solvent model and consideration of the solvent accessible surface area. MM-GBSA is employed to predict the binding free energy and decompose the free energy contributions to the binding free energy of a protein–protein complex on a per-residue basis. The binding free energy with MM-GBSA was computed by averaging the results of computations over 1000 samples from the equilibrium ensembles.

First, the computational protocol must be selected between the "single-trajectory" (one trajectory of the complex) or "separate-trajectory" (three separate trajectories of the complex, receptor, and ligand). To reduce the noise in the calculations, it is common that each term is evaluated on frames from the trajectory of the bound complex. In this study, we choose the "single-trajectory" protocol, because it is less noisy due to the cancellation of intermolecular energy contributions. This protocol applies to cases where significant structural changes upon binding are not expected. Entropy calculations typically dominate the computational cost of the MM-GBSA estimates. However, for the absolute affinities, the entropy term is needed, owing to the loss of translational and rotational freedom when the ligand binds. In this study, the entropy contribution was not included in the calculations of binding free energies of the complexes because the entropic differences in estimates of relative binding affinities are expected to be small owing to small mutational changes and preservation of the conformational dynamics [69,70]. MM-GBSA energies were evaluated with the MMPBSA.py script in the AmberTools21 package [71].

2.4. Graph-Based Dynamic Network Analysis of KRAS-DARPIN Complexes

To analyze protein structures, we employed a graph-based representation where residues are modeled as network nodes, and non-covalent interactions between residue side-chains define the edges. This approach captures the spatial and functional relationships between residues, providing insights into the protein's structural and dynamic properties. The graph-based framework allows for the integration of both structural and evolutionary information, enabling a comprehensive analysis of residue interactions. The Residue Interaction Network Generator (RING) program [72–75] was used to generate the initial residue interaction networks from the crystal structures of the KRAS-DARPIN protein complexes. The residue interaction networks for conformational ensembles were constructed by defining edges based on non-covalent interactions between residue side-chains [76,77]. The weights of these edges were determined using two key metrics: (a) dynamic residue cross-correlations derived from MD simulations where these correlations quantify the coordinated motions of residue pairs [78] and (b) coevolutionary couplings measured using mutual information scores, where these couplings reflect evolutionary constraints and residue-residue dependencies [79]. The edge lengths (weights) between nodes ii and jj were computed using generalized correlation coefficients, which integrate both dynamic correlations and coevolutionary mutual information. Only residue pairs observed in at least one independent simulation were included in the network. The matrix of communication distances was constructed using generalized correlations between composite variables that describe both the dynamic positions of residues and their coevolutionary mutual information. This matrix provides a quantitative measure of communication efficiency between residues, reflecting both their physical proximity and evolutionary relationships.

Network graph calculations were performed using the Python NetworkX 3.4.2. version 3.13. package NetworkX [80,81]. This included the computation of key network parameters, such as shortest paths and betweenness centrality, to identify residues critical for communication within the protein structure. The short path betweenness (SPC) of residue i is defined to be the sum of the fraction of shortest paths between all pairs of residues that pass through residue i:

$$C_b(n_{i)} = \sum_{j<k}^{N} \frac{g_{jk}(i)}{g_{jk}} \tag{6}$$

where g_{jk} denotes the number of shortest geodesics paths connecting j and k, and $g_{jk}(i)$ is the number of shortest paths between residues j and k passing through the node n_i. Residues with high occurrence in the shortest paths connecting all residue pairs have higher betweenness values. For each node n, the betweenness value is normalized by the number of node pairs excluding n given as (N−1)(N−2)/2, where N is the total number of nodes in the connected component that node n belongs to.

$$C_b(n_i) = \frac{1}{(N-1)(N-2)} \sum_{\substack{j<k \\ j \neq i \neq k}}^{N} \frac{g_{jk}(i)}{g_{jk}} \quad (7)$$

To account for differences in network size, the betweenness centrality of each residue ii was normalized by the number of node pairs excluding ii. The normalized short path betweenness of residue i can be expressed as:g_{jk} is the number of shortest paths between residues j and k; $g_{jk}(i)$ is the fraction of these shortest paths that pass through residue i. Residues with high normalized betweenness centrality values were identified as key mediators of communication within the protein structure network.

2.5. Network-Based Mutational Profiling of Allosteric Residue Centrality

Through mutation-based perturbations of protein residues, we compute dynamic couplings of residues and changes in the short path betweenness centrality (SPC) averaged over all possible modifications in a given position.

The change of SPC upon mutational changes of each node is reminiscent of the calculation of residue centralities by systematically removing nodes from the network.

$$\Delta L_i = \langle \mid \left| \Delta L_i^{node}(j) \right| \mid^2 \rangle \quad (8)$$

where i is a given site, j is a mutation, and $\langle \ldots \rangle$ denotes averaging over mutations. $\Delta L_i^{node}(j)$ describes the change of SPC parameters upon mutation j in a residue node i. ΔL_i is the average change of ASPL triggered by mutational changes in position i. Z-score is then calculated for each node as follows:

$$Z_i = \frac{\Delta L_i - \langle \Delta L \rangle}{\sigma} \quad (9)$$

$\langle \Delta L \rangle$ is the change in the SPC network parameter under mutational scanning, averaged over all protein residues, and σ is the corresponding standard deviation. The ensemble-average Z score changes are computed from network analysis of the conformational ensembles of KRAS-DARPIN complexes using 1000 snapshots of the simulation trajectory.

3. Results

3.1. MD Simulations of the KRAS–Protein Complexes Reveal Distinct Dynamic Signatures

To investigate the conformational dynamics and functional flexibility of KRAS in complexes with DARPIN binders, we employed crystal structures of bound DARPIN K27 (pdb id 5O2S) K55 (pdb id 52T), K13 (pdb id 6H46), and K19 (pdb id 6H47) (Figure 1). Using these structures, we conducted microsecond-scale molecular dynamics (MD) simulations of the KRAS–DARPIN complexes. These DARPINs engage KRAS through distinct modes. K27 targets the GDP-bound inactive conformation, acting as a conformational blocker that prevents nucleotide exchange by stabilizing the P-loop and Switch I regions (Figure 1A). K55 mimics natural effectors like RAF1, binding to the GTP-bound active state and engaging both Switch I and II loops (Figure 1B). K13 and K19 recognize a novel allosteric lobe, centered on residues around helix α3–loop–α4, including H95, Y96, L133, S136, and Y137, which are KRAS-specific and not conserved across RAS isoforms (Figure 1C,D).

We used MD simulations of the KRAS–RAF1 complex (PDB ID: 6VJJ) as a representative reference system for understanding how protein–protein interactions modulate KRAS dynamics. MD simulations for KRAS-RAF1 were initiated from the closed conformation of switch I (state II), where GTP is stably bound and coordinated by Mg^{2+} ions, T35, and S17. Y32 remains positioned over the nucleotide, stabilizing the active state conformation (Figure 2A). The key structural regions of KRAS are the P-loop (residues 10–17), Switch I (residues 25–40), and Switch II regions (residues 60–76), reflecting their functional coupling during nucleotide binding and hydrolysis (Figure 1). The α-helices (α-helix 1: 15–24; α-helix 2: 67–73; α-helix 3: 87–104; α-helix 4: 127–136; α-helix 5: 148–166) and β-sheets (β-strand 1: 3–9; β-strand 2: 38–44; β-strand 5: 109–115; β-strand 6: 139–143) showed moderate correlations with the switch regions, while anti-correlated motions are observed between β-sheets (β-strand 3: 51–57; β-strand 4: 77–84) and the allosteric lobe (residues 87–166) of KRAS (Figure 1).

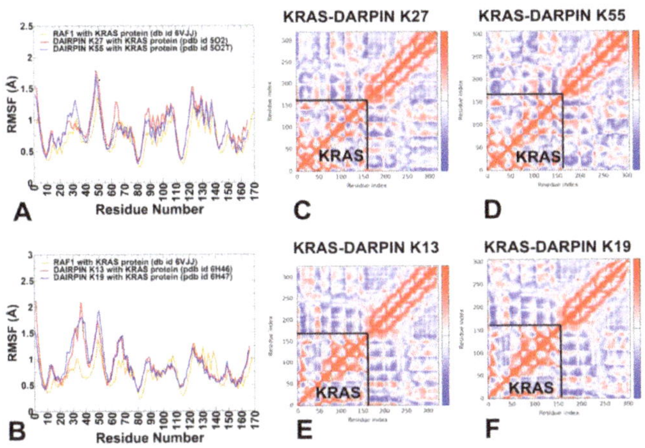

Figure 2. Conformational dynamics of KRAS-DARPin complexes. (**A**) The RMSF profiles of the KRAS protein residues obtained from MD simulations of the KRAS complexes bound with RBD of RAF1, pdb id 6VJJ (in orange lines), KRAS-K27 complex, pdb id 5OS (in red lines), and KRAS-K55 complex, pdb id 5O2T (in blue lines). (**B**) The RMSF profiles for the KRAS residues from MD simulations of the crystal structure of wild-type KRAS (bound with the RBD of RAF1, pdb id 6VJJ (in orange lines), KRAS-K13, pdb id 6H46 (in red lines), KRAS-K19, pdb id 6H47 (in blue lines). The DCCM maps for the KRAS residues in the KRAS complex with K27 (**C**), K55 (**D**), K13 (**E**), and K19 (**F**). DCC measures the correlation between the motions of pairs of residues over the course of the MD simulation. Positive correlations indicate that residues move in the same direction, while negative correlations indicate anti-correlated motions. The KRAS protein residues (residues 1–170) are highlighted on the DCCM maps.

To assess the equilibration and stability of the KRAS-DARPin complexes, we calculated the root mean square deviation (RMSD) of the protein backbone over the course of the microsecond-scale MD simulations (Supporting Information, Figure S2). The RMSD plots reveal that all systems reached stable equilibration within the first 200–400 ns, with fluctuations thereafter remaining within acceptable limits (±1–2 Å). In the KRAS-K27 complex, the RMSD stabilized around 1.5 Å after an initial equilibration period of ∼500 ns, indicating a tightly constrained inactive conformation. The KRAS-K55 complex showed slightly higher RMSD values (∼2.0 Å), reflecting the increased rigidity imposed by effector-like binding. For the allosteric binders K13 and K19, the RMSD values fluctuated more broadly (∼2.5 Å), consistent with their role in modulating KRAS dynamics rather than fully immobilizing the protein (Supporting Information, Figure S2).

The root mean square fluctuation (RMSF) analysis provided a detailed view of the residue-wise flexibility of KRAS upon binding to DARPINs K27, K55, K13, and K19

(Figure 2). Using RMSF profiles from all-atom MD simulations, we can identify how each DARPin affects the flexibility of key functional motifs (P-loop, Switch I, Switch II), stabilization of interface regions, and differential impact on dynamics between effector-mimicking and allosteric binders. DARPin K27 binds to the GDP-bound inactive conformation of KRAS and functions as a conformational restrainer without inducing large-scale structural changes. In the complex formed between KRAS and DARPin K27, RMSF profiling showed that P-loop (residues 10–17) and Switch I (24–40) had reduced backbone fluctuations, particularly at T35, I36, Y40, and Q70 (Figure 2A). KRAS residues Y32, D33, T35, I36, Y40, and Q70, which form the primary interaction patch, are among the most rigid segments of the protein, suggesting that K27 sequesters KRAS in a closed, inactive conformation without inducing global structural changes (Figure 2A). Switch II (residues 60–70), although engaged to some extent, remains more flexible than Switch I, indicating that K27 does not fully restrict switch motion, but rather modulates dynamic equilibrium toward the GDP-bound state (Figure 2A). The RMSF profile for the KRAS–K55 complex reflects a stronger resemblance to natural effector engagement, with significant immobilization of both Switch I and Switch II loops (Figure 2A). RMSF analysis showed that both Switch I and Switch II become more rigid upon K55 binding, indicating a dual-lock mechanism that enhances effector-like stabilization of the switch regions. Switch II residues Y64 and M67 also show substantial rigidity, indicating that K55 binds to an effector-like orientation, locking these residues into a signaling-ready geometry. The P-loop and upstream β-strands (e.g., β1 and β2) retain moderate flexibility, suggesting that K55 allows some degree of nucleotide sampling (Figure 2A).

Unlike K27 and K55, K13 and K19 bind to a novel allosteric site located at the junction of helix α3–loop–α4, with minimal involvement of the canonical effector interface (Figure 1C,D). This leads to a different pattern of RMSF modulation. For the KRAS-K13 complex, allosteric lobe residues H95 and Y96 show reduced mobility, confirming their role as central anchors in this system (Figure 2B). Residues in β-strand 5 and C-terminal loop regions (e.g., L133, S136, Y137) also display lower RMSF values, indicating that K13 stabilizes the structural scaffold supporting the allosteric pocket. Notably, Switch I and Switch II become more mobile, suggesting that K13 does not block functional transitions, but instead alters the conformational ensemble. In the KRASB-K19 complex, H95 and R102 showed strong reductions in fluctuation compared to free KRAS or effector-bound KRAS-RAF1 states (Figure 2B). As with K13, the Switch I and Switch II loops remain relatively flexible, indicating that K19 modulates function indirectly, possibly by altering signal propagation across the central β-sheet. Hence, K13 and K19 act as dynamic modulators, engaging pockets that are structurally distant from the effector interface, yet capable of influencing long-range communication within KRAS (Figure 2B). The RMSF profiles of KRAS–DARPin complexes clearly distinguish four molecular recognition strategies. K27 locks KRAS into a closed conformation, reducing Switch I dynamics. K55 stabilizes both Switch I and II, resembling natural effectors. K13 and K19 target a KRAS-specific allosteric motif, leading to selective stabilization of the α3–loop–α4 region, while allowing the switch regions to remain partially dynamic.

To understand how different binding partners influence the intrinsic conformational dynamics and residue–residue communication in KRAS, we performed dynamic cross-correlation matrix (DCCM) analysis using MD simulations. This approach maps correlated motions between residues across the protein. The KRAS–K27 complex shows that the P-loop and Switch I (residues 10–40) remain tightly coupled, consistent with K27's preference for the inactive conformation. However, correlations between Switch I and Switch II are reduced, suggesting that K27 may partly suppress the cooperative motion typically seen in active-state systems (Figure 2C). Positive correlations are reinforced within the

P-loop and Switch I. Helix 2 (67–73) and β-strands 5–6 (109–143) —which form part of the central β-sheet —show moderate negative correlations with Switch II, implying that its engagement may indirectly dampen signaling transitions. The KRAS–K55 complex, which mimics natural effectors like RAF1, displays a correlation profile similar to KRAS–K27 with some key distinctions (Figure 2D). Strong positive correlations are maintained between Switch I (25–40) and Switch II (60–76), reflecting K55's ability to stabilize both regions simultaneously. Helix 3 (87–104) and β-strand 4 (77–84) exhibit positive correlations with Switch II, suggesting that K55 binding enhances long-range communication through the central β-sheet. Negative correlations between Switch II and upstream β-strands (e.g., β3 and β4) are diminished, indicating that K55 binding stabilizes a more rigid, effector-like network. The KRAS–K13 complex, which engages the allosteric lobe (helix α3–loop–helix α4, residues 87–104), shows a distinct pattern of correlated motion (Figure 2E). Enhanced correlations emerge between the allosteric lobe (residues 87–104) and β-strand 3–4 (51–84), suggesting that K13 modulates signal propagation through the central β-sheet. P-loop and Switch I retain moderate internal correlations, but show reduced coupling with Switch II, indicating that K13 alters the canonical switch–switch communication without fully immobilizing either loop. Residues H95 and Y96, central to K13 binding, become hub residues in the correlation map, showing strong positive correlations with distal regions, including Loop 7 (residues 130–140) and helix 5 (residues 148–166), suggesting that K13 introduces new allosteric pathways that extend beyond the effector interface. This pattern indicates that K13 functions not by blocking switch motion, but by modulating the global allosteric network. Similar to K13, K19 targets the allosteric lobe, but with greater involvement of polar interactions (Figure 2F). The H95–D108 motif becomes a central hub, showing strong correlations with both β-strand 4 (77–84) and helix 5 (148–166). A correlation between Switch I and Switch II is preserved, though slightly weakened, indicating that K19 allows for partial switch dynamics. P-loop and helix 1 (15–24) show increased independence from the rest of the structure, suggesting that K19 reduces the global coherence of KRAS motions, favoring localized stabilization over broad conformational restriction.

Structural mapping of the conformational mobility profiles of the KRAS-K27 complex emphasized rigidification of P-loop and Switch I (Supporting Information, Figure S3A). Switch I residues (Y32, D33, T35, I36, Y40) show strongly suppressed mobility, especially around T35 and Y40, which form hydrophobic clusters and hydrogen bonds with the DARPin scaffold. Switch II residues (M67, Q70, Y64) remain moderately flexible, suggesting that K27 does not fully immobilize this loop but instead allows some degree of motion. The core β-sheets (β1–β6) remain highly rigid, consistent with their role in maintaining overall fold integrity. The helices α2 and α3 show moderate stabilization, while helices α4 and α5 become slightly more mobile, reflecting allosteric modulation of distal regions due to K27 binding (Supporting Information, Figure S3A). Structural mapping of dynamics profiles for the KRAS–K55 complex highlighted greater stabilization of Switch I and Switch II (Supporting Information, Figure S3B). The Y40–M67–F79 aromatic cluster becomes a central rigid domain, showing minimal deviation across simulation time. Structural mapping of dynamic profiles clearly demonstrates that K55 causes more widespread immobilization of KRAS than K27, even in regions far from the immediate binding interface. This is due to several key factors, including effector-mimicking engagement of both Switch I and II and strong hydrophobic and electrostatic interactions centered around Y40, Y64, M67, and R41. In addition, the results suggest enhanced network coupling between functional motifs, including the central β-sheet and helix α5, leading to global rigidification. In contrast, K27 stabilizes the inactive state by focusing on the P-loop and Switch I, with minimal impact on Switch II dynamics, resulting in localized stabilization rather than broad conformational restriction. The structural map of the KRAS–K13 complex showed stabilization of the allosteric pocket while pointing to the enhanced switch mobility. K13 acts as a dynamic modulator, where it locally stabilizes the allosteric lobe, but does not restrict global switch motion (Supporting Information, Figure S3C). K19 targets residues H95, Y96, and R102, leading to strong suppression of fluctuations in the allosteric lobe. Structural mapping showed that K19 also induces moderate-to-strong immobilization in regions far from the binding site, such as helix α5 (residues 148–166) and parts of the central β-sheet, suggesting that K19 binding alters global signal transmission rather than just local dynamics (Supporting Information, Figure S3D). The widespread immobilization of KRAS regions observed in the presence of K55 and K19, even in areas far from the binding interface, underscores the integrated nature of KRAS dynamics.

3.2. Mutational Scanning of KRAS-DARPin Complexes

Using conformational ensembles obtained from MD simulations, we performed systematic mutational scanning of the KRAS residues in the KRAS-DARPin complexes (Figure 3). Mutational scanning is a computational approach used to evaluate the impact of mutations on protein stability and binding affinity, often leveraging statistical potentials or knowledge-based energy models such as the BeAtMuSiC approach [62–64]. In silico mutational scanning was performed by averaging the binding free energy changes over the equilibrium ensembles and allows for predictions of the mutation-induced changes of the binding interactions and the stability of the complex. To provide a systematic comparison, we constructed mutational heatmaps for the KRAS interface residues. We used the BeatMusic approach to identify binding interface residues. Residues are considered part of the interface if they are within a defined cutoff distance (typically 5 Å) from atoms in the binding partner.

In the KRAS–K27 complex, binding hotspot residues include Y32, D33, T35, I36, Y40, R41, M67, Q70, and Y71, which contribute to its modular, backbone-mediated binding strategy (Figure 3A,C). These residues are primarily located in the Switch I and II regions, forming a modular hydrogen bond network that stabilizes the GDP-bound inactive conformation of KRAS. K27 uses a broad, modular recognition strategy where multiple weakly interacting residues collectively stabilize the complex. This enhances its mutation tolerance and makes it particularly effective against variants that stabilize the GDP-bound form. The most critical residue is R41, which forms strong electrostatic interactions with conserved acidic motifs on the DARPin. Substitutions at this site (e.g., R41C/H/N) severely impair binding (Figure 3A,C). These complexes are extremely sensitive to mutations affecting T35 and Y40, which play central roles in hydrophobic stabilization and nucleotide coordination.

KRAS shows some resilience to conservative mutations, especially those retaining an aromatic character (e.g., Y40F/H) (Figure 3A,C). K27 binding is somewhat less affected by mutations at Switch II residues (e.g., Y64, M67), indicating that it does not rely heavily on this region (Figure 3A,C). Binding of KRAS with DARPin K55 suggests a mechanism of effector mimicry through hydrophobic anchoring and electrostatic stabilization. K55 engages both Switch I and Switch II regions, forming a tight hydrophobic core centered around Y40, M67, and F79, while also establishing electrostatic networks involving D38 and R41. The key binding hotspots include I36, S39, Y40, Y64, M67, R41, and D38 (Figure 3B,D), where positions I36, S39, Y40, Y64, and M67 emerged as the least mutation-tolerant. Structural mapping shows that K55 recognizes conserved motifs in Switch I/II, making it ideal for blocking signal propagation rather than nucleotide cycling (Figure 3D). Mutational scanning revealed sensitivity to mutation positions that disrupt aromatic stacking, such as Y64G and M67L, which lead to significant destabilization. T35A/N/V substitutions have minimal impact on K55 binding while E37 and D38 substitutions moderately affect binding, reflecting their role in interfacial hydrogen bonding and salt bridge formation (Figure 3B).

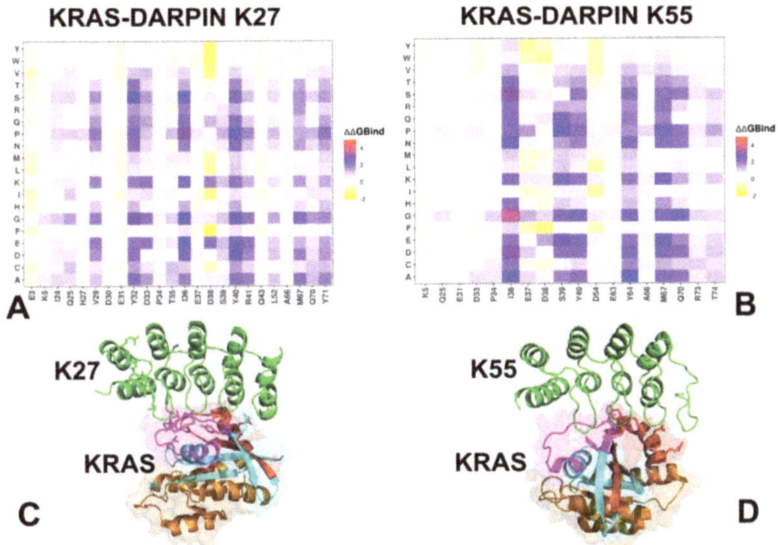

Figure 3. The ensemble-based mutational scanning of binding for the KRAS complexes with DARPin K27 and K55 proteins. The mutational scanning heatmap for the KRAS binding epitope residues in the KRAS complex with DARPin K27 (**A**) and the mutational scanning heatmap for the KRAS binding epitope residues in the KRAS complex with DARPin K55 (**B**). The binding energy hotspots correspond to residues with high mutational sensitivity. The heatmaps show the computed binding free energy changes for 20 single mutations on the sites of the variants. The squares on the heatmap are colored using a 3-colored scale, yellow–white–blue–red, with blue color indicating the appreciable unfavorable effects on stability and red pointing to very large destabilizing effects of mutations. The horizontal axis represents the binding epitope residues. Residues are considered part of the interface if they are within a defined cutoff distance of 5 Å from atoms in the binding partner. The Y-axis depicts all possible substitutions of a given binding epitope residue, denoting mutations to letters using a single letter annotation of the amino acid residues. Structural mapping of the binding energy hotspots in the KRAS complexes with DARPin K27 (**C**) and DARPin K55 (**D**). The binding epitope hotspot positions for K27 (Q25, V29, Y32, D33, T35, I36, S39, Y40, R41. M67, Q70, Y71) and hotspots for K55 (I36, S39, Y40, Y64, M67, Q70, R73, T74) are shown in sticks. The Switch I region (residue 24–40) is depicted in magenta, and the Switch II region (residues 60–76) is in red. The allosteric lobe (residues 87–166) is shown in orange. The DARPin molecules are shown in green ribbons.

However, T35 mutations severely disrupt K27 binding, highlighting differences in epitope dependency. Substitutions at Y40 and M67 affect both DARPins, consistent with their central role in hydrophobic stabilization. Similarly, Y64 mutations strongly impair K55 binding due to its reliance on aromatic stacking and side-chain interactions, whereas K27 remains largely unaffected, likely due to its focus on upstream P loop and Switch I residues. This analysis reinforced the notion that K55 mimics natural effectors like RAF1, relying on localized, high-affinity contacts within the switch regions, which makes it highly potent but also more vulnerable to mutations that affect aromatic or charged residues.

Despite targeting partially overlapping regions on the KRAS surface, only a subset of residues contributes to both K27 and K55 binding. These residues, I36, S39, Y40, and M67, emerge as shared anchors between the two DARPins, playing roles in stabilizing the interaction. These findings align with experimental observations [32] showing that engineered protein binders can achieve inhibition through complementary strategies.

Both DARPins K13 and K19 engage a novel allosteric lobe on KRAS, located in the α3–loop–α4 motif, distinct from the classical effector interface. Despite this shared location, they differ in their residue-specific contributions and polar vs hydrophobic dependency. For KRAS–K13 interactions, the binding hotspots are K88, F90, H94, H95, Y96 R102, L133, S136, Y137, and G138 (Figure 4A,C). This set of residues forms a central hub in the allosteric lobe, where H95 and Y96 contribute the most to binding via π-stacking and hydrophobic packing with tryptophan residues in the DARPins (W35 and W37) (Figure 4C). In the KRAS-K19 complex, the binding hotspots are F90, H95, E107, D108, L133, S136, and Y137 (Figure 4B). K19 shares many key residues with K13 but introduces additional polar interaction networks, particularly at D108 (Figure 4B,D). Mutational scanning confirms that hotspot substitutions severely impair binding in both systems. H94D/E/N and H95D/E/P/V/G lead to strong destabilization in both complexes due to loss of aromatic stacking. R102K/R/D substitutions have a stronger impact on K13, where this residue contributes more directly to polar anchoring. E98 mutations (E98A/D/K/P) cause significant destabilization in K19, confirming its importance in maintaining interface compatibility in this system. Conservative mutations, such as S136T/A, Y137F/H, and L133V/I, are generally better tolerated, supporting their structural compatibility rather than direct functional inhibition.

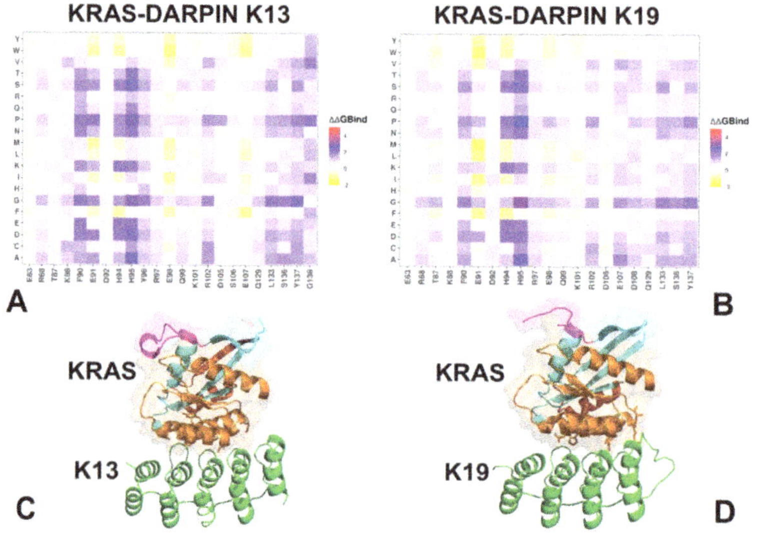

Figure 4. The ensemble-based mutational scanning of binding for the KRAS complexes with DARPin K13 and K19 proteins. The mutational scanning heatmap for the KRAS binding epitope residues in the KRAS complex with DARPin K13 (**A**) and the mutational scanning heatmap for the KRAS binding epitope residues in the KRAS complex with DARPin K19 (**B**). The binding energy hotspots correspond to residues with high mutational sensitivity. The heatmaps show the computed binding free energy changes for 20 single mutations on the sites of the variants. The squares on the heatmap are colored using a 3–colored scale, yellow–white–blue–red, with blue color indicating the appreciable unfavorable effects on stability and red pointing to very large destabilizing effects of mutations. The horizontal axis represents the binding epitope residues. Residues are considered part of the interface if they are within a defined cutoff distance of 5 Å from atoms in the binding partner. The Y-axis depicts all possible substitutions of a given binding epitope residue, denoting mutations to letters using a single-letter annotation of the amino acid residues. Structural mapping of the binding energy hotspots in the KRAS complexes with DARPin K13 (**C**) and DARPin K19 (**D**). The binding epitope hotspot positions for K13 (K88, F90, H94, H95, R102, L133, S136, Y138, G138) and hotspots for K19 (F90, H94, H95, E107, D108, L133, S136, Y137) are shown in sticks. The Switch I region (residue 24–40) is depicted in magenta, and the Switch II region (residues 60–76) is in red. The allosteric lobe (residues 87–166) is shown in orange. The DARPin molecules are shown in green ribbons.

Several KRAS residues are engaged by multiple DARPINs, albeit to different degrees. I36 is important for hydrophobic core integrity in K27 and K55. Y40 is a dominant anchor in K27 and K55, contributing to both hydrophobic and electrostatic interactions. R41 is a central residue in K27 and K55, though K27 shows greater reliance on its side chain. H95 is engaged exclusively by K13 and K19, reinforcing their KRAS-specificity. These findings indicate that certain residues, particularly I36 and Y40, are evolutionarily constrained interaction hubs, making them ideal targets for broad-spectrum inhibition strategies. The analysis reveals how each binder engages KRAS through distinct mechanisms with both shared and unique energetic hotspots. The presence of unique hotspots suggests that combining these DARPINs could raise the genetic barrier to resistance, since escape would require simultaneous mutation of structurally distant and functionally divergent residues.

These findings highlight that mutational scanning is highly effective in identifying major binding hotspots and predicting the impact of loss-of-function mutations. However, they also reveal limitations in capturing subtle specificity shifts, particularly for polar and charged residues at the periphery of the binding interface. The resolution of the simplified knowledge-based energy function used in our analysis appears sufficient for detecting large-effect mutations (e.g., those involving hydrophobic collapse or charge reversal), but may be somewhat less sensitive to fine-grained changes in electrostatic networks or hydrogen bond rearrangements that may influence binding selectivity. Importantly, the observed conservation of key hotspot residues across natural effectors and engineered binders underscores the presence of conservative mechanisms in KRAS recognition.

3.3. MM-GBSA Analysis of the Binding Energetics Provides Quantitative Characterization of Thermodynamic Drivers of DARPin Interactions

Using the conformational equilibrium ensembles obtained from MD simulations of the KRAS-DARPin complexes, we computed the binding free energies for these complexes and analyzed the residue-specific contributions to binding (Figures 5 and 6). The MM-GBSA residue-based decomposition reveals that R41 is the dominant hotspot in the KRAS–K27 interface, contributing ΔG = −16.34 kcal/mol, primarily through electrostatic interactions (ΔGelec = −137.18 kcal/mol) (Figure 5A). This residue acts as a central anchor, more so than any other in the system. Other notable contributors include Y32, T35, I36, Y40, M67, Q70, and Y71, all of which show favorable van der Waals contributions, reinforcing the hydrophobic nature of the K27 interface (Figure 5B). D33 and D38 contribute via polar contacts and hydrogen bonds, playing supportive roles in maintaining overall interface compatibility (Figure 5C). K55 derives its strongest stabilization from van der Waals interactions and engages a tighter, more localized interface with strong contributions from M67 (ΔG_vdw = −4.07 kcal/mol), Y64 (ΔG_vdw = −3.3 kcal/mol), Y40 (ΔG_vdw = −3.45 kcal/mol), I36 (ΔG_vdw = −2.48 kcal/mol) (Figure 5D,E). These residues form a central hydrophobic cluster, especially involving Y40–M67–F79, that mimics natural effectors. This difference highlights how K27 uses a modular, distributed energy model, while K55 exploits a concentrated, high-affinity hotspot network. Electrostatic contributions are also significant, particularly at D33 (ΔG_elec = −50.42 kcal/mol), which forms long-range salt bridges with R41 and R73 in K55 (Figure 5F). E37 and D54 also contribute to polar networks that stabilize Switch I geometry (Figure 5F).

Unlike K27, K55 does not rely heavily on R41, though this residue still plays a supportive role in shaping the electrostatic landscape. Both K27 and K55 exploit hydrophobic interactions to stabilize the complex. Y40 contributes significantly via aromatic stacking, and I36 and M67 provide favorable van der Waals contributions, forming part of a hydrophobic core that anchors the interface (Figure 5B,E). This reflects the broader theme seen across multiple effectors and inhibitors: hydrophobic clusters play a dominant role in maintaining interface compatibility, especially when they involve conserved aromatic or aliphatic side chains.

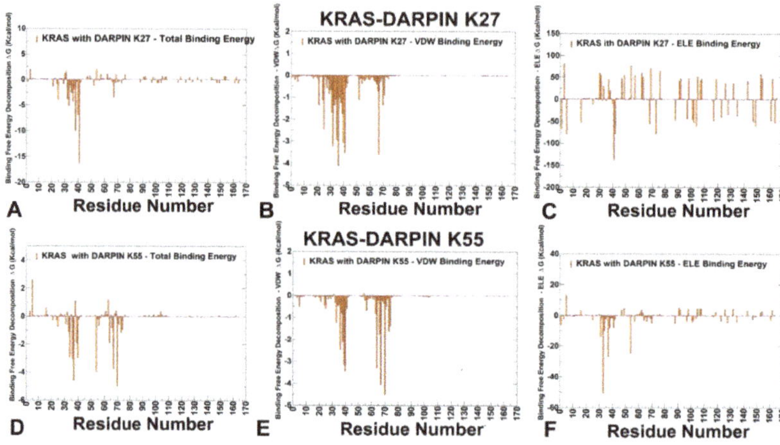

Figure 5. The residue-based decomposition of the total binding MM–GBSA energies for the KRAS residues in the KRAS complex with DARPin K27 (**A–C**) and DARPin K55 (**D–F**). The residue-based decomposition of the total binding energy (**A**), van der Waals contribution (**B**), and electrostatic contribution to the total MM-GBSA binding energy (**C**) for the KRAS residues in the KRAS complex with K27. The residue-based decomposition of the total binding energy (**D**), van der Waals contribution (**E**), and electrostatic contribution to the total MM-GBSA binding energy (**F**) for the KRAS residues in the KRAS complex with K55. The MM-GBSA contributions are evaluated using 1000 samples from the equilibrium MD simulations of respective KRAS complexes.

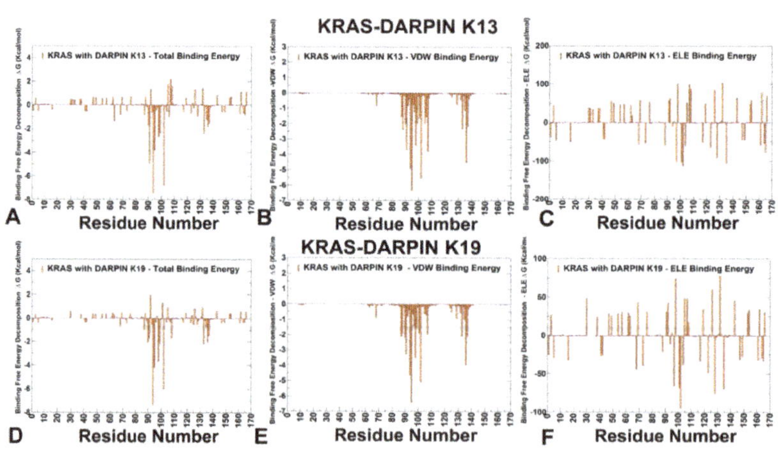

Figure 6. The residue-based decomposition of the total binding MM-GBSA energies for the KRAS residues in the KRAS complex with DARPin K13 (**A–C**) and DARPin K19 (**D–F**). The residue-based decomposition of the total binding energy (**A**), van der Waals contribution (**B**), and electrostatic contribution to the total MM-GBSA binding energy (**C**) for the KRAS residues in the KRAS complex with K13. The residue-based decomposition of the total binding energy (**D**), van der Waals contribution (**E**), and electrostatic contribution to the total MM-GBSA binding energy (**F**) for the KRAS residues in the KRAS complex with K19. The MM-GBSA contributions are evaluated using 1000 samples from the equilibrium MD simulations of the respective KRAS complexes.

The MM-GBSA results suggested that the electrostatic contributions support the interface integrity (Figure 5C,F). While not the primary driver in all cases, electrostatic interactions contribute to binding stability in both systems. In K27, key contributors include R41, D33, and D38, where long-range charge complementarity enhances binding affinity (Figure 5C). In K55, the residues D33, E37, and D54 participate in salt bridges and hydrogen bonds with basic motifs in the DARPin, reinforcing interface compatibility (Figure 5F). Despite differences in how each binder utilizes these interactions, electrostatic networks remain essential for stabilizing the switch regions. The MM-GBSA results clearly distinguish K27 and K55 as two functionally divergent classes of engineered binders. K27 emphasizes electrostatic anchoring and modular hydrogen bonding, stabilizing the GDP-bound inactive conformation. K55 exploits hydrophobic clusters and effector-like electrostatic networks, locking KRAS into a signaling-ready state that blocks downstream effector access. Despite targeting partially overlapping surfaces, these DARPINs differ significantly in their energetic dependencies and sensitivity to mutations. MM-GBSA computations suggested that K27 binding is driven by a dominant electrostatic hotspot at R41, which anchors the interface through salt bridge formation. Hydrophobic contributions from Y32, T35, I36, Y40, M67, Q70, and Y71 form a modular hydrogen bond network that sequesters KRAS in an inactivation-prone orientation. We argue that the mechanism of K27 binding is determined by modular hydrophobic and electrostatic anchoring. K55 binding is primarily driven by a central hydrophobic core involving Y40, M67, and F79, reinforced by electrostatic networks at D38 and S39. As a result, the K55 binding mechanism resembles effector mimicry through compact hydrophobic clustering. These findings support the concept of multi-binder strategies, where combinations of conformational blockers (K27) and effector mimics (K55) can synergistically reduce oncogenic KRAS activity.

DARPins K13 and K19 bind to an evolutionary divergent but structurally conserved allosteric pocket involving H94, H95, Y96, L133, S136, and Y137 (Figure 6). These residues form part of a hydrophobic core that supports structural compatibility between KRAS and the DARPIN scaffold, even though the pocket is not involved in effector recognition or nucleotide exchange. Both systems exploit conserved hydrophobic motifs —especially involving H95 and Y96 —which serve as central anchors for stabilizing the interaction. In both complexes, van der Waals interactions dominate the binding mechanism as H95 contributes significantly through aromatic stacking with tryptophan residues in the DARPin scaffold. L133 and Y137 reinforce shape complementarity, reducing conformational freedom in Loop 7 and helix α5. These findings are consistent with experimental observations showing that H95 is the central hotspot of both K13 and K19 [33]. Despite their shared allosteric engagement, K13 and K19 differ significantly in their electrostatic contributions and mutation sensitivity, leading to distinct mechanisms of stabilization and isoform selectivity.

K13 relies heavily on H94 and R102, forming a dense hydrophobic cluster that stabilizes the α3–loop–α4 interface. It engages E91 and L133 via moderate electrostatic and van der Waals forces. The top contributors to binding stability are H94 > R102 > E91 > H95 (Figure 6A–C). This ranking highlights that K13 relies strongly on E91, which forms polar interactions and hydrogen bonds that support structural compatibility between the DARPin and KRAS. The analysis showed that K13 forms a dense hydrophobic core involving H94 and H95, supported by polar networks formed with E91 and R102 (Figure 6B). The electrostatic contribution of R102 also plays a critical anchoring role (Figure 6C). In contrast, K19 shows a different residue dependency, with the following hotspot ranking: H94 > R102 > H95 > E98 (Figure 6D–F). This shift reflects the role of E98, which engages in backbone hydrogen bonding and water-mediated interactions that contribute significantly to interface stability. While E98 does not form direct side-chain contacts with the DARPin, its presence helps maintain local structure and hydration compatibility, enhancing overall binding fidelity. Additionally, H94 shows improved electrostatic compatibility in K19 (Figure 6F), indicating that it participates in enhanced hydrogen bond formation compared to K13. The MM-GBSA analysis of the KRAS–K13 and KRAS–K19 complexes reveals that both DARPINs engage a structurally stable allosteric lobe but do so with different energetic strategies. K13 emphasizes E91 and R102, forming moderate electrostatic networks that support interface integrity. K19 leverages E98 and H94, introducing additional polar constraints that enhance KRAS specificity. K19 shares many of the same hydrophobic anchors as K13, especially at H94 and H95, but introduces new polar constraints that increase KRAS selectivity. In particular, D108 emerged as an additional hotpot for K19, and E98 appeared to be a significant contributor to binding stability. This enhanced electrostatic dependency allows K19 to fine-tune the signaling output of KRAS.

We also examined the consistency and agreement between mutational scanning predictions derived from the BeAtMuSiC approach and rigorous MM-GBSA calculations on equilibrium ensembles obtained from microsecond-scale MD simulations. The residue-based decomposition of binding free energies confirmed the critical roles of hotspot residues identified by BeAtMuSiC. In particular, in the KRAS-K27 complex, R41 emerged as the dominant hotspot ($\Delta G = -16.34$ kcal/mol), consistent with BeAtMuSiC predictions. Other notable contributors, including Y32, T35, I36, Y40, and M67, also showed favorable van der Waals contributions, reinforcing the hydrophobic nature of the K27 interface. In the KRAS-K55 complex, the hydrophobic residues Y40, M67, and F79 were identified as major contributors to binding stability, aligning with the BeAtMuSiC results. Electrostatic interactions at D38 and R41 further stabilized the interface, supporting the effector-mimetic signature of K55. For the allosteric binders K13 and K19, residues H95, Y96, L133, and S136 were validated as central anchors, corroborating BeAtMuSiC predictions. Specifically, H95 contributed significantly through aromatic stacking with tryptophan residues in the DARPin scaffold, while L133 and Y137 reinforced shape complementarity. Our computational findings are in good agreement with the available experimental data, underscoring the reliability of our predictions. Indeed, the KRAS residues I36, S39, Y40, and M67 were identified as shared anchors between K27 and K55, playing critical roles in stabilizing the interaction interface. These results align with experimental observations [32], which demonstrate that engineered protein binders can achieve inhibition through complementary strategies. Furthermore, the computational predictions from both mutational scanning and MM-GBSA computations of allosteric binding by K13 and K19 showed stabilization of KRAS-specific pocket in the α3–loop–α4 motif, leveraging conserved hydrophobic motifs centered on H95 and Y96. These results are consistent with experimental mutagenesis studies revealing that H95A/V/N severely impair binding for both K13 and K19, validating its role as a KRAS-specific hotspot [33]. Together, these comparisons highlight the strong agreement between our computational predictions and experimental observations, reinforcing the validity of our mutational scanning analysis.

3.4. Network Analysis of Allosteric Communication in KRAS–DARPin Complexes

We used the ensemble-based network centrality analysis and the network-based mutational profiling of allosteric residue propensities to characterize the global network of allosteric communications. We constructed residue interaction graphs using the ensemble-averaged contact maps from MD simulations of KRAS complexes with DARPINs K27, K55, K13, and K19. Each node represents a KRAS residue, and edges are defined based on proximity ($\leq$5 Å) in the conformational ensemble. The Shortest Path Betweenness (SPC) centrality was computed for each residue, measuring how often it lies on the shortest paths connecting distant parts of the protein. High-SPC residues function as central routers in the network, playing critical roles in information transfer even if they are not directly involved in binding. To assess how mutations perturb this network, we calculated the Z-score of SPC changes across an ensemble of mutation-induced structures. A high positive Z-score indicates that a mutation disrupts communication pathways, suggesting that the wild-type residue contributes significantly to network integrity and signaling robustness Allosteric hotspots are identified as residues in which mutations incur significant perturbations of the global residue interaction network that disrupt the network connectivity and cause a significant impairment of global network communications and compromise signaling.

Network analysis of KRAS-K27 revealed a number of high-SPC residues including positions in the interacting KRAS regions (Y40, R41, D38, S39, Y71, V14, K16, L23, I24, F78, F82, F90) as well as residues in the allosteric lobe (L113, S136, L159) (Figure 7A). In the KRAS–K27 complex, SPC analysis also identified several clusters of high-centrality residues: P-loop (residues 10–17): L6, G15, S17, I18; Switch I (residues 24–40): Y32, P34, T35, I36, E37, D38, S39; α-helix 1 (residues 80–85): M80, A83; α-helix 3 (residues 87–104): R97, H95, Y96; and β-strand 4 (residues 77–84): L77, V81, L83. Y40 and R41 emerged as the dominant hubs, showing the highest SPC value. This residue acts as a central router in the network, mediating long-range communication between the P-loop and C-terminal helices. T35 and Y40 also showed elevated SPC, reinforcing their role in maintaining nucleotide coordination geometry. The central β-sheet, especially β-strand 2 (E37–D38–S39) and β-strand 3 (I55–L56–D57), exhibited increased network centrality, indicating that K27 indirectly stabilizes these motifs, even though it does not engage them directly (Figure 7A). K55 interface includes key aromatic anchors, such as Y40, M67, and F79, which form a hydrophobic cluster that locks the switch loops into a signaling-ready configuration. The highest SPC positions in the network centrality profile for KRAS-K55 complex include residues Q70, Y71, Y64, M67, E37, D38, T74, F78, F82, I93, V114, N116, L133, and Y137 (Figure 7B). Although the overall shape and patterns of the SPC profiles for K27 and K55 complexes are generally similar, as may be expected, there are noticeable differences in the relative ranking of the top SPC residues. For the KRAS-K55 complex, the SPC profile highlighted the importance of Switch II residues Q70, Y71, Y64, and M67, which form a hydrophobic cluster that locks the switch loops into a signaling-ready configuration. The network mediating role of these Switch II residues reflected their critical roles in effector mimicry and hydrophobic stabilization (Figure 7B). E37 and D38, known for their involvement in salt bridge formation with RAF1, also showed high SPC values, suggesting they could function as local hubs in the effector lobe. Importantly, among SPC profile peaks are residues F78, L80, and F82 from β-strand 4; I93 from α-helix 3; V114 and N116 from β-strand 5, and Y157, L159, I163, and R164 from α-helix 5 of the allosteric lobe (Figure 7A,B). Hence, the SPC analysis highlighted the presence of allosteric mediating centers from different functional regions, including positions located away from the binding interface. The results also highlight the importance of allosteric communication in KRAS, where distant residues influence the binding interface with RAF1 through long-range interactions. This communication is mediated by a network of residues that function as bridges or bottlenecks in the allosteric network.

Unlike K27 and K55, which target the effector lobe, K13 and K19 bind to a novel allosteric pocket involving helix 3–loop–helix 4, centered around H95, Y96, and surrounding motifs. SPC analysis revealed elevated centrality values for F78, H94, F90, F82, N116, I93, Y137, L159, L113, I142, H95, and Y96 (Figure 7C). H95 and Y96 form the core of the allosteric hub while L133, S136, Y137 provide allosteric communication centers For KRAS-K19, the major allosteric centers are F82, F90, V114, S136, E91, T144, M72, H95, and Y96 (Figure 7D). These residues do not show strong overlap with those identified in effector systems, suggesting that K13 and K19 engage a structurally and dynamically distinct region of the network, offering a modular route for isoform-specific targeting. Importantly, all

KRAS-DARPin complexes share a number of common allosteric hotspots, including F78, F82 (β-strand 4), F90, I93 (α-helix 3), V114, and N116 (β-strand 5), that are located in different functional regions of KRAS (Figure 7). Structural mapping showed that many of these residues form communication routes via the central β-sheet, linking distal pockets to the active-site lobe (Supporting Information, Figure S4). The predicted allosteric hotspots are in strong agreement with experimental data [34], which independently identified residues in β-strand 4 (F78, F82) and β-strand 5 (V112, V114, N116) as critical mediators of long-range communication, linking distant functional sites and stabilizing specific conformations. These hotspots, which align strongly with experimental data, underscore a conserved allosteric infrastructure that is universally important for KRAS signaling. The findings highlight a conserved allosteric infrastructure that can be exploited for therapeutic design.

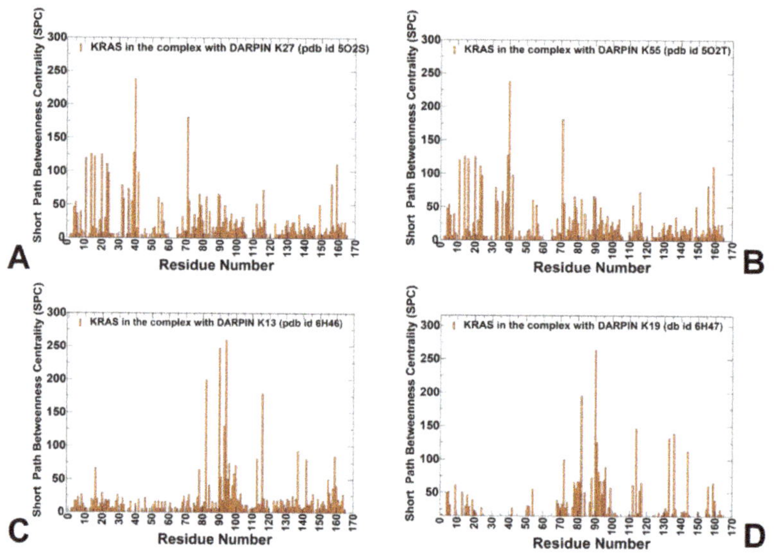

Figure 7. Network centrality analysis of the KRAS complexes with DARPIN proteins. The ensemble-averaged SPC centrality for the KRAS residues in the KRAS-WT complex with DARPIN K27 (pdb id 5O2S) (**A**), KRAS-WT complex with DARPIN K55 (**B**), KRAS-WT complex with DARPIN K13 (**C**), and KRAS-WT complex with DARPIN K19 (**D**). The profiles are shown in orange-colored filled bars.

By comparing the top-ranked residues based on their Z-scores, we can gain deeper insight into the distinct mechanisms by which these DARPINs engage KRAS. While mutational scanning identified hotspot residues such as R41, Y40, T35, I36, and M67 as major contributors to interface stability, the Z-score SPC analysis revealed additional mutation-sensitive control points that are not part of the direct DARPin interface yet have profound effects on network integrity. Network-based mutational scanning revealed that certain residues, although not part of the direct interface, strongly affect binding when mutated. Among the top Z-score profile residues are Y40, Y71, T20, K16, V14, L23, F78, S39, and N116 (Figure 8A). These findings suggest that K27's mechanism extends beyond localized stabilization. For K55, the top Z-score SPC residues identified are quite different and include Q70, Y71, and T74, along with residues in the allosteric lobe, Y137, L133, R97, and A134. (Figure 8B). These residues highlight regions that are central to signal propagation and network stability, even though they may not directly interact with the DARPin. Mutations at these network-embedded residues, even if they do not contact the

DARPin directly, can severely disrupt information flow, leading to reduced binding affinity or altered functional outcomes.

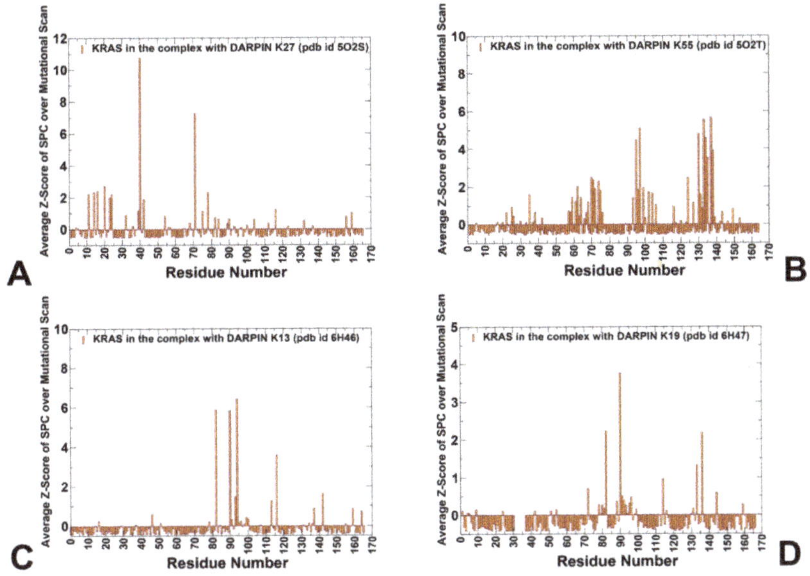

Figure 8. Network centrality analysis of the KRAS complexes with DARPIN proteins. The average Z-score of the ASPL over the mutational scan is obtained for each node based on the change of the characteristic path length under node removal, averaged over all protein residues. The Z-scores are then computed over all mutational changes in a given position. The network metric profile corresponding to the average Z-score of the ASPL over mutational scan for the KRAS residues in the KRAS-WT complex with DARPIN K27 (pdb id 5O2S) (**A**), KRAS-WT complex with DARPIN K55 (**B**), KRAS-WT complex with DARPIN K13 (**C**), and KRAS-WT complex with DARPIN K19 (**D**). The profiles are shown in orange-colored filled bars.

Despite both targeting the effector lobe, the Z-score profiles show that K27 engages a broader, distributed network, where mutations at structurally distant residues can severely disrupt long-range signaling, even when they are not directly involved in binding. K55 focuses on a more compact, effector-like topology, where disruption of key Switch II and allosteric lobe motifs has disproportionate effects on network efficiency, indicating a higher sensitivity to aromatic loss and local destabilization. Unlike K27, K55's Z-score RCA peaks are concentrated around Switch II and the allosteric lobe, suggesting that it relies on localized signal reinforcement rather than broad network modulation. The fact that mutations at Y137, L133, and R97 significantly impair network efficiency indicates that K55 exploits not only the effector interface but also regions previously thought to be functionally remote, such as Loop 7 and helix α4. For K13, the major mutation-sensitive cont5rol points are M78, F90, H94, F82, L113, N116, I142, Y137, and L159 (Figure 8C). For K19 the hubs are generally similar but not identical: F90, F82, S136, L133, V114, M72, V114, and T144 (Figure 8D). The important finding of this analysis is that the predicted allosteric hotspots are consistent with the seminal experimental studies of KRAS allostery, showing that allosteric effects in KRAS are mediated by the central β-strand of KRAS, which acts as a hub for transmitting conformational changes linking distant functional sites [34]. According to DMS analysis of KRAS allostery, KRAS positions F78, F82 from β-strand 4, and residues V112, V114, and N116 from β-strand 5 can act as allosteric centers affecting binding with effector proteins through long-range interactions. The important revelation of our predictions is that the allosteric binding hotspots correspond to conserved positions that could control allosteric binding with a diverse range of binding partners, including DARPin proteins.

The important revelation of our predictions is that the allosteric binding hotspots correspond to conserved positions that could control allosteric binding with a diverse range of binding partners targeting different pockets on KRAS. In addition, a group of residues from the α-helix 3 (I93, H95, and Y96) is critical for linking the binding interface region with the central β-strand and subsequently to the allosteric lobe. Finally, the predicted allosterically important positions (Y157, L159, I163, and R164 from α-helix 5) were found in the experimentally validated allosteric pocket located in the C-terminal lobe of the protein (this pocket is formed by residues 97, 101, 107–111, 136–140, and 161–166) and is the most distant pocket from the binding interface [34]. Our analysis identifies these sites as major hotspots of allosteric communication that affect interaction paths from remote parts of KRAS to the binding interface.

4. Discussion

KRAS has long stood as a paradigm of functional complexity—its conformational transitions between GDP- and GTP-bound states, its reliance on switch regions for effector engagement, and its integration into signaling networks have made it both an essential regulator and a formidable therapeutic target. The emergence of engineered binders like DARPINs K27, K55, K13, and K19 has opened new avenues for targeting this oncogenic hub with precision and selectivity. These systems do not simply block KRAS function; they engage it through diverse molecular strategies that span effector mimicry, conformational restriction, and allosteric modulation. By combining mutational scanning, MM-GBSA decomposition, RMSF profiling, and network-based analysis, we uncover a layered understanding of how these DARPINs exploit the structural and energetic architecture of KRAS to modulate its activity. At the heart of our findings is the realization that KRAS recognition by engineered proteins is not monolithic. Instead, it reflects a spectrum of binding mechanisms, each with unique implications for mutation sensitivity and functional interference. The comparison between K27 and K55 reveals a striking dichotomy in the way engineered proteins can engage KRAS. While both target the classical effector interface involving Switch I and II loops, their mechanisms diverge significantly. K27 binds to the inactive, GDP-bound state, stabilizing the nucleotide-binding pocket without inducing major structural rearrangements. Key residues such as R41, T35, and Y40 contribute through modular hydrogen bonding and electrostatic anchoring, forming a broad interaction scaffold rather than relying on localized high-affinity contacts. In contrast, K55 mimics natural effectors like RAF1 and PI3Kγ, locking KRAS into a signaling-ready geometry through a compact hydrophobic cluster centered on Y40, M67, and F79, supported by electrostatic interactions at D38 and R41. This system acts more like a molecular trap, interfering with downstream signal propagation by restricting motion in both switch regions and altering the electrostatic environment around the effector lobe.

Mutational scanning offers a powerful lens through which to evaluate residue-level contributions to complex stability, revealing a core set of evolutionarily constrained hotspots that are critical for recognition across all systems. Notably, I36 and Y40 emerge as universal hubs, contributing significantly to binding in both natural effectors and engineered proteins. However, mutational effects diverge between DARPINs. For instance, R41 becomes the dominant hotspot in K27, where mutations such as R41C/H/N/D cause severe destabilization due to loss of salt bridge formation and backbone-mediated hydrogen bonding. In contrast, Y64 and M67 are uniquely vulnerable in K55, with substitutions like Y64G or M67L/V/A leading to dramatic reductions in binding affinity, likely due to disruption of aromatic stacking and shape complementarity. These results underscore that while certain residues serve as central anchors across multiple binders, others define specificity and escape vulnerability, highlighting the importance of tailoring therapeutic strategies to match the energetic and mutational context of each system.

The MM-GBSA residue-wise decomposition deepens this mechanistic picture, showing that hydrophobic forces dominate interface stabilization, even when electrostatic contributions vary significantly between systems. In K27, the electrostatic environment around R41 plays a defining role, with a total contribution of $\Delta G = -16.34$ kcal/mol, driven primarily by favorable polar interactions. This contrasts sharply with K55, where van der Waals contacts at Y40 and M67 form the backbone of binding stability. Despite their shared engagement of the effector lobe, the energetic dependency of these two DARPINs reflects their distinct functional roles: K27 acts as a conformational restrainer, preventing SOS1-mediated activation, while K55 functions as an effector mimic, blocking downstream signal propagation by locking KRAS into a rigid, signaling-ready orientation. This divergence in thermodynamic drivers provides a rationale for combining these agents in therapeutic regimens, as their overlapping yet mechanistically distinct modes of action raise the genetic barrier to resistance. Beyond the classical effector interface, K13 and K19 offered a different strategy in KRAS targeting, engaging a structurally remote allosteric lobe centered on the α3–loop–α4 motif. This pocket includes H95, Y96, L133, S136, and Y137, with D108 emerging as a defining feature of K19, where it forms salt bridges and hydrogen bonds that enhance selectivity for KRAS over HRAS and NRAS. Unlike K27 and K55, K13 and K19 do not fully restrict switch loop motion, but instead alter the conformational equilibrium of KRAS, subtly shifting the balance toward functionally constrained states. This dynamic modulation aligns with experimental observations that these DARPINs allow for partial nucleotide exchange and effector engagement, distinguishing them from strict pathway blockers and positioning them as fine-tuners of KRAS activity rather than complete suppressors.

Network-based analysis reveals that the impact of DARPin binding extends beyond the immediate interface. High Z-score peaks in K27-bound KRAS include Y40, Y71, T20, K16, V14, L23, F78, S39, and N116, many of which are not directly involved in binding but function as bottlenecks in long-range communication. These residues contribute to maintaining the structural integrity of the central β-sheet, suggesting that K27 indirectly reshapes the global network topology, favoring inactive-state communication routes. In contrast, K55 shows elevated Z-scores at Switch II residues such as Q70, Y71, T74, Y137, L133, R97, A130, and A134, indicating that its mechanism relies on localized rigidity and enhanced effector-like connectivity. Mutations at these positions severely disrupt network efficiency, supporting the idea that K55 reinforces the active-state residue interaction map, making it overly sensitive to aromatic loss but less affected by upstream P-loop variations. The comprehensive analysis of KRAS binding with engineered DARPINs—K27, K55, K13, and K19—reveals a strikingly consistent theme: despite their distinct mechanisms of recognition, all systems engage a unifying allosteric architecture that spans multiple functional motifs. This architecture is not only preserved across complexes but also mirrors the intrinsic communication framework of KRAS itself, where specific residues function as central hubs, transmitting conformational changes across the protein. Notably, this includes positions such as H94, H95, Y96, L133, S136, Y137, and D108 (in K19) —residues located within the allosteric lobe formed by helix α3–loop–α4, which has been experimentally validated as a key regulatory domain for modulating RAS function.

This study reveals that KRAS binding with DARPINs and small molecule inhibitors follows a common architectural logic, governed by a conserved network of residues that span both the effector interface and the allosteric lobe. The Z-score RCA and SPC profiles highlight how mutations at seemingly remote sites—such as N116, F78, and H95—can exert profound effects on binding stability and signaling output, reinforcing the idea that KRAS functions as an integrated communication system, where local perturbations reshape global dynamics. Moreover, the overlap between allosteric centers recognized by DARPINs and those targeted by small molecule inhibitors like MTX-1106 and Sotorasib [46] suggests that this remote lobe is not only structurally conserved but also evolutionarily privileged for therapeutic intervention. The presence of H95 and Y96 as central hubs further supports the notion that KRAS-specific allostery can be harnessed for precision oncology. Ultimately, our findings suggest that the future of KRAS inhibition lies in exploiting this universal architecture, where modular hydrogen bonding, aromatic stacking, and network-level restructuring combine to deliver potent, durable, and mutation-resilient targeting of one of cancer's most persistent drivers.

A key strength of this study lies in the rigorous validation of BeAtMuSiC predictions through complementary computational and experimental approaches. First, MM-GBSA calculations provided an independent assessment of residue-specific contributions to binding stability, confirming the energetic roles of hotspot residues identified by BeAtMuSiC. For example, the dominant contribution of R41 in the KRAS-K27 interface and the hydrophobic cluster centered on Y40, M67, and F79 in the KRAS-K55 complex were consistently identified across both methods. Similarly, residues H95, Y96, L133, and S136 were validated as central anchors for allosteric binders K13 and K19, highlighting the reliability of our mutational scanning approach.

Furthermore, our computational findings align closely with experimental observations. For instance, the identification of I36, S39, Y40, and M67 as shared anchors between K27 and K55 is consistent with experimental studies showing that engineered protein binders can achieve inhibition through complementary strategies [32]. Likewise, the critical role of H95 in allosteric modulation, as predicted computationally, has been experimentally validated for both K13 and K19 [33].

The convergence of computational predictions and experimental data highlights a universal architectural logic governing KRAS recognition. Key residues such as I36, Y40, M67, and H95 emerge as evolutionarily constrained hotspots that govern both local binding events and long-range signal propagation. This conserved network topology underscores the potential for designing multi-targeted inhibitors that exploit both canonical and non-canonical binding sites on KRAS. For example, combining effector mimics like K55 with allosteric modulators like K13 or K19 could raise the genetic barrier to resistance by targeting structurally distant and functionally divergent residues. Additionally, the identification of H95 and Y96 as central hubs in KRAS-specific allostery supports the notion that these residues represent privileged therapeutic targets for isoform-specific modulation. By integrating dynamic, energetic, and network-level insights, this study lays the groundwork for a systems-level approach to KRAS inhibition, where residue interaction networks and hinge point redistribution may help to guide the design of next-generation inhibitors.

5. Conclusions

This study presents a comprehensive and integrative computational analysis of the binding of KRAS with engineered DARPINs, combining conformational dynamics, mutational scanning, MM-GBSA residue decomposition, and network-based centrality profiling to uncover the molecular mechanisms that define binding specificity, mutation sensitivity, and allosteric control. These findings provide a detailed mechanistic framework for understanding how distinct classes of binders—effector mimics, conformational blockers, and allosteric modulators—exploit both local and global features of KRAS to achieve targeted inhibition. MD simulations reveal that KRAS exhibits significant conformational plasticity, even under bound conditions, with key functional regions such as the P-loop, Switch I, and Switch II displaying distinct flexibility patterns depending on the nature of the interaction partner. Notably, K27 and K55 stabilize the effector lobe through different dynamic strategies. K27 restricts the P-loop and Switch I motion, locking KRAS into an inactive-state conformation without inducing large-scale structural rearrangements. K55 immobilizes both switch loops simultaneously, mimicking natural effectors like RAF1. These results highlight that functional outcomes are not solely determined by static structures, but rather by how each binder reshapes the intrinsic flexibility landscape of KRAS.

Mutational scanning across all four DARPIN systems identifies a core set of evolutionarily constrained residues that function as universal hotspots in KRAS recognition. KRAS residues I36, Y40, M67, and H95 consistently emerge as critical contributors to binding stability. Importantly, mutational effects extend beyond direct contact points. Residues not directly involved in binding, such as R41, F78, and N116, also show high sensitivity to mutation, indicating their role in supporting long-range communication pathways and modulating network efficiency. This emphasizes that the KRAS function is governed by distributed interactions, where mutations at structurally distant sites can have profound consequences on signaling output. The MM-GBSA residue-based energy decomposition provides a quantitative view of these interactions, showing that K27 relies heavily on electrostatic contributions, especially from R41, which forms salt bridges and hydrogen bonds that anchor the interface. K55 exploits a dense hydrophobic cluster involving Y40, M67, and F79, with electrostatic networks at D38 and E37 enhancing its effector-mimetic signature. K13 and K19 engage a novel allosteric pocket centered on helix α3–loop–α4, where H95, Y96, L133, and D108 (in K19) contribute strongly to binding stability via aromatic stacking and polar interactions. These energetic profiles confirm that while hydrophobic forces dominate interface stabilization, electrostatic contributions play a defining role in determining specificity.

Network-based analysis reveals that DARPin binding restructures the global communication architecture of KRAS, even when engaging structurally remote pockets. We found that KRAS allostery involves not only localized conformational changes but also global network restructuring, reinforcing the idea that effective targeting requires consideration of both direct and indirect effects on protein communication. This work establishes that KRAS operates within a highly coordinated communication infrastructure, where certain residues function as central hubs governing both local binding and global conformational transitions. What emerges from our findings is a universal network topology underlying KRAS allostery, where certain positions are conserved across diverse binders —both natural effectors and synthetic inhibitors—and play critical roles in shaping the energetic, dynamic, and network-level properties of the protein. This suggests that KRAS does not simply fold into isolated functional units; rather, it operates as an integrated signaling module, with evolutionarily constrained hotspots that govern both local binding events and long-range signal propagation. Notably, several of these hotspot residues, including H94, H95, and Y96, are engaged by both small-molecule inhibitors and protein-based binders, suggesting a universally conserved allosteric architecture. This convergence across modalities underscores the evolutionary importance of this region and supports its use as a therapeutic nexus for isoform-specific modulation. By integrating dynamic, energetic, and network-level insights, we demonstrate that effective KRAS targeting must move beyond traditional views of local binding motifs and instead embrace a systems-level approach, where residue interaction networks and hinge point redistribution guide the design of next-generation inhibitors.

Supplementary Materials: The following supporting information can be downloaded at: https://www.mdpi.com/article/10.3390/biom15060819/s1, Figure S1 presents a structural representation of KRAS in its open (state I) and closed (state II) conformations, highlighting the dynamic equilibrium between these states in the GTP-bound form. Figure S2 presents the RMSD profiles for K27-KRAS and K13 -KRAS complexes. Figure S3 presents structural mapping of conformational mobility profiles of KRAS-DARPin complexes obtained from MD simulations. Figure S4 presents structural mapping of allosteric communication hotspots in KRAS complexes with DARPin proteins. Table S1 presents a summary of the systems used in this study. Table S2 details the simulation setup for KRAS-DARPin complexes analyzed in this study.

Author Contributions: Conceptualization, G.V.; methodology, M.A., V.P., B.F. and G.V.; software, M.A., V.P., B.F. and G.V.; validation, G.V.; formal analysis, G.V., M.A., V.P. and B.F.; investigation, G.V.; resources, V.P., M.A, and G.V.; data curation, M.A., V.P. and G.V.; writing—original draft preparation, G.V.; writing—review and editing, G.V.; visualization, V.P., M.A. and G.V.; supervision G.V.; project administration, G.V.; funding acquisition, G.V. All authors have read and agreed to the published version of the manuscript.

Funding: This research was funded by the National Institutes of Health under Award 1R01AI181600-01 and Subaward 6069-SC24-11 to G.V.

Institutional Review Board Statement: Not applicable.

Informed Consent Statement: Not applicable.

Data Availability Statement: The original contributions presented in this study are included in the article/Supplementary Material. Crystal structures were obtained and downloaded from the Protein Data Bank (http://www.rcsb.org) (accessed on 10 April 2025). The rendering of protein structures was performed with the interactive visualization program UCSF ChimeraX package (https://www.rbvi.ucsf.edu/chimerax/) (accessed on 15 April 2025) and Pymol (https://pymol.org/2/) (accessed on 27 April 2025). All mutational heatmaps were produced using the developed software that is freely available at (https://alshahrani.shinyapps.io/HeatMapViewerApp/) (accessed on 16 April 2025).

Acknowledgments: The authors acknowledge support from Schmid College of Science and Technology at Chapman University for providing computing resources at the Keck Center for Science and Engineering.

Conflicts of Interest: The authors declare no conflicts of interest. The funders had no role in the design of the study; in the collection, analyses, or interpretation of data; in the writing of the manuscript; or in the decision to publish the results.

References

1. Saperstein, R.; Goel, S.; Maitra, R. Noncoding RNA Profile in Reovirus Treated KRAS-Mutated Colorectal Cancer Patients. *Diseases* **2023**, *11*, 142. [CrossRef]
2. Spencer-Smith, R. The RAF Cysteine-Rich Domain: Structure, Function, and Role in Disease. *Adv. Cancer Res.* **2024**, *164*, 69–91. [CrossRef]
3. Spencer-Smith, R.; Morrison, D.K. Regulation of RAF Family Kinases: New Insights from Recent Structural and Biochemical Studies. *Biochem. Soc. Trans.* **2024**, *52*, 1061–1069. [CrossRef]
4. Jeon, H.; Tkacik, E.; Eck, M.J. Signaling from RAS to RAF: The Molecules and Their Mechanisms. *Annu. Rev. Biochem.* **2024**, *93*, 289–316. [CrossRef] [PubMed]
5. Fang, Z.; Lee, K.-Y.; Huo, K.-G.; Gasmi-Seabrook, G.; Zheng, L.; Moghal, N.; Tsao, M.-S.; Ikura, M.; Marshall, C.B. Multivalent Assembly of KRAS with the RAS-Binding and Cysteine-Rich Domains of CRAF on the Membrane. *Proc. Natl. Acad. Sci. USA* **2020**, *117*, 12101–12108. [CrossRef] [PubMed]
6. Spencer-Smith, R.; Terrell, E.M.; Insinna, C.; Agamasu, C.; Wagner, M.E.; Ritt, D.A.; Stauffer, J.; Stephen, A.G.; Morrison, D.K. RASopathy Mutations Provide Functional Insight into the BRAF Cysteine-Rich Domain and Reveal the Importance of Autoinhibition in BRAF Regulation. *Mol. Cell* **2022**, *82*, 4262–4276.e5. [CrossRef]

7. Fernandez-Medarde, A.; Santos, E. Ras in Cancer and Developmental Diseases. *Genes Cancer* **2011**, *2*, 344–358. [CrossRef] [PubMed]
8. Ghimessy, A.; Radeczky, P.; Laszlo, V.; Hegedus, B.; Renyi-Vamos, F.; Fillinger, J.; Klepetko, W.; Lang, C.; Dome, B.; Megyesfalvi, Z. Current Therapy of KRAS-Mutant Lung Cancer. *Cancer Metastasis Rev.* **2020**, *39*, 1159–1177. [CrossRef]
9. Vigil, D.; Martin, T.D.; Williams, F.; Yeh, J.J.; Campbell, S.L.; Der, C.J. Aberrant Overexpression of the Rgl2 Ral Small GTPase-Specific Guanine Nucleotide Exchange Factor Promotes Pancreatic Cancer Growth through Ral-Dependent and Ral-Independent Mechanisms. *J. Biol. Chem.* **2010**, *285*, 34729–34740. [CrossRef]
10. Lu, S.; Jang, H.; Gu, S.; Zhang, J.; Nussinov, R. Drugging Ras GTPase: A Comprehensive Mechanistic and Signaling Structural View. *Chem. Soc. Rev.* **2016**, *45*, 4929–4952. [CrossRef]
11. Lu, S.; Jang, H.; Nussinov, R.; Zhang, J. The Structural Basis of Oncogenic Mutations G12, G13 and Q61 in Small GTPase K-Ras4B. *Sci. Rep.* **2016**, *6*, 21949. [CrossRef] [PubMed]
12. Vatansever, S.; Erman, B.; Gümüş, Z.H. Oncogenic G12D Mutation Alters Local Conformations and Dynamics of K-Ras. *Sci. Rep.* **2019**, *9*, 11730. [CrossRef] [PubMed]
13. Baussand, J.; Kleinjung, J. Specific Conformational States of Ras GTPase upon Effector Binding. *J. Chem. Theory Comput.* **2013**, *9*, 738–749. [CrossRef] [PubMed]
14. Spoerner, M.; Herrmann, C.; Vetter, I.R.; Kalbitzer, H.R.; Wittinghofer, A. Dynamic Properties of the Ras Switch I Region and Its Importance for Binding to Effectors. *Proc. Natl. Acad. Sci. USA* **2001**, *98*, 4944–4949. [CrossRef]
15. Spoerner, M.; Nuehs, A.; Ganser, P.; Herrmann, C.; Wittinghofer, A.; Kalbitzer, H.R. Conformational States of Ras Complexed with the GTP Analogue GppNHp or GppCH2p: Implications for the Interaction with Effector Proteins. *Biochemistry* **2005**, *44*, 2225–2236. [CrossRef]
16. Spoerner, M.; Hozsa, C.; Poetzl, J.A.; Reiss, K.; Ganser, P.; Geyer, M.; Kalbitzer, H.R. Conformational States of Human Rat Sarcoma (Ras) Protein Complexed with Its Natural Ligand GTP and Their Role for Effector Interaction and GTP Hydrolysis. *J. Biol. Chem.* **2010**, *285*, 39768–39778. [CrossRef]
17. Chao, F.-A.; Dharmaiah, S.; Taylor, T.; Messing, S.; Gillette, W.; Esposito, D.; Nissley, D.V.; McCormick, F.; Byrd, R.A.; Simanshu, D.K.; et al. Insights into the Cross Talk between Effector and Allosteric Lobes of KRAS from Methyl Conformational Dynamics. *J. Am. Chem. Soc.* **2022**, *144*, 4196–4205. [CrossRef]
18. Ostrem, J.M.; Peters, U.; Sos, M.L.; Wells, J.A.; Shokat, K.M. K-Ras(G12C) Inhibitors Allosterically Control GTP Affinity and Effector Interactions. *Nature* **2013**, *503*, 548–551. [CrossRef]
19. Janes, M.R.; Zhang, J.; Li, L.-S.; Hansen, R.; Peters, U.; Guo, X.; Chen, Y.; Babbar, A.; Firdaus, S.J.; Darjania, L.; et al. Targeting KRAS Mutant Cancers with a Covalent G12C-Specific Inhibitor. *Cell* **2018**, *172*, 578–589.e17. [CrossRef]
20. Canon, J.; Rex, K.; Saiki, A.Y.; Mohr, C.; Cooke, K.; Bagal, D.; Gaida, K.; Holt, T.; Knutson, C.G.; Koppada, N.; et al. The Clinical KRAS(G12C) Inhibitor AMG 510 Drives Anti-Tumour Immunity. *Nature* **2019**, *575*, 217–223. [CrossRef]
21. Hallin, J.; Engstrom, L.D.; Hargis, L.; Calinisan, A.; Aranda, R.; Briere, D.M.; Sudhakar, N.; Bowcut, V.; Baer, B.R.; Ballard, J.A.; et al. The KRASG12C Inhibitor MRTX849 Provides Insight toward Therapeutic Susceptibility of KRAS-Mutant Cancers in Mouse Models and Patients. *Cancer Discov.* **2020**, *10*, 54–71. [CrossRef]
22. Murugan, A.K.; Grieco, M.; Tsuchida, N. RAS Mutations in Human Cancers: Roles in Precision Medicine. *Semin. Cancer Biol.* **2019**, *59*, 23–35. [CrossRef] [PubMed]
23. Fetics, S.K.; Guterres, H.; Kearney, B.M.; Buhrman, G.; Ma, B.; Nussinov, R.; Mattos, C. Allosteric Effects of the Oncogenic RasQ61L Mutant on Raf-RBD. *Structure* **2015**, *23*, 505–516. [CrossRef]
24. Tran, T.H.; Chan, A.H.; Young, L.C.; Bindu, L.; Neale, C.; Messing, S.; Dharmaiah, S.; Taylor, T.; Denson, J.-P.; Esposito, D.; et al. KRAS Interaction with RAF1 RAS-Binding Domain and Cysteine-Rich Domain Provides Insights into RAS-Mediated RAF Activation. *Nat. Commun.* **2021**, *12*, 1176. [CrossRef] [PubMed]
25. Zhang, M.; Jang, H.; Li, Z.; Sacks, D.B.; Nussinov, R. B-Raf Autoinhibition in the Presence and Absence of 14-3-3. *Structure* **2021**, *29*, 768–777.e2. [CrossRef]
26. Park, E.; Rawson, S.; Li, K.; Kim, B.W.; Ficarro, S.B.; Pino, G.G.; Sharif, H.; Marto, J.A.; Jeon, H.; Eck, M.J. Architecture of Autoinhibited and Active BRAF–MEK1–14-3-3 Complexes. *Nature* **2019**, *575*, 545–550. [CrossRef]
27. Martinez Fiesco, J.A.; Durrant, D.E.; Morrison, D.K.; Zhang, P. Structural Insights into the BRAF Monomer-to-Dimer Transition Mediated by RAS Binding. *Nat. Commun.* **2022**, *13*, 486. [CrossRef] [PubMed]
28. Park, E.; Rawson, S.; Schmoker, A.; Kim, B.W.; Oh, S.; Song, K.; Jeon, H.; Eck, M.J. Cryo-EM Structure of an RAS/RAF Recruitment Complex. *Nat. Commun.* **2023**, *14*, 4580. [CrossRef]
29. Lee, S.Y.; Eun, H.J.; Lee, K.Y. Effector Binding Sequentially Alters KRAS Dimerization on the Membrane: New Insights into RAS-Mediated RAF Activation. *Adv. Sci.* **2024**, *11*, e2401530. [CrossRef]
30. Zhou, Z.W.; Ambrogio, C.; Bera, A.K.; Li, Q.; Li, X.X.; Li, L.; Son, J.; Gondi, S.; Li, J.; Campbell, E.; et al. KRASQ61H Preferentially Signals through MAPK in a RAF Dimer-Dependent Manner in Non-Small Cell Lung Cancer. *Cancer Res.* **2020**, *80*, 3719–3731. [CrossRef]

31. Kwon, J.J.; Dilly, J.; Liu, S.; Kim, E.; Bian, Y.; Dharmaiah, S.; Tran, T.H.; Kapner, K.S.; Ly, S.H.; Yang, X.; et al. Comprehensive Structure-Function Analysis Reveals Gain- and Loss-of-Function Mechanisms Impacting Oncogenic KRAS Activity. *bioRxiv* **2024**, 2024.10.22.618529. [CrossRef] [PubMed] [PubMed Central]
32. Guillard, S.; Kolasinska-Zwierz, P.; Debreczeni, J.; Breed, J.; Zhang, J.; Bery, N.; Marwood, R.; Tart, J.; Overman, R.; Stocki, P.; et al. Structural and Functional Characterization of a DARPin Which Inhibits Ras Nucleotide Exchange. *Nat. Commun.* **2017**, *8*, 16111. [CrossRef]
33. Bery, N.; Legg, S.; Debreczeni, J.; Breed, J.; Embrey, K.; Stubbs, C.; Kolasinska-Zwierz, P.; Barrett, N.; Marwood, R.; Watson, J.; et al. KRAS-Specific Inhibition Using a DARPin Binding to a Site in the Allosteric Lobe. *Nat. Commun.* **2019**, *10*, 2607. [CrossRef]
34. Weng, C.; Faure, A.J.; Escobedo, A.; Lehner, B. The Energetic and Allosteric Landscape for KRAS Inhibition. *Nature* **2024**, *626*, 643–652. [CrossRef] [PubMed]
35. Chen, C.C.; Er, T.K.; Liu, Y.Y.; Hwang, J.K.; Barrio, M.J.; Rodrigo, M.; Garcia-Toro, E.; Herreros-Villanueva, M. Computational Analysis of KRAS Mutations: Implications for Different Effects on the KRAS p.G12D and p.G13D Mutations. *PLoS ONE* **2013**, *8*, e55793. [CrossRef] [PubMed]
36. Vatansever, S.; Erman, B.; Gümüş, Z.H. Comparative Effects of Oncogenic Mutations G12C, G12V, G13D, and Q61H on Local Conformations and Dynamics of K-Ras. *Comput. Struct. Biotechnol. J.* **2020**, *18*, 1000–1011. [CrossRef]
37. Nguyen, K.; López, C.A.; Neale, C.; Van, Q.N.; Carpenter, T.S.; Di Natale, F.; Travers, T.; Tran, T.H.; Chan, A.H.; Bhatia, H.; et al. Exploring CRD Mobility during RAS/RAF Engagement at the Membrane. *Biophys. J.* **2022**, *121*, 3630–3650. [CrossRef]
38. Zhou, Y.; Prakash, P.; Liang, H.; Cho, K.J.; Gorfe, A.A.; Hancock, J.F. Lipid-Sorting Specificity Encoded in K-Ras Membrane Anchor Regulates Signal Output. *Cell* **2017**, *168*, 239–251.e16. [CrossRef]
39. Sayyed-Ahmad, A.; Prakash, P.; Gorfe, A.A. Distinct Dynamics and Interaction Patterns in H- and K-Ras Oncogenic P-Loop Mutants. *Proteins* **2017**, *85*, 1618–1632. [CrossRef]
40. Li, Z.L.; Prakash, P.; Buck, M. A "Tug of War" Maintains a Dynamic Protein-Membrane Complex: Molecular Dynamics Simulations of C-Raf RBD-CRD Bound to K-Ras4B at an Anionic Membrane. *ACS Cent. Sci.* **2018**, *4*, 298–305. [CrossRef]
41. Ancy, I.; Penislusshiyan, S.; Ameen, F.; Chitra, L. Microsecond Molecular Dynamics Simulation to Gain Insight into the Binding of MRTX1133 and Trametinib with KRASG12D Mutant Protein for Drug Repurposing. *J. Mol. Recognit.* **2024**, *37*, e3103. [CrossRef] [PubMed]
42. Tu, G.; Gong, Y.; Yao, X.; Liu, Q.; Xue, W.; Zhang, R. Pathways and Mechanism of MRTX1133 Binding to KRAS G12D Elucidated by Molecular Dynamics Simulations and Markov State Models. *Int. J. Biol. Macromol.* **2024**, *274*, 133374. [CrossRef] [PubMed]
43. Zhao, Z.; Bohidar, N.; Bourne, P.E. Analysis of KRAS-Ligand Interaction Modes and Flexibilities Reveals the Binding Characteristics. *J. Chem. Inf. Model.* **2023**, *63*, 1362–1370. [CrossRef] [PubMed]
44. Li, Y.; Han, L.; Zhang, Z. Understanding the Influence of AMG 510 on the Structure of KRASG12C Empowered by Molecular Dynamics Simulation. *Comput. Struct. Biotechnol. J.* **2022**, *20*, 1056–1067. [CrossRef]
45. Mir, S.A.; Nayak, B.; Aljarba, N.H.; Kumarasamy, V.; Subramaniyan, V.; Dhara, B. Exploring KRas Protein Dynamics: An Integrated Molecular Dynamics Analysis of KRas Wild and Mutant Variants. *ACS Omega* **2024**, *9*, 30665–30674. [CrossRef]
46. Pandey, D.; Roy, K.K. Decoding KRAS Dynamics: Exploring the Impact of Mutations and Inhibitor Binding. *Arch. Biochem. Biophys.* **2025**, *764*, 110279. [CrossRef]
47. Pantsar, T.; Rissanen, S.; Dauch, D.; Laitinen, T.; Vattulainen, I.; Poso, A. Assessment of Mutation Probabilities of KRAS G12 Missense Mutants and Their Long-Timescale Dynamics by Atomistic Molecular Simulations and Markov State Modeling. *PLoS Comput. Biol.* **2018**, *14*, e1006458. [CrossRef]
48. Shi, S.; Zheng, L.; Ren, Y.; Wang, Z. Impacts of Mutations in the P-Loop on Conformational Alterations of KRAS Investigated with Gaussian Accelerated Molecular Dynamics Simulations. *Molecules* **2023**, *28*, 2886. [CrossRef]
49. Chen, J.; Wang, J.; Yang, W.; Zhao, L.; Hu, G. Conformations of KRAS4B Affected by Its Partner Binding and G12C Mutation: Insights from GaMD Trajectory-Image Transformation-Based Deep Learning. *J. Chem. Inf. Model.* **2024**, *64*, 6880–6898. [CrossRef]
50. Chen, J.; Wang, J.; Yang, W.; Zhao, L.; Zhao, J.; Hu, G. Molecular Mechanism of Phosphorylation-Mediated Impacts on the Conformation Dynamics of GTP-Bound KRAS Probed by GaMD Trajectory-Based Deep Learning. *Molecules* **2024**, *29*, 2317. [CrossRef]
51. Rose, P.W.; Prlic, A.; Altunkaya, A.; Bi, C.; Bradley, A.R.; Christie, C.H.; Costanzo, L.D.; Duarte, J.M.; Dutta, S.; Feng, Z.; et al. The RCSB Protein Data Bank: Integrative View of Protein, Gene and 3D Structural Information. *Nucleic Acids Res.* **2017**, *45*, D271–D281. [CrossRef] [PubMed]
52. Webb, B.; Sali, A. Comparative Protein Structure Modeling Using MODELLER. *Curr. Protoc. Bioinform.* **2016**, *54*, 5.6.1–5.6.37. [CrossRef]
53. Fernandez-Fuentes, N.; Zhai, J.; Fiser, A. ArchPRED: A Template Based Loop Structure Prediction Server. *Nucleic Acids Res.* **2006**, *34*, W173–W176. [CrossRef] [PubMed]
54. Krivov, G.G.; Shapovalov, M.V.; Dunbrack, R.L., Jr. Improved Prediction of Protein Side-Chain Conformations with SCWRL4. *Proteins* **2009**, *77*, 778–795. [CrossRef]

55. Søndergaard, C.R.; Olsson, M.H.; Rostkowski, M.; Jensen, J.H. Improved Treatment of Ligands and Coupling Effects in Empirical Calculation and Rationalization of pKa Values. *J. Chem. Theory Comput.* **2011**, *7*, 2284–2295. [CrossRef]
56. Olsson, M.H.; Søndergaard, C.R.; Rostkowski, M.; Jensen, J.H. PROPKA3: Consistent Treatment of Internal and Surface Residues in Empirical pKa Predictions. *J. Chem. Theory Comput.* **2011**, *7*, 525–537. [CrossRef]
57. Maier, J.A.; Martinez, C.; Kasavajhala, K.; Wickstrom, L.; Hauser, K.E.; Simmerling, C. ff14SB: Improving the Accuracy of Protein Side Chain and Backbone Parameters from ff99SB. *J. Chem. Theory Comput.* **2015**, *11*, 3696–3713. [CrossRef] [PubMed]
58. Tian, C.; Kasavajhala, K.; Belfon, K.A.A.; Raguette, L.; Huang, H.; Migues, A.N.; Bickel, J.; Wang, Y.; Pincay, J.; Wu, Q.; et al. ff19SB: Amino-Acid-Specific Protein Backbone Parameters Trained against Quantum Mechanics Energy Surfaces in Solution. *J. Chem. Theory Comput.* **2020**, *16*, 528–552. [CrossRef]
59. Roe, D.R.; Cheatham, T.E., III. PTRAJ and CPPTRAJ: Software for Processing and Analysis of Molecular Dynamics Trajectory Data. *J. Chem. Theory Comput.* **2013**, *9*, 3084–3095. [CrossRef]
60. Fernandes, H.S.; Sousa, S.F.; Cerqueira, N.M.F.S.A. VMD Store—A VMD Plugin to Browse, Discover, and Install VMD Extensions. *J. Chem. Inf. Model.* **2019**, *59*, 4519–4523. [CrossRef]
61. Grant, B.J.; Skjaerven, L.; Yao, X.Q. The Bio3D Packages for Structural Bioinformatics. *Protein Sci.* **2021**, *30*, 20–30. [CrossRef] [PubMed]
62. Dehouck, Y.; Kwasigroch, J.M.; Rooman, M.; Gilis, D. BeAtMuSiC: Prediction of Changes in Protein-Protein Binding Affinity on Mutations. *Nucleic Acids Res.* **2013**, *41*, W333–W339. [CrossRef]
63. Dehouck, Y.; Gilis, D.; Rooman, M. A New Generation of Statistical Potentials for Proteins. *Biophys. J.* **2006**, *90*, 4010–4017. [CrossRef] [PubMed]
64. Dehouck, Y.; Grosfils, A.; Folch, B.; Gilis, D.; Bogaerts, P.; Rooman, M. Fast and Accurate Predictions of Protein Stability Changes upon Mutations Using Statistical Potentials and Neural Networks: PoPMuSiC-2.0. *Bioinformatics* **2009**, *25*, 2537–2543. [CrossRef]
65. Srinivasan, J.; Cheatham, T.E.; Cieplak, P.; Kollman, P.A.; Case, D.A. Continuum Solvent Studies of the Stability of DNA, RNA, and Phosphoramidate−DNA Helices. *J. Am. Chem. Soc.* **1998**, *120*, 9401–9409. [CrossRef]
66. Kollman, P.A.; Massova, I.; Reyes, C.; Kuhn, B.; Huo, S.; Chong, L.; Lee, M.; Lee, T.; Duan, Y.; Wang, W.; et al. Calculating Structures and Free Energies of Complex Molecules: Combining Molecular Mechanics and Continuum Models. *Acc. Chem. Res.* **2000**, *33*, 889–897. [CrossRef]
67. Hou, T.; Wang, J.; Li, Y.; Wang, W. Assessing the Performance of the MM/PBSA and MM/GBSA Methods. 1. The Accuracy of Binding Free Energy Calculations Based on Molecular Dynamics Simulations. *J. Chem. Inf. Model.* **2011**, *51*, 69–82. [CrossRef] [PubMed]
68. Weng, G.; Wang, E.; Wang, Z.; Liu, H.; Zhu, F.; Li, D.; Hou, T. HawkDock: A Web Server to Predict and Analyze the Protein–Protein Complex Based on Computational Docking and MM/GBSA. *Nucleic Acids Res.* **2019**, *47*, W322–W330. [CrossRef]
69. Sun, H.; Duan, L.; Chen, F.; Liu, H.; Wang, Z.; Pan, P.; Zhu, F.; Zhang, J.Z.H.; Hou, T. Assessing the Performance of MM/PBSA and MM/GBSA Methods. 7. Entropy Effects on the Performance of End-Point Binding Free Energy Calculation Approaches. *Phys. Chem. Chem. Phys.* **2018**, *20*, 14450–14460. [CrossRef]
70. Weng, G.; Wang, E.; Chen, F.; Sun, H.; Wang, Z.; Hou, T. Assessing the Performance of MM/PBSA and MM/GBSA Methods. 9. Prediction Reliability of Binding Affinities and Binding Poses for Protein-Peptide Complexes. *Phys. Chem. Chem. Phys.* **2019**, *21*, 10135–10145. [CrossRef]
71. Miller, B.R., III; McGee, T.D., Jr.; Swails, J.M.; Homeyer, N.; Gohlke, H.; Roitberg, A.E. MMPBSA.Py: An Efficient Program for End-State Free Energy Calculations. *J. Chem. Theory Comput.* **2012**, *8*, 3314–3321. [CrossRef]
72. Brinda, K.V.; Vishveshwara, S. A Network Representation of Protein Structures: Implications for Protein Stability. *Biophys. J.* **2005**, *89*, 4159–4170. [CrossRef]
73. Vijayabaskar, M.S.; Vishveshwara, S. Interaction Energy Based Protein Structure Networks. *Biophys. J.* **2010**, *99*, 3704–3715. [CrossRef] [PubMed]
74. Sethi, A.; Eargle, J.; Black, A.A.; Luthey-Schulten, Z. Dynamical Networks in tRNA:Protein Complexes. *Proc. Natl. Acad. Sci. USA* **2009**, *106*, 6620–6625. [CrossRef]
75. Stetz, G.; Verkhivker, G.M. Computational Analysis of Residue Interaction Networks and Coevolutionary Relationships in the Hsp70 Chaperones: A Community-Hopping Model of Allosteric Regulation and Communication. *PLoS Comput. Biol.* **2017**, *13*, e1005299. [CrossRef] [PubMed]
76. Martin, A.J.; Vidotto, M.; Boscariol, F.; Di Domenico, T.; Walsh, I.; Tosatto, S.C. RING: Networking Interacting Residues, Evolutionary Information and Energetics in Protein Structures. *Bioinformatics* **2011**, *27*, 2003–2005. [CrossRef] [PubMed]
77. Piovesan, D.; Minervini, G.; Tosatto, S.C. The RING 2.0 Web Server for High Quality Residue Interaction Networks. *Nucleic Acids Res.* **2016**, *44*, W367–W374. [CrossRef]
78. Clementel, D.; Del Conte, A.; Monzon, A.M.; Camagni, G.F.; Minervini, G.; Piovesan, D.; Tosatto, S.C.E. RING 3.0: Fast Generation of Probabilistic Residue Interaction Networks from Structural Ensembles. *Nucleic Acids Res.* **2022**, *50*, W651–W656. [CrossRef]

79. Del Conte, A.; Camagni, G.F.; Clementel, D.; Minervini, G.; Monzon, A.M.; Ferrari, C.; Piovesan, D.; Tosatto, S.C.E. RING 4.0: Faster Residue Interaction Networks with Novel Interaction Types across over 35,000 Different Chemical Structures. *Nucleic Acids Res.* **2024**, *52*, W306–W312. [CrossRef]
80. Floyd, R.W. Algorithm 97: Shortest Path. *Commun. ACM* **1962**, *5*, 345. [CrossRef]
81. Hagberg, A.A.; Schult, D.A.; Swart, P.J. Exploring Network Structure, Dynamics, and Function Using NetworkX. In Proceedings of the 7th Python in Science Conference (SciPy2008), Pasadena, CA, USA, 19–24 August 2008; pp. 11–15.

biomolecules

Article

Affinity Tag-Free Purification of SARS-CoV-2 N Protein and Its Crystal Structure in Complex with ssDNA

Atanu Maiti * and Hiroshi Matsuo *

Cancer Innovation Laboratory, Frederick National Laboratory for Cancer Research, Frederick, MD 21702, USA
* Correspondence: atanu.maiti@nih.gov (A.M.); hiroshi.matsuo@nih.gov (H.M.); Tel.: +1-301-846-6826 (A.M.); +1-301-228-4375 (H.M.)

Abstract: The nucleocapsid (N) protein is one of the four structural proteins in SARS-CoV-2, playing key roles in viral assembly, immune evasion, and stability. One of its primary functions is to protect viral RNA by forming the nucleocapsid. However, the precise mechanisms by which the N protein interacts with viral RNA and assembles into a nucleocapsid remain unclear. Compared to other SARS-CoV-2 components, targeting the N protein has several advantages: it exhibits higher sequence conservation, lower mutation rates, and stronger immunogenicity, making it an attractive target for antiviral drug development and diagnostics. Therefore, a detailed understanding of the N protein's structure is essential for deciphering its role in viral assembly and developing effective therapeutics. In this study, we report the expression and purification of a soluble recombinant N protein, along with a 1.55 Å resolution crystal structure of its nucleic acid-binding domain (N-NTD) in complex with ssDNA. Our structure revealed new insights into the conformation and interaction of the flexible N-arm, which could aid in understanding nucleocapsid assembly. Additionally, we identified residues that are critical for ssDNA interaction.

Keywords: COVID-19; SARS-CoV-2; nucleocapsid protein; protein expression; protein purification; mass photometry; N-terminal domain; X-ray crystallography; atomic-resolution structure; protein–DNA interactions

1. Introduction

The COVID-19 pandemic, caused by Severe Acute Respiratory Syndrome Coronavirus-2 (SARS-CoV-2), is one of the most devastating health and socioeconomic crises in modern history [1–3]. According to the World Health Organization (WHO), approximately 777 million people have been infected, and 7.1 million people have died due to SARS-CoV-2 infection to date (https://data.who.int/dashboards/covid19/cases). Although the timely development and worldwide implementation of vaccines have helped contain the pandemic, newly emerging variants still pose threats to human health and present risks for future pandemics [4,5]. Therefore, efforts to better understand SARS-CoV-2 and to develop antiviral agents have been ongoing since the pandemic began [6–8].

SARS-CoV-2 is a positive-sense single-stranded RNA (ssRNA) virus that belongs to the betacoronavirus family and shares genetic and structural homology with SARS-CoV (which caused the 2002 SARS epidemic) and MERS-CoV (which emerged in 2012) [9,10]. The large RNA genome (approximately 30 kb) of SARS-CoV-2 encodes 16 non-structural proteins (NSPs) and at least six accessory proteins. It also encodes four structural proteins—Spike (S), Nucleocapsid (N), Membrane (M), and Envelope (E) proteins [11] (Figure 1a)—which play important roles in the viral life cycle, host infection, and pathogenicity [12,13]. While the S, M, and E proteins mainly contribute to the outer shell structure of the virus particle, the N protein forms the viral capsid by packaging the viral RNA genome into a helical ribonucleoprotein (RNP) complex [13,14]. Thus, the N protein plays a pivotal role in protecting the viral genome by forming a nucleocapsid within the viral particle, making it less prone to mutation [15,16]. The N protein also participates in other processes such as viral genome transcription, replication, and immune regulation by interacting with viral and host proteins. The N protein is highly immunogenic, the most abundant viral protein during infection, and highly conserved among other coronaviruses. These specific features make the N protein a potential target for antiviral drug development and diagnosis, and hence an attractive subject for structural studies [12,17–23].

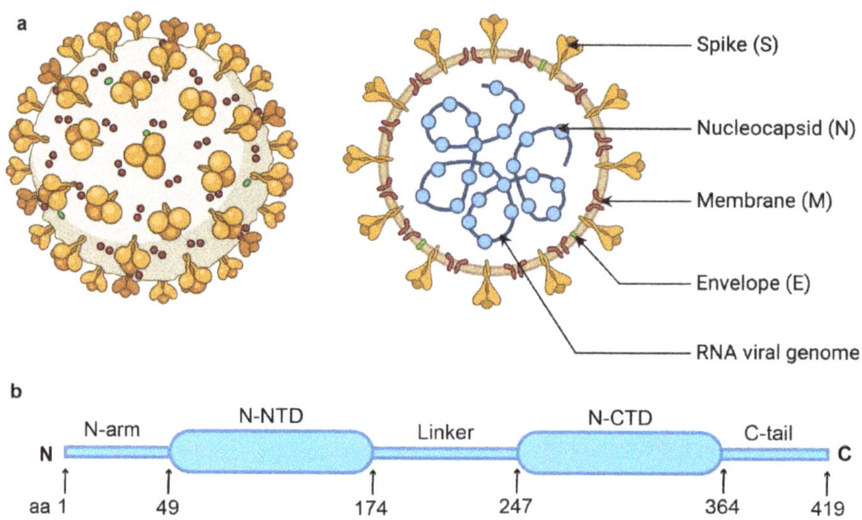

Figure 1. (**a**) Schematic representation of SARS-CoV-2 virus and its structural proteins (S, N, M, and E) with genomic RNA [created with BioRender.com]. (**b**) The N protein is composed of several distinct regions: the N-terminal domain (N-NTD) and the C-terminal domain (N-CTD) connected by a flexible linker, and intrinsically disordered regions like the N-arm and the C-tail.

Numerous efforts have been made to understand the structure and function of coronaviruses' N proteins. The SARS-CoV-2 N protein consists of 419 amino acids and comprises two structural domains: the N-terminal domain (N-NTD; residues 49–174) and the C-terminal domain (N-CTD; residues 247–364). These two domains are connected by a highly flexible linker region (residues 175–246), and at either end of these domains are intrinsically disordered regions, namely the N-arm (at the N-terminus; residues 1–48) and the C-tail (at the C-terminus; residues 365–419) [24] (Figure 1b). While the N protein interacts with RNA through the N-NTD, the N-CTD is involved in the self-dimerization of full-length N protein and RNA binding [25–32]. The flexible linker region (residues 175–246) has been shown to regulate the interaction of the N protein with RNA as well as with other viral and host proteins [33,34]. Though it is believed that the disordered C-tail plays a role in interaction with viral M protein and the packaging signal [22,35], the role of the N-arm remains unclear. Many structures of the N-NTD and N-CTD domains of the SARS-CoV-2 N protein have been solved separately [23,25,28–32,36–44]. Although the nucleocapsid (N) protein is a key component of the SARS-CoV-2 virus and an important target for therapeutic development, no high-resolution structure of the full-length N protein is currently available. Recently, structures of the full-length SARS-CoV-2 N protein were resolved by Cryo-EM at 6.00 Å and

4.57 Å resolutions as a dimer and monomer, respectively [45], but these were insufficient in obtaining detailed, precise information about the molecule. Thus, a high-resolution structure of the full-length SARS-CoV-2 N protein is still needed.

Escalating research on the N protein requires the isolation and purification of high-quality N protein in large quantities, which is generally achieved by overexpressing the recombinant protein in bacterial cells. Several groups have reported the purification of recombinant N protein from *E. coli* using affinity tags [31,46–49]. Although affinity tags simplify the protein purification process, retention or incomplete removal of the tags is not always ideal for downstream studies with the protein [50–55]. Furthermore, the method used for purification can also impact the quality and molecular properties of the N protein [56]. Therefore, developing a procedure to purify high-quality, full-length SARS-CoV-2 N protein without an affinity tag is valuable. Here, we report an affinity tag-free purification method for recombinant, full-length SARS-CoV-2 N protein in a large quantity and with high purity. Our purified N protein is highly soluble, homogeneous, free of nucleic acids, and suitable for biochemical, biophysical, and structural studies.

Although the main role of the SARS-CoV-2 N protein in the viral life cycle is to form the viral nucleoprotein complex (RNP) by interacting with genomic RNA and assembling the nucleoprotein complex, the detailed mechanism of this process is not well understood. Recent studies have also highlighted the involvement of the SARS-CoV-2 N protein in interactions with ssDNA and dsDNA. The ability of the SARS-CoV-2 N protein to unwind dsDNA as well as to anneal ssDNA may significantly interfere with host cell DNA replication, repair, and recombination processes. Additionally, the inhibition of SARS-CoV-2 replication by DNA aptamers shows promise for designing DNA-based inhibitors targeting the N protein [57–60]. To advance our understanding of nucleocapsid assembly formation and nucleic acid interactions, we aimed to crystallize the full-length N protein in complex with nucleic acids (ssDNA). Our crystallization efforts resulted in a high-resolution (1.55 Å) structure of the N-NTD with a small part of the N-arm in complex with ssDNA. The crystal packing of our structure suggests a possible role for the flexible N-arm in RNP assembly formation. Our structure showed that ssDNA binds to the N-NTD at the same RNA binding pocket, but the mode of interaction differs from that of RNA. Additionally, our structure revealed the amino acid residues important for ssDNA binding. As the inhibition of nucleocapsid assembly is a major pathway for antiviral development, our structure could be valuable for the structure-guided design of inhibitors.

2. Results and Discussion

2.1. N Protein Production

Considering the high demand for high-quality full-length SARS-CoV-2 N protein, we developed an *E. coli* overexpression vector for the SARS-CoV-2 N gene without any affinity tag. We used a polyethyleneimine (PEI) precipitation method to separate nucleic acids from proteins. The supernatant was treated with ion-exchange chromatography (SP Sepharose) followed by size-exclusion chromatography (Superdex-75), which allowed us to purify full-length N protein. Size-exclusion chromatography and SDS-PAGE analysis (Figure 2a) revealed the eluted protein was a single, major species with high purity. UV absorption analysis confirmed the absence of nucleic acid contamination in the purified N protein. As reported by other groups [16,22], our mass photometry (MP) analysis showed that the N protein remained as a homogeneous dimer at a concentration of 50 to 100 nM (Figure 2b). Overall, our purified N protein was highly soluble, homogeneous, nucleic acid-free, and suitable for biochemical, biophysical, and structural studies.

2.2. Crystal Formation and Structure of N-NTD with Part of Flexible N-Arm

We aimed to capture the oligomeric structure of the full-length N protein in complex with nucleic acids. As reported by Zhao et al. [61], dimeric N protein's tendency to form higher-order oligomers depends on the nucleic acid length and shows a strong binding affinity with 10 nt poly-T (T-10) and 20 nt poly-T (T-20), forming higher-order oligomers. Accordingly, we attempted to crystallize the full-length N protein in complex with T-10 (10 nt T) and T-20 (20 nt T) ssDNA. After approximately a year, crystals appeared in a single condition (F4 condition from PEG/pH screen, Hampton Research) at 4 °C, in a well with N protein and T-20 ssDNA (Supplementary Figure S1a). A cryoprotected crystal diffracted at high resolution (Supplementary Figure S1b). The crystal belonged to the I 41 space group, with unit cell parameters (a = b = 90.89 Å, c = 36.34 Å, α = β = γ = 90°) (Table 1). Cell content analysis using the CCP4 program suit suggested that the unit cell volume was insufficient to accommodate the full-length N protein but could contain half of it.

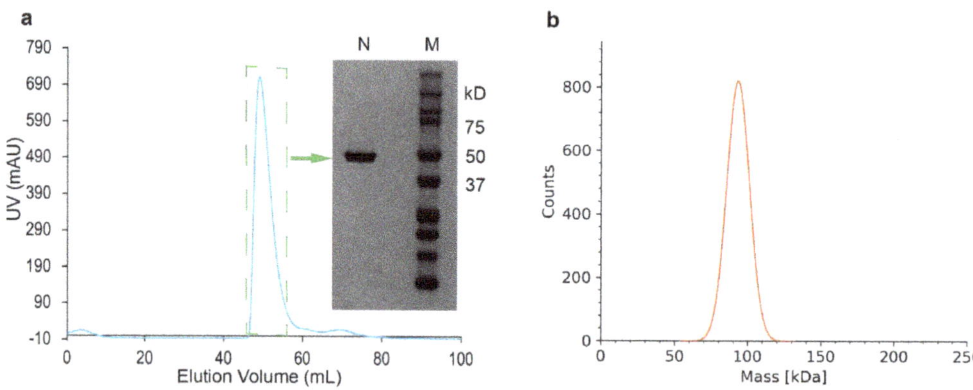

Figure 2. SARS-CoV-2 N protein purified by polyethyleneimine precipitation followed by SP Sepharose ion-exchange and finally size-exclusion chromatography. (**a**) FPLC size-exclusion chromatography and SDS-PAGE analysis of the purified N protein. The blue line represents the UV absorption spectrum of the eluted protein. The dashed green box highlights the absorption of the collected pure N protein. 'N' indicates the purified full-length N protein, while 'M' represents the protein marker. (**b**) Mass photometry analysis revealed that the N protein appeared as a monodisperse dimer (~93 kD) in-solution.

Molecular replacement using the available structural coordinates of N-NTD (PDB ID:7N0R) and N-CTD (PDB ID:6ZCO) together (for the full-length protein) or using only N-CTD (for the C-terminal domain) failed to produce a solution. However, molecular replacement with only N-NTD (for the N-terminal domain) successfully provided a solution, resulting in a single-protein molecule per asymmetric unit and final structure determination at 1.55 Å resolution. The crystallization of only the N-terminal part of the full-length N protein, after an extended period, led us to hypothesize that the full-length N protein might have been digested in the crystallization condition, with only the N-terminal part crystallizing under the given condition.

To confirm this, we collected the residual crystallization sample (RS) after crystal formation in the N+T-20 crystallization well, and the crystallization sample (CS) from the N+T-10 crystallization well without crystal formation, and examined them by SDS-PAGE. The SDS-PAGE analysis confirmed that the full-length protein was indeed digested under the crystallization condition, producing a significant fragment with an approximate molecular weight of 20 kD (Supplementary Figure S1d,e). Supporting previous reports, our findings indicate that the crystallization of full-length N protein is challenging due to its intrinsic flexibility and protease sensitivity [16,23,62,63].

Although we could not determine the exact length of the crystallized fragment of the N protein, we were able to build a structural model based on the electron density observed after phase determination via molecular replacement. Like previously reported N-NTD structures, strong electron density was observed for the N-NTD structured region (residues 49–173) along with a part of the flexible N-arm (residues 39–48), allowing us to confidently build the structural model. No electron density was observed for the rest of the protein, and it was therefore not modeled in the final crystal structure. Overall, we solved the crystal structure of the N protein from residues 39 to 173, covering part of the flexible N-arm and the entire N-NTD (Supplementary Figure S1c,f and Figure 3a).

Table 1. Data collection and refinement statistics.

Data Collection	
Space group	I 41
Cell dimensions a, b, c (Å) α, β, γ (°)	 90.89, 90.89, 36.34 90.00, 90.00, 90.00
Resolution (Å)	50.00–1.55 (1.61–1.55) *
R_{merge} (%)	7.0 (65.3)
R_{pim} (%)	1.9 (18.6)
I/σI	34.42 (2.94)
CC1/2	0.998 (0.945)
Completeness (%)	99.8 (97.0)
Redundancy	12.9 (11.6)
Refinement	
Resolution (Å)	33.74–1.55 (1.61–1.55)
No. of reflections	20,986 (1666)
R_{work}/R_{free} (%)	17.80/20.20
No. of atoms	2193
Protein	1972
DNA	44
Water	177
B-factor	
Average B-factors ($Å^2$)	21.0
Protein/DNA	19.2
Water	30.1
Ramachandran plot (%)	
Favored regions	98.5
Allowed regions	1.5
Outliers	0
r.m.s deviations	
Bond lengths (Å)	0.009
Bond angles (°)	1.26

* Values in parentheses are for highest-resolution shell.

We also observed strong electron density for a nucleotide along with additionally strong density for two adjoining phosphates, but weak density for the corresponding base. Since our crystallization sample contained 20 nt T, the electron density observed here could represent a stable portion of the poly-T interacting with N-NTD, while the remaining poly-T remains flexible and invisible. Therefore, we modeled the trinucleotide as TTT (Supplementary Figure S1c and Figure 3a).

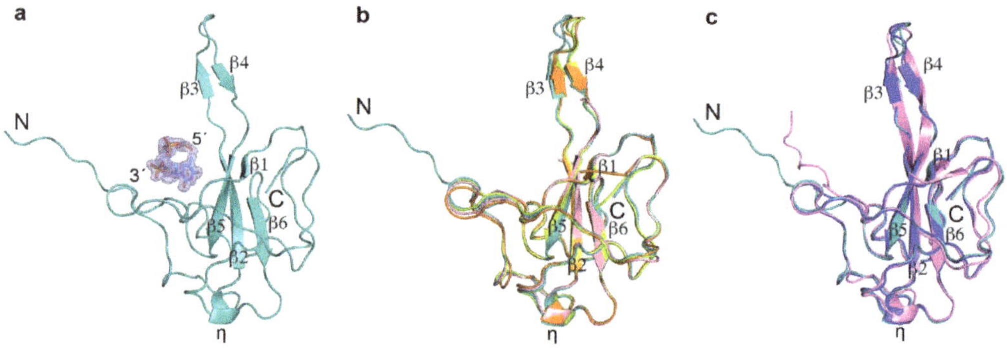

Figure 3. Comparisons of coronavirus N protein structures. (**a**) Our N protein structure (digested full-length N) showing N-NTD with part of the flexible N-arm (PDB ID:9CJ6; teal color, cartoon) in complex with ssDNA (blue, stick). A 2Fo–Fc electron density map contoured at 1σ is shown in light blue around the ssDNA. (**b**) Superimposition of our structure with other published SARS-CoV-2 N-NTD structures (PDB ID:7XWZ; light pink color, PDB ID:7CDZ; orange color, PDB ID:7N0R; limon color). (**c**) Superimposition of our structure with closely related SARS-CoV N-NTD (PDB ID: 2OFZ; navy blue color) and MERS-CoV- N-NTD (PDB ID: 4UD1; purple color) structures. N and C indicate the N- and C-terminal ends of the protein, respectively. The 5′ and 3′ represent the 5′ end and 3′ end of ssDNA. ß and η represent the ß-strand and short 3_{10} helix, respectively.

2.3. Comparison with Published N-NTD Structures

Since the COVID-19 pandemic, several structures of SARS-CoV-2 N-NTD have been published (Supplementary Table S1): some in complex with dsRNA or ssRNA [28,30], some with antibodies or proteins [37,38,42,44], some with small molecules, and some in the apo form [23,25,29,30,44]. Among all the crystal structures deposited in the PDB, the apo N-NTD has been crystallized as a trimer (PDB ID: 6VYO, 6WKP, 7UW3, 6M3M, 7CDZ, 7XX1, and 7VBD); N-NTD in complex with dsRNA has been crystallized as a dimer (PDB ID: 7XWZ); and N-NTD in complex with antibodies or proteins has been crystallized as a monomer (PDB ID: 7STR, 7STS, 7SUE, 7N3C, 7N3D, 7CR5, 7N0R, 7R98, and 7VNU) in the asymmetric unit. Our N-NTD in complex with ssDNA crystallized as a monomer in the asymmetric unit (Figure 3a) and maintained good topological agreement with other published crystal structures of SARS-CoV-2 N-NTD (Figure 3b), as well as with the closely related SARS-CoV (PDB ID: 2OFZ) and MERS-CoV N-NTD (PDB ID: 4UD1) structures (Figure 3c).

The N-NTD resembles a right-handed fist, containing a core of four-stranded antiparallel ß-sheets (ß1-ß5-ß2-ß6), which is sandwiched between two loops, a short 3_{10} helix (η), and a protruding ß-hairpin (ß3 and ß4) between ß2 and ß5 (Figure 3a–c). As previously reported, the protruding ß-hairpin (ß3 and ß4) region is flexible, exhibiting a range of dynamic conformations across different structures. Most previous N-NTD structures were determined using the stable N-NTD construct spanning residues 48 to174. However, in our structure (a digested full-length N), in addition to the stable N-NTD, we determined the structure of a small part of the flexible N-arm at high resolution. The N-arm extends outward from the N-NTD core, like what has been observed in the MERS-CoV N-NTD structure (Figure 3c) [64].

2.4. Putative Role of N-Arm in Nucleocapsid Assembly

In the viral life cycle, the primary role of the N protein is to bind viral genomic RNA to form the viral nucleoprotein complex (RNP) and assemble this complex within the virus particle. Previous studies have indicated that the assembly of the RNP complex involves multiple regions of the N protein that mediate protein–RNA and protein–protein interactions [17,22,65–67]. However, the mechanism of RNP complex assembly formation remains elusive.

It is known that intrinsically disordered proteins (IDPs) or intrinsically disordered regions (IDRs) of a protein lack a defined structure yet play important roles in biological processes, especially in macromolecular interactions [68–71]. Due to its intrinsically disordered nature, the structural information and biological role of the N-arm of the SARS-CoV-2 N protein are still unclear.

Our crystal structure revealed that the N-NTD, along with part of the flexible N-arm, arranges in a specific pattern where the N-arm extends to neighboring molecules and anchors between two molecules through significant hydrogen bonding (Figure 4). Specifically, the mainchain amine of L45 forms a direct hydrogen bond with the sidechain carbonyl of Q160 in loop 2. The mainchain amine and carbonyl of Q43 form hydrogen bonds with the sidechain carbonyl of N75 and the mainchain amine of T76, respectively, in loop 1. The mainchain carbonyl of R41 forms a hydrogen bond with the sidechain amine of N77 in loop 1. Additionally, the sidechain nitrogen of R41 forms a hydrogen bond with the sidechain NH_2 of R93 (near ß3) of another neighboring molecule. The crystal packing of our structure suggested a possible role for the flexible N-arm in RNP assembly formation. However, further experiments are required to support this hypothesis.

2.5. ssDNA Interacts at the RNA Binding Pocket of N-NTD with Different Interaction Modes

Despite the main role of the N protein in packaging the viral RNA genome, new findings highlight several biological processes that require the interaction of the N protein with ssDNA or dsDNA [57–59]. Previous computational and structural studies have provided significant insights into how the N protein interacts with dsRNA and ssRNA [28,30]. However, structural information on how the N protein interacts with dsDNA or ssDNA is lacking. To address this gap, we aimed to crystallize poly-T (T-10 and T-20) in complex with the N protein.

In our structure, we were able to model a small segment (TTT) of the T-20 ssDNA based on well-defined electron density. The absence of electron density for other parts of the ssDNA may have been due to the dynamic nature of ssDNA or possible degradation under the crystallization condition [72–74]. As we were uncertain about which segment of the TTT nucleotide stretched stably bound to the N protein and got crystallized, we numbered the T with well-defined electron density as 0, with −1 and +1 indicating the positions toward the 5′ and 3′ ends, respectively. Nevertheless, our structure provides valuable insights into the interaction between the N protein and ssDNA.

Superimposing our structure with previous structures of N-NTD in complex with dsRNA and ssRNA (PDB ID: 7XWZ and 7ACT) showed that ssDNA binds to the N-NTD at the same RNA binding pocket, but in a different mode (Figure 5a,b). The Watson–Crick face of ssDNA interacts with the protein through direct and water-mediated hydrogen bonds as well as π–π stacking (Figure 5c). The Watson–Crick face of T_0 forms two direct and one water-mediated interaction with the protein. The 2-carbonyl group forms a hydrogen bond with the mainchain amino proton of S51, the N3 atom forms a hydrogen bond with the sidechain –OH group of Y111, and the 4-carbonyl group forms a water-mediated hydrogen bond with the guanidino group of R88. T_0 also stacks with Y109 through π–π stacking. Additionally, the deoxyribose O3′ of T_0 forms a direct hydrogen bond with the guanidino group of R149 (Figure 5c and Supplementary Figure S3). The Watson–Crick face of T_{-1} interacts with the protein through one direct and one water-mediated hydrogen bond. The 2-carbonyl group forms a hydrogen bond with the sidechain amino group of N47, and the N3 atom forms a water-mediated hydrogen bond with the mainchain carbonyl group of T49. Additionally, the 5′-phosphate group of T_{+1} is supported by one direct hydrogen bond with the guanidino group of R149 and one water-mediated hydrogen bond with the mainchain carbonyl group of P151 (Figure 5c and Supplementary Figure S3). In contrast, RNA interacts with the N protein through its phosphate backbone (Figure 5d) [28,30]. Residues R88, Y111, Y109, and R149 are involved in interactions with both DNA and RNA.

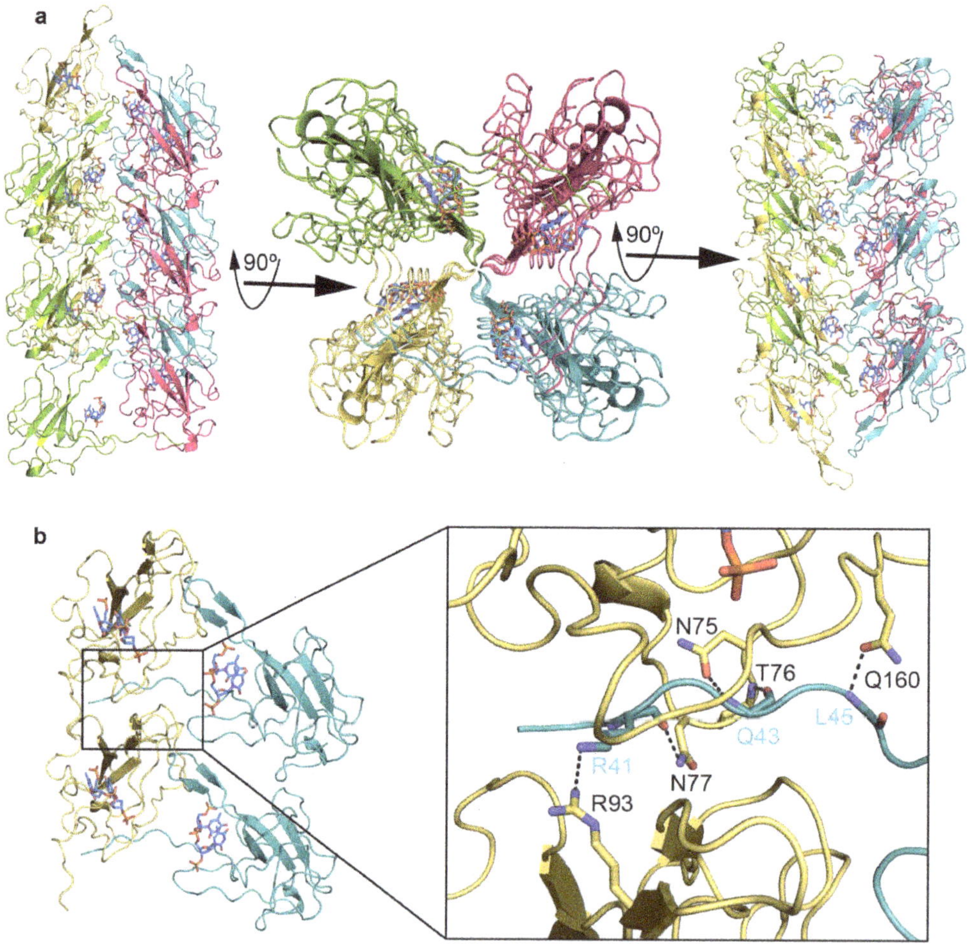

Figure 4. Crystal packaging of digested N protein. (**a**) In the crystal, four neighboring protein molecules are colored as teal, hot pink, split pea, and yellow-orange, showing crystal packing at different 90° angles. (**b**) The N-arm extends to neighboring molecules and is anchored between two molecules through hydrogen bonding. Gray dashed lines represent the hydrogen bonds between amino acids.

Recent studies have highlighted the use of oligonucleotides as potential inhibitors of protein function [75]. A designed ssDNA aptamer has shown promise as an antiviral therapy against COVID-19 by disrupting the liquid–liquid phase separation (LLPS) mediated by the N protein [59]. Our structure, along with previous structures, will aid in designing more effective nucleic acid-based therapeutics against SARS-CoV-2.

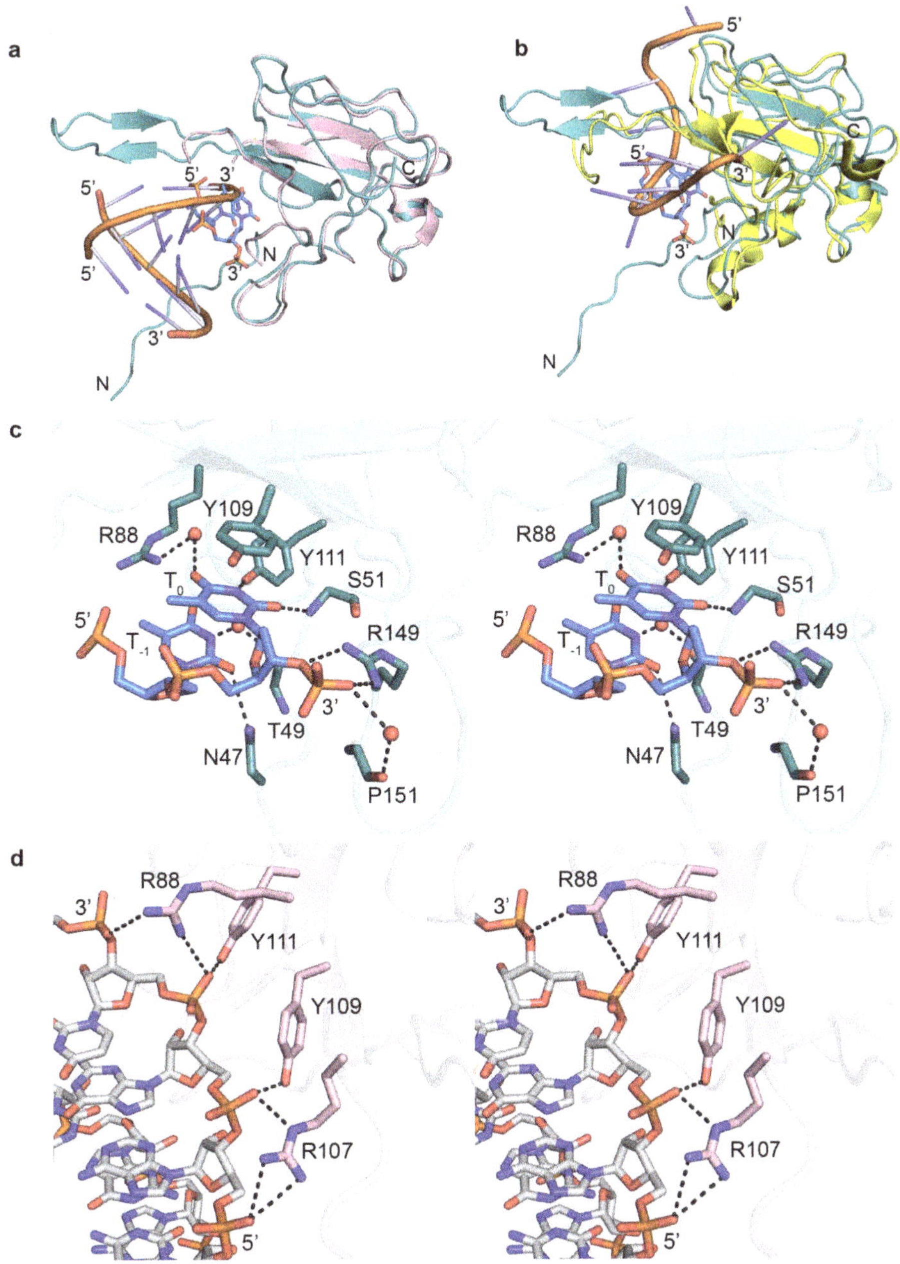

Figure 5. Interactions of N protein with nucleic acids. (**a**) Superimposition of our structure (teal, PDB ID: 9CJ6) with the crystal structure of the N-NTD complex with dsRNA (light pink, PDB ID: 7XWZ).

(**b**) Superimposition of our structure (teal, PDB ID: 9CJ6) with the NMR structure of the N-NTD complex with ssRNA (yellow, PDB ID: 7ACT). RNA is shown as a cartoon and ssDNA as a stick. ssDNA binds to N-NTD at the RNA binding pocket. (**c**) Stereo-view of detailed interaction of ssDNA with N-NTD (teal, our structure). (**d**) Stereo view of the detailed interaction of dsRNA with N-NTD (light pink, PDB ID: 7XWZ). ssDNA and RNA are represented as sticks, and water is represented as the red sphere. Gray dashed lines represent the hydrogen bonds. The nucleic acid bases of ssDNA interact with protein through direct and water-mediated hydrogen bonds. The RNA interacts with protein through hydrogen bonds in the phosphate backbone.

3. Materials and Methods

3.1. Plasmid Generation and Protein Overexpression and Purification

The protein expression vector was generated by cloning the SARS-CoV-2 N gene (*Gene ID: 43740575*) into the pET-11a vector (GenScript) using cloning sites 5′ NdeI and 3′ BamHI, without any affinity tag. The *pET-11a* expression vector carrying the SARS-CoV-2 N gene was transformed into BL21 (DE3) cells (Invitrogen). Cells were grown in LB medium at 37 °C until they reached an optical density of 0.5–0.6 at 600 nm. The temperature was then reduced to 17 °C, and protein overexpression was induced by adding 0.2 mM IPTG at an optical density of 0.6–0.8. Cells were further grown overnight at 17 °C. All protein purification steps were performed at 4 °C unless otherwise specified.

E. coli cells from a 1 L culture were harvested by centrifugation and resuspended in 25 mL of lysis buffer (20 mM Tris-HCl, pH 7.5, 200 mM NaCl, 1 mM DTT, 5% glycerol) with protease inhibitors (Roche, Basel, Switzerland). The suspended cells were disrupted by sonication, and cell debris was removed by centrifugation at 12,000 rpm for 45 min.

The supernatant was treated with a gradual addition of 5% polyethyleneimine to reach a final concentration of 0.5% polyethyleneimine in solution and stirred for 1 h to allow maximum precipitation. The precipitate was separated by centrifugation at 10,000 rpm for 10 min, and the supernatant containing the N protein was retained.

The supernatant was added to 6 mL of SP Sepharose beads (pre-equilibrated with lysis buffer) and agitated overnight. The N protein bound to the SP Sepharose beads was passed through a column under gravity. After washing the beads with 40 mL of lysis buffer, the protein was eluted with 10 mL of elution buffer in fractions E1–E6 (lysis buffer with varying salt concentrations: E1 = 0.3 M, E2 = 0.4 M, E3 = 0.5 M, E4 = 0.6 M, E5 = 0.8 M, and E6 = 1 M NaCl). The N protein was completely eluted in fractions E2 (0.4 M NaCl) and E3 (0.5 M NaCl), as confirmed by SDS-PAGE electrophoresis (Supplementary Figure S2a).

The eluted protein was further purified using a Superdex 75 size-exclusion column (GE Healthcare Life Sciences) with FPLC buffer (20 mM Tris-HCl, pH 7.5, 100 mM NaCl, 1 mM DTT, 1% glycerol) on an AKTA FPLC system (Supplementary Figure S2b). The purity and concentration of the proteins were measured by gel electrophoresis and UV spectroscopy.

3.2. Mass Photometry

Mass photometry experiments were performed using a TwoMP mass photometer (Refeyn Ltd., Oxford, UK). High-precision cover glasses (24 × 50 mm, thickness No. 1.5H, Paul Marienfeld GmbH and Co KG, Lauda-Königshofen, Germany) and six-well pre-cut CultureWell gaskets (Grace Bio-Labs, Bend, OR, USA) were used. Coverslips were rinsed sequentially with deionized water and isopropyl alcohol (IPA) in the following order, water–IPA–water–IPA–water, and then dried with a stream of ultrapure nitrogen. A drop of Immersol 518F immersion oil (Carl Zeiss Jena, Oberkochen, Germany) was placed on the microscope lens before placing the coverslips on the mass photometer stage.

Protein samples were prepared at concentrations of 50 nM and 100 nM in sample buffer (20 mM Tris-HCl, pH 7.5, 100 mM NaCl, 1 mM DTT). The instrument was calibrated with a mix of beta-amylase from sweet potato and thyroglobulin from bovine thyroid (Sigma-Aldrich), both dissolved in sample buffer. Autofocusing was performed in buffer-free mode, and 15 μL of protein solution was placed in a well on the coverslip. Movies were recorded for 60 s using the regular view setting in AcquireMP (version 2023 R1.1) software. The experiments were performed in triplicate with freshly prepared samples. Data analysis was conducted in histogram mode with the manual selection of observed peaks, followed by Gaussian fitting using DiscoverMP software (version 2023 R1.2).

3.3. Crystal Growth and Data Collection

We aimed to crystallize the full-length SARS-CoV-2 N protein in complex with ssDNA. For this purpose, we used two different ssDNAs, T-10 (TTTTTTTTTT) and T-20 (TTTTTTTTTTTTTTTTTTTT), as reported by Zhao et al. [61]. T-10 and T-20, at 10% molar excess, were each mixed separately with purified SARS-CoV-2 N protein at a concentration of 30 μM. The mixtures were then concentrated to approximately 900 μM protein concentration using Amicon Ultra-4 (Merck Millipore). Crystallization screening was performed using various commercially available crystallization screens and the sitting drop vapor diffusion method at 4 °C and 20 °C. Crystal drops were prepared by mixing 0.3 μL of the sample with 0.3 μL of reservoir solution in a 2-well sitting drop crystallization plate (Molecular Dimension) using a Mosquito Crystal robot (TTP Labtech). Crystals appeared after approximately one year under a condition containing 0.2 M HEPES (pH 7.4) and 20% *w/v* PEG 4000 (F4 condition from PEG/pH screen, Hampton Research). Crystals were cryoprotected using the reservoir solution supplemented with 15% *v/v* glycerol and flash-frozen in liquid nitrogen. X-ray diffraction data were collected at the Southeast Regional Collaborative Access Team (SER-CAT) 22-ID beam line at the Advanced Photon Source, Argon National Laboratory.

3.4. Structural Determination and Analysis

The collected diffraction data were indexed, integrated, and scaled using the HKL2000 program [76]. The crystals belong to the space group I 41, with unit cell parameters (a = b = 90.89 Å, c = 36.34 Å, $\alpha = \beta = \gamma = 90°$) (Table 1). Cell content analysis was performed using Mathew's coefficient [77] in the CCP4 program suit. The structure was solved at 1.55 Å resolution by molecular replacement using the program Phaser [78] and the N-NTD structure (PDB ID:7N0R) as the search model. Model building of the protein and bound DNA was manually performed using the program Coot [79]. The structural model was refined with Phenix refinement [80,81], and the final model was validated using the PDB validation tool (https://validate.wwpdb.org) and MolProbity [82]. Structural figures were created using the software PyMOL. Data collection statistics and refinement parameters of the crystal structure are provided in Table 1.

4. Conclusions

In this study, we developed an optimized protocol for the affinity tag-free purification of the SARS-CoV-2 N protein with a high yield. Our purification method involved polyethyleneimine precipitation, SP Sepharose chromatography, and size-exclusion chromatography, resulting in the isolation of highly pure N protein. The purified N protein was highly soluble, homogeneous, nucleic acid-free, and suitable for structural studies. Mass photometry analysis confirmed that the purified N protein dimerized in solution, consistent with previous reports. Our purification method provides a reliable source of N protein for the further exploration of its biochemical and biophysical properties, particularly its interactions with nucleic acids.

Despite our efforts to crystallize the full-length N protein in complex with ssDNA, we encountered the inherent challenges posed by its flexibility and protease sensitivity, leading to the crystallization of only the N-terminal domain (N-NTD). This result also underscores the possibility of the spontaneous cleavage of the SARS-CoV-2 N protein. As previously observed, under certain conditions, the N-terminal sides of Ser and Thr residues are prone to spontaneous cleavage [83,84]. Notably, the flexible linker region of the SARS-CoV-2 N protein contains several Ser and Thr residues, making it highly susceptible to such cleavage. Our findings provide an important insight for researchers, emphasizing the need to carefully assess the stability of full-length N protein during biochemical and biophysical experiments, as well as during crystallization trials.

Our structural analysis revealed that the N-NTD, along with the flexible N-arm, formed a specific arrangement that facilitated higher-order oligomerization and suggests a potential contribution to the assembly of the viral ribonucleoprotein complex (RNP). Notably, the crystallized N-NTD displayed a monomeric form in complex with a trinucleotide, suggesting a distinct mode of interaction with ssDNA compared to RNA. These structural insights, combined with previous findings, expand our understanding of the N protein's role in viral assembly and its potential as a target for nucleic acid-based therapeutics. The unique binding mode of ssDNA observed in our structure highlights key residues involved in this interaction, providing a foundation for future therapeutic development. Given the challenges of crystallizing the full-length N protein, further experiments are required to fully elucidate its role in RNP assembly and its interactions with different nucleic acids. These findings open new avenues for designing inhibitors targeting the N protein, with the potential to disrupt SARS-CoV-2 replication and pathogenesis.

Supplementary Materials: The following supporting information can be downloaded at: https://www.mdpi.com/article/10.3390/biom14121538/s1. Figure S1: Crystallization of SARS-CoV-2 N protein with Poly-T (T-20) ssDNA; Figure S2: SDS-PAGE analysis of SARS-CoV-2 N protein expression and purification; Figure S3: Summary of the interactions between SARS-CoV-2 N-NTD and nucleotides TTT; Table S1: Structures of SARS-CoV-2 N-NTD reported in Protein Data Bank (PDB).

Author Contributions: A.M. designed and performed all the experiments, structural analysis, and interpretation; A.M. prepared all the figures; A.M. and H.M. wrote the paper; A.M. conceived the project; H.M. secured funding. All authors have read and agreed to the published version of the manuscript.

Funding: This work was supported by federal funds from the National Cancer Institute, National Institutes of Health, under contract 75N91019D00024. The content of this publication does not necessarily reflect the views or policies of the Department of Health and Human Services, nor does mention of trade names, commercial products, or organizations imply endorsement by the U.S. Government.

Institutional Review Board Statement: Not applicable.

Informed Consent Statement: Not applicable.

Data Availability Statement: Atomic coordinates and structural factors were deposited in the Protein Data Bank with accession code 9CJ6 (https://www.rcsb.org/structure/9CJ6). Data are available from A.M. and H.M upon request (e-mail: atanu.maiti@nih.gov and hiroshi.matsuo@nih.gov).

Acknowledgments: The authors would like to acknowledge Sergey G. Tarasov and Marzena A. Dyba at the Biophysics Resources of CCR/NIH for providing instruments and technical support. The schematic representation of the SARS-CoV-2 virus and its structural proteins (S, N, M, and E) with genomic RNA was created with BioRender.com.

Conflicts of Interest: The authors declare no conflicts of interest.

References

1. Atar, S.; Atar, I. An invited commentary on "The socio-economic implications of the coronavirus and COVID-19 pandemic: A review". *Int. J. Surg.* **2020**, *78*, 122. [CrossRef] [PubMed]
2. Li, Y.; Niu, Y. A commentary on "The socio-economic implications of the coronavirus pandemic (COVID-19): A review". *Int. J. Surg.* **2021**, *95*, 106048. [CrossRef] [PubMed]
3. Nicola, M.; Alsafi, Z.; Sohrabi, C.; Kerwan, A.; Al-Jabir, A.; Iosifidis, C.; Agha, M.; Agha, R. The socio-economic implications of the coronavirus pandemic (COVID-19): A review. *Int. J. Surg.* **2020**, *78*, 185–193. [CrossRef]
4. Chen, Y.; Liu, Q.; Zhou, L.; Zhou, Y.; Yan, H.; Lan, K. Emerging SARS-CoV-2 variants: Why, how, and what's next? *Cell Insight* **2022**, *1*, 100029. [CrossRef]
5. Aleem, A.; Akbar Samad, A.B.; Vaqar, S. Emerging Variants of SARS-CoV-2 and Novel Therapeutics Against Coronavirus (COVID-19). In *StatPearls*; StatPearls Publishing: Treasure Island, FL, USA, 2024.
6. Ahmad, I. The race to treat COVID-19: Potential therapeutic agents for the prevention and treatment of SARS-CoV-2. *Eur. J. Med. Chem.* **2021**, *213*, 113157. [CrossRef] [PubMed]
7. Liu, C.; Zhou, Q.; Li, Y.; Garner, L.V.; Watkins, S.P.; Carter, L.J.; Smoot, J.; Gregg, A.C.; Daniels, A.D.; Jervey, S.; et al. Research and Development on Therapeutic Agents and Vaccines for COVID-19 and Related Human Coronavirus Diseases. *ACS Cent. Sci.* **2020**, *6*, 315–331. [CrossRef]
8. Li, G.; Hilgenfeld, R.; Whitley, R.; De Clercq, E. Therapeutic strategies for COVID-19: Progress and lessons learned. *Nat. Rev. Drug Discov.* **2023**, *22*, 449–475. [CrossRef]
9. Zhu, Z.; Lian, X.; Su, X.; Wu, W.; Marraro, G.A.; Zeng, Y. From SARS and MERS to COVID-19: A brief summary and comparison of severe acute respiratory infections caused by three highly pathogenic human coronaviruses. *Respir. Res.* **2020**, *21*, 224. [CrossRef]
10. Krishnamoorthy, S.; Swain, B.; Verma, R.S.; Gunthe, S.S. SARS-CoV, MERS-CoV, and 2019-nCoV viruses: An overview of origin, evolution, and genetic variations. *Virusdisease* **2020**, *31*, 411–423. [CrossRef]
11. Helmy, Y.A.; Fawzy, M.; Elaswad, A.; Sobieh, A.; Kenney, S.P.; Shehata, A.A. The COVID-19 Pandemic: A Comprehensive Review of Taxonomy, Genetics, Epidemiology, Diagnosis, Treatment, and Control. *J. Clin. Med.* **2020**, *9*, 1225. [CrossRef]
12. Yan, W.; Zheng, Y.; Zeng, X.; He, B.; Cheng, W. Structural biology of SARS-CoV-2: Open the door for novel therapies. *Signal Transduct. Target. Ther.* **2022**, *7*, 26. [CrossRef] [PubMed]
13. Arya, R.; Kumari, S.; Pandey, B.; Mistry, H.; Bihani, S.C.; Das, A.; Prashar, V.; Gupta, G.D.; Panicker, L.; Kumar, M. Structural insights into SARS-CoV-2 proteins. *J. Mol. Biol.* **2021**, *433*, 166725. [CrossRef] [PubMed]
14. Yao, H.; Song, Y.; Chen, Y.; Wu, N.; Xu, J.; Sun, C.; Zhang, J.; Weng, T.; Zhang, Z.; Wu, Z.; et al. Molecular Architecture of the SARS-CoV-2 Virus. *Cell* **2020**, *183*, 730–738.e713. [CrossRef]
15. Masters, P.S.; Sturman, L.S. Background paper. Functions of the coronavirus nucleocapsid protein. *Adv. Exp. Med. Biol.* **1990**, *276*, 235–238. [CrossRef] [PubMed]
16. Cong, Y.; Kriegenburg, F.; de Haan, C.A.M.; Reggiori, F. Coronavirus nucleocapsid proteins assemble constitutively in high molecular oligomers. *Sci. Rep.* **2017**, *7*, 5740. [CrossRef] [PubMed]
17. McBride, R.; van Zyl, M.; Fielding, B.C. The coronavirus nucleocapsid is a multifunctional protein. *Viruses* **2014**, *6*, 2991–3018. [CrossRef]
18. Chang, C.K.; Sue, S.C.; Yu, T.H.; Hsieh, C.M.; Tsai, C.K.; Chiang, Y.C.; Lee, S.J.; Hsiao, H.H.; Wu, W.J.; Chang, W.L.; et al. Modular organization of SARS coronavirus nucleocapsid protein. *J. Biomed. Sci.* **2006**, *13*, 59–72. [CrossRef]
19. Tilocca, B.; Soggiu, A.; Sanguinetti, M.; Musella, V.; Britti, D.; Bonizzi, L.; Urbani, A.; Roncada, P. Comparative computational analysis of SARS-CoV-2 nucleocapsid protein epitopes in taxonomically related coronaviruses. *Microbes Infect.* **2020**, *22*, 188–194. [CrossRef]
20. Kopecky-Bromberg, S.A.; Martinez-Sobrido, L.; Frieman, M.; Baric, R.A.; Palese, P. Severe acute respiratory syndrome coronavirus open reading frame (ORF) 3b, ORF 6, and nucleocapsid proteins function as interferon antagonists. *J. Virol.* **2007**, *81*, 548–557. [CrossRef]
21. Surjit, M.; Liu, B.; Chow, V.T.; Lal, S.K. The nucleocapsid protein of severe acute respiratory syndrome-coronavirus inhibits the activity of cyclin-cyclin-dependent kinase complex and blocks S phase progression in mammalian cells. *J. Biol. Chem.* **2006**, *281*, 10669–10681. [CrossRef]
22. Wu, W.; Cheng, Y.; Zhou, H.; Sun, C.; Zhang, S. The SARS-CoV-2 nucleocapsid protein: Its role in the viral life cycle, structure and functions, and use as a potential target in the development of vaccines and diagnostics. *Virol. J.* **2023**, *20*, 6. [CrossRef] [PubMed]
23. Peng, Y.; Du, N.; Lei, Y.; Dorje, S.; Qi, J.; Luo, T.; Gao, G.F.; Song, H. Structures of the SARS-CoV-2 nucleocapsid and their perspectives for drug design. *EMBO J.* **2020**, *39*, e105938. [CrossRef] [PubMed]
24. Eltayeb, A.; Al-Sarraj, F.; Alharbi, M.; Albiheyri, R.; Mattar, E.; Abu Zeid, I.M.; Bouback, T.A.; Bamagoos, A.; Aljohny, B.O.; Uversky, V.N.; et al. Overview of the SARS-CoV-2 nucleocapsid protein. *Int. J. Biol. Macromol.* **2024**, *260*, 129523. [CrossRef]

25. Kang, S.; Yang, M.; Hong, Z.; Zhang, L.; Huang, Z.; Chen, X.; He, S.; Zhou, Z.; Zhou, Z.; Chen, Q.; et al. Crystal structure of SARS-CoV-2 nucleocapsid protein RNA binding domain reveals potential unique drug targeting sites. *Acta Pharm. Sin. B* **2020**, *10*, 1228–1238. [CrossRef]
26. Jayaram, H.; Fan, H.; Bowman, B.R.; Ooi, A.; Jayaram, J.; Collisson, E.W.; Lescar, J.; Prasad, B.V. X-ray structures of the N- and C-terminal domains of a coronavirus nucleocapsid protein: Implications for nucleocapsid formation. *J. Virol.* **2006**, *80*, 6612–6620. [CrossRef]
27. Grossoehme, N.E.; Li, L.; Keane, S.C.; Liu, P.; Dann, C.E., 3rd; Leibowitz, J.L.; Giedroc, D.P. Coronavirus N protein N-terminal domain (NTD) specifically binds the transcriptional regulatory sequence (TRS) and melts TRS-cTRS RNA duplexes. *J. Mol. Biol.* **2009**, *394*, 544–557. [CrossRef]
28. Luan, X.; Li, X.; Li, Y.; Su, G.; Yin, W.; Jiang, Y.; Xu, N.; Wang, F.; Cheng, W.; Jin, Y.; et al. Antiviral drug design based on structural insights into the N-terminal domain and C-terminal domain of the SARS-CoV-2 nucleocapsid protein. *Sci. Bull.* **2022**, *67*, 2327–2335. [CrossRef] [PubMed]
29. Sarkar, S.; Runge, B.; Russell, R.W.; Movellan, K.T.; Calero, D.; Zeinalilathori, S.; Quinn, C.M.; Lu, M.; Calero, G.; Gronenborn, A.M.; et al. Atomic-Resolution Structure of SARS-CoV-2 Nucleocapsid Protein N-Terminal Domain. *J. Am. Chem. Soc.* **2022**, *144*, 10543–10555. [CrossRef]
30. Dinesh, D.C.; Chalupska, D.; Silhan, J.; Koutna, E.; Nencka, R.; Veverka, V.; Boura, E. Structural basis of RNA recognition by the SARS-CoV-2 nucleocapsid phosphoprotein. *PLoS Pathog.* **2020**, *16*, e1009100. [CrossRef]
31. Ye, Q.; West, A.M.V.; Silletti, S.; Corbett, K.D. Architecture and self-assembly of the SARS-CoV-2 nucleocapsid protein. *Protein Sci.* **2020**, *29*, 1890–1901. [CrossRef]
32. Zinzula, L.; Basquin, J.; Bohn, S.; Beck, F.; Klumpe, S.; Pfeifer, G.; Nagy, I.; Bracher, A.; Hartl, F.U.; Baumeister, W. High-resolution structure and biophysical characterization of the nucleocapsid phosphoprotein dimerization domain from the COVID-19 severe acute respiratory syndrome coronavirus 2. *Biochem. Biophys. Res. Commun.* **2021**, *538*, 54–62. [CrossRef] [PubMed]
33. Peng, T.Y.; Lee, K.R.; Tarn, W.Y. Phosphorylation of the arginine/serine dipeptide-rich motif of the severe acute respiratory syndrome coronavirus nucleocapsid protein modulates its multimerization, translation inhibitory activity and cellular localization. *FEBS J.* **2008**, *275*, 4152–4163. [CrossRef]
34. Carlson, C.R.; Asfaha, J.B.; Ghent, C.M.; Howard, C.J.; Hartooni, N.; Safari, M.; Frankel, A.D.; Morgan, D.O. Phosphoregulation of Phase Separation by the SARS-CoV-2 N Protein Suggests a Biophysical Basis for its Dual Functions. *Mol. Cell* **2020**, *80*, 1092–1103.e4. [CrossRef]
35. Masters, P.S. Coronavirus genomic RNA packaging. *Virology* **2019**, *537*, 198–207. [CrossRef]
36. Yang, M.; He, S.; Chen, X.; Huang, Z.; Zhou, Z.; Zhou, Z.; Chen, Q.; Chen, S.; Kang, S. Structural Insight Into the SARS-CoV-2 Nucleocapsid Protein C-Terminal Domain Reveals a Novel Recognition Mechanism for Viral Transcriptional Regulatory Sequences. *Front. Chem.* **2020**, *8*, 624765. [CrossRef] [PubMed]
37. Kang, S.; Yang, M.; He, S.; Wang, Y.; Chen, X.; Chen, Y.Q.; Hong, Z.; Liu, J.; Jiang, G.; Chen, Q.; et al. A SARS-CoV-2 antibody curbs viral nucleocapsid protein-induced complement hyperactivation. *Nat. Commun.* **2021**, *12*, 2697. [CrossRef] [PubMed]
38. Ye, Q.; Lu, S.; Corbett, K.D. Structural Basis for SARS-CoV-2 Nucleocapsid Protein Recognition by Single-Domain Antibodies. *Front. Immunol.* **2021**, *12*, 719037. [CrossRef] [PubMed]
39. Jia, Z.; Liu, C.; Chen, Y.; Jiang, H.; Wang, Z.; Yao, J.; Yang, J.; Zhu, J.; Zhang, B.; Yuchi, Z. Crystal structures of the SARS-CoV-2 nucleocapsid protein C-terminal domain and development of nucleocapsid-targeting nanobodies. *FEBS J.* **2022**, *289*, 3813–3825. [CrossRef]
40. Rafael Ciges-Tomas, J.; Franco, M.L.; Vilar, M. Identification of a guanine-specific pocket in the protein N of SARS-CoV-2. *Commun. Biol.* **2022**, *5*, 711. [CrossRef]
41. Mercaldi, G.F.; Bezerra, E.H.S.; Batista, F.A.H.; Tonoli, C.C.C.; Soprano, A.S.; Shimizu, J.F.; Nagai, A.; da Silva, J.C.; Filho, H.V.R.; do Nascimento Faria, J.; et al. Discovery and structural characterization of chicoric acid as a SARS-CoV-2 nucleocapsid protein ligand and RNA binding disruptor. *Sci. Rep.* **2022**, *12*, 18500. [CrossRef]
42. Ni, X.; Han, Y.; Zhou, R.; Zhou, Y.; Lei, J. Structural insights into ribonucleoprotein dissociation by nucleocapsid protein interacting with non-structural protein 3 in SARS-CoV-2. *Commun. Biol.* **2023**, *6*, 193. [CrossRef] [PubMed]
43. Vandervaart, J.P.; Inniss, N.L.; Ling-Hu, T.; Minasov, G.; Wiersum, G.; Rosas-Lemus, M.; Shuvalova, L.; Achenbach, C.J.; Hultquist, J.F.; Satchell, K.J.F.; et al. Serodominant SARS-CoV-2 Nucleocapsid Peptides Map to Unstructured Protein Regions. *Microbiol. Spectr.* **2023**, *11*, e0032423. [CrossRef]
44. Kim, Y.; Maltseva, N.; Tesar, C.; Jedrzejczak, R.; Endres, M.; Ma, H.; Dugan, H.L.; Stamper, C.T.; Chang, C.; Li, L.; et al. Epitopes recognition of SARS-CoV-2 nucleocapsid RNA binding domain by human monoclonal antibodies. *iScience* **2024**, *27*, 108976. [CrossRef]
45. Casasanta, M.A.; Jonaid, G.M.; Kaylor, L.; Luqiu, W.Y.; DiCecco, L.A.; Solares, M.J.; Berry, S.; Dearnaley, W.J.; Kelly, D.F. Structural Insights of the SARS-CoV-2 Nucleocapsid Protein: Implications for the Inner-workings of Rapid Antigen Tests. *Microsc. Microanal.* **2023**, *29*, 649–657. [CrossRef]

46. Li, G.; Li, W.; Fang, X.; Song, X.; Teng, S.; Ren, Z.; Hu, D.; Zhou, S.; Wu, G.; Li, K. Expression and purification of recombinant SARS-CoV-2 nucleocapsid protein in inclusion bodies and its application in serological detection. *Protein Expr. Purif.* **2021**, *186*, 105908. [CrossRef] [PubMed]
47. Zeng, W.; Liu, G.; Ma, H.; Zhao, D.; Yang, Y.; Liu, M.; Mohammed, A.; Zhao, C.; Yang, Y.; Xie, J.; et al. Biochemical characterization of SARS-CoV-2 nucleocapsid protein. *Biochem. Biophys. Res. Commun.* **2020**, *527*, 618–623. [CrossRef]
48. Djukic, T.; Mladenovic, M.; Stanic-Vucinic, D.; Radosavljevic, J.; Smiljanic, K.; Sabljic, L.; Devic, M.; Cujic, D.; Vasovic, T.; Simovic, A.; et al. Expression, purification and immunological characterization of recombinant nucleocapsid protein fragment from SARS-CoV-2. *Virology* **2021**, *557*, 15–22. [CrossRef] [PubMed]
49. Di, D.; Dileepan, M.; Ahmed, S.; Liang, Y.; Ly, H. Recombinant SARS-CoV-2 Nucleocapsid Protein: Expression, Purification, and Its Biochemical Characterization and Utility in Serological Assay Development to Assess Immunological Responses to SARS-CoV-2 Infection. *Pathogens* **2021**, *10*, 1039. [CrossRef] [PubMed]
50. Uhlen, M.; Forsberg, G.; Moks, T.; Hartmanis, M.; Nilsson, B. Fusion proteins in biotechnology. *Curr. Opin. Biotechnol.* **1992**, *3*, 363–369. [CrossRef]
51. Sergeev, N.V.; Glukhova, N.S.; Nazimov, I.V.; Guliaev, V.A.; Donetskii, I.A.; Miroshnikov, A.I. Analytical biotechnology of recombinant peptides and proteins. II. Primary structure of the fusion protein containing human proinsulin and optimization of its proteolysis by trypsin. *Bioorg. Khim* **2000**, *26*, 516–521. [CrossRef]
52. Wingfield, P.T. Overview of the purification of recombinant proteins. *Curr. Protoc. Protein Sci.* **2015**, *80*, 6.1.1–6.1.35. [CrossRef] [PubMed]
53. Wingfield, P.T. Overview of the purification of recombinant proteins produced in *Escherichia coli*. *Curr. Protoc. Protein Sci.* **2003**, *30*, 6.1.1–6.1.37. [CrossRef]
54. Kimple, M.E.; Brill, A.L.; Pasker, R.L. Overview of affinity tags for protein purification. *Curr. Protoc. Protein Sci.* **2013**, *73*, 9.9.1–9.9.23. [CrossRef] [PubMed]
55. Waugh, D.S. An overview of enzymatic reagents for the removal of affinity tags. *Protein Expr. Purif.* **2011**, *80*, 283–293. [CrossRef]
56. Tarczewska, A.; Kolonko-Adamska, M.; Zarebski, M.; Dobrucki, J.; Ozyhar, A.; Greb-Markiewicz, B. The method utilized to purify the SARS-CoV-2 N protein can affect its molecular properties. *Int. J. Biol. Macromol.* **2021**, *188*, 391–403. [CrossRef]
57. Zhang, B.; Xie, Y.; Lan, Z.; Li, D.; Tian, J.; Zhang, Q.; Tian, H.; Yang, J.; Zhou, X.; Qiu, S.; et al. SARS-CoV-2 Nucleocapsid Protein Has DNA-Melting and Strand-Annealing Activities With Different Properties From SARS-CoV-2 Nsp13. *Front. Microbiol.* **2022**, *13*, 851202. [CrossRef]
58. Morse, M.; Sefcikova, J.; Rouzina, I.; Beuning, P.J.; Williams, M.C. Structural domains of SARS-CoV-2 nucleocapsid protein coordinate to compact long nucleic acid substrates. *Nucleic Acids Res.* **2023**, *51*, 290–303. [CrossRef]
59. Huang, Y.; Huang, C.; Chen, J.; Chen, S.; Li, B.; Li, J.; Jin, Z.; Zhang, Q.; Pan, P.; Du, W.; et al. Inhibition of SARS-CoV-2 replication by a ssDNA aptamer targeting the nucleocapsid protein. *Microbiol. Spectr.* **2024**, *12*, e0341023. [CrossRef]
60. Esler, M.A.; Belica, C.A.; Rollie, J.A.; Brown, W.L.; Moghadasi, S.A.; Shi, K.; Harki, D.A.; Harris, R.S.; Aihara, H. A compact stem-loop DNA aptamer targets a uracil-binding pocket in the SARS-CoV-2 nucleocapsid RNA-binding domain. *Nucleic Acids Res.* **2024**, gkae874. [CrossRef]
61. Zhao, H.; Wu, D.; Nguyen, A.; Li, Y.; Adao, R.C.; Valkov, E.; Patterson, G.H.; Piszczek, G.; Schuck, P. Energetic and structural features of SARS-CoV-2 N-protein co-assemblies with nucleic acids. *iScience* **2021**, *24*, 102523. [CrossRef]
62. Kemp, G.; Webster, A.; Russell, W.C. Proteolysis is a key process in virus replication. *Essays Biochem.* **1992**, *27*, 1–16. [PubMed]
63. Lutomski, C.A.; El-Baba, T.J.; Bolla, J.R.; Robinson, C.V. Multiple Roles of SARS-CoV-2 N Protein Facilitated by Proteoform-Specific Interactions with RNA, Host Proteins, and Convalescent Antibodies. *JACS Au* **2021**, *1*, 1147–1157. [CrossRef] [PubMed]
64. Papageorgiou, N.; Lichiere, J.; Baklouti, A.; Ferron, F.; Sevajol, M.; Canard, B.; Coutard, B. Structural characterization of the N-terminal part of the MERS-CoV nucleocapsid by X-ray diffraction and small-angle X-ray scattering. *Acta Crystallogr. D Struct. Biol.* **2016**, *72*, 192–202. [CrossRef]
65. Surjit, M.; Liu, B.; Kumar, P.; Chow, V.T.; Lal, S.K. The nucleocapsid protein of the SARS coronavirus is capable of self-association through a C-terminal 209 amino acid interaction domain. *Biochem. Biophys. Res. Commun.* **2004**, *317*, 1030–1036. [CrossRef] [PubMed]
66. Chang, C.K.; Hou, M.H.; Chang, C.F.; Hsiao, C.D.; Huang, T.H. The SARS coronavirus nucleocapsid protein--forms and functions. *Antiviral Res.* **2014**, *103*, 39–50. [CrossRef]
67. Malik, Y.A. Properties of Coronavirus and SARS-CoV-2. *Malays. J. Pathol.* **2020**, *42*, 3–11.
68. Trivedi, R.; Nagarajaram, H.A. Intrinsically Disordered Proteins: An Overview. *Int. J. Mol. Sci.* **2022**, *23*, 14050. [CrossRef]
69. Berlow, R.B.; Dyson, H.J.; Wright, P.E. Expanding the Paradigm: Intrinsically Disordered Proteins and Allosteric Regulation. *J. Mol. Biol.* **2018**, *430*, 2309–2320. [CrossRef]
70. Dyson, H.J.; Wright, P.E. Intrinsically unstructured proteins and their functions. *Nat. Rev. Mol. Cell Biol.* **2005**, *6*, 197–208. [CrossRef]

71. Xie, H.; Vucetic, S.; Iakoucheva, L.M.; Oldfield, C.J.; Dunker, A.K.; Uversky, V.N.; Obradovic, Z. Functional anthology of intrinsic disorder. 1. Biological processes and functions of proteins with long disordered regions. *J. Proteome Res.* **2007**, *6*, 1882–1898. [CrossRef]
72. Kouno, T.; Silvas, T.V.; Hilbert, B.J.; Shandilya, S.M.D.; Bohn, M.F.; Kelch, B.A.; Royer, W.E.; Somasundaran, M.; Kurt Yilmaz, N.; Matsuo, H.; et al. Crystal structure of APOBEC3A bound to single-stranded DNA reveals structural basis for cytidine deamination and specificity. *Nat. Commun.* **2017**, *8*, 15024. [CrossRef]
73. Fang, Y.; Xiao, X.; Li, S.X.; Wolfe, A.; Chen, X.S. Molecular Interactions of a DNA Modifying Enzyme APOBEC3F Catalytic Domain with a Single-Stranded DNA. *J. Mol. Biol.* **2018**, *430*, 87–101. [CrossRef] [PubMed]
74. Xiao, X.; Li, S.X.; Yang, H.; Chen, X.S. Crystal structures of APOBEC3G N-domain alone and its complex with DNA. *Nat. Commun.* **2016**, *7*, 12193. [CrossRef]
75. Maiti, A.; Hedger, A.K.; Myint, W.; Balachandran, V.; Watts, J.K.; Schiffer, C.A.; Matsuo, H. Structure of the catalytically active APOBEC3G bound to a DNA oligonucleotide inhibitor reveals tetrahedral geometry of the transition state. *Nat. Commun.* **2022**, *13*, 7117. [CrossRef] [PubMed]
76. Otwinowski, Z.; Minor, W. Processing of X-ray diffraction data collected in oscillation mode. *Methods Enzymol.* **1997**, *276*, 307–326. [CrossRef]
77. Matthews, B.W. Solvent content of protein crystals. *J. Mol. Biol.* **1968**, *33*, 491–497. [CrossRef] [PubMed]
78. Bunkoczi, G.; Echols, N.; McCoy, A.J.; Oeffner, R.D.; Adams, P.D.; Read, R.J. Phaser. MRage: Automated molecular replacement. *Acta Crystallogr. D Biol. Crystallogr.* **2013**, *69*, 2276–2286. [CrossRef]
79. Emsley, P.; Cowtan, K. Coot: Model-building tools for molecular graphics. *Acta Crystallogr. D Biol. Crystallogr.* **2004**, *60*, 2126–2132. [CrossRef]
80. Adams, P.D.; Afonine, P.V.; Bunkoczi, G.; Chen, V.B.; Echols, N.; Headd, J.J.; Hung, L.W.; Jain, S.; Kapral, G.J.; Grosse Kunstleve, R.W.; et al. The Phenix software for automated determination of macromolecular structures. *Methods* **2011**, *55*, 94–106. [CrossRef]
81. Echols, N.; Grosse-Kunstleve, R.W.; Afonine, P.V.; Bunkoczi, G.; Chen, V.B.; Headd, J.J.; McCoy, A.J.; Moriarty, N.W.; Read, R.J.; Richardson, D.C.; et al. Graphical tools for macromolecular crystallography in PHENIX. *J. Appl. Crystallogr.* **2012**, *45*, 581–586. [CrossRef]
82. Chen, V.B.; Arendall, W.B., 3rd; Headd, J.J.; Keedy, D.A.; Immormino, R.M.; Kapral, G.J.; Murray, L.W.; Richardson, J.S.; Richardson, D.C. MolProbity: All-atom structure validation for macromolecular crystallography. *Acta Crystallogr. D Biol. Crystallogr.* **2010**, *66*, 12–21. [CrossRef] [PubMed]
83. Lyons, B.; Kwan, A.H.; Truscott, R.J. Spontaneous cleavage of proteins at serine and threonine is facilitated by zinc. *Aging Cell* **2016**, *15*, 237–244. [CrossRef] [PubMed]
84. Butre, C.I.; Buhler, S.; Sforza, S.; Gruppen, H.; Wierenga, P.A. Spontaneous, non-enzymatic breakdown of peptides during enzymatic protein hydrolysis. *Biochim. Biophys. Acta* **2015**, *1854*, 987–994. [CrossRef] [PubMed]

biomolecules

MDPI

Article

Mechanistic Insights into the Mechanism of Allosteric Inhibition of Ubiquitin-Specific Protease 7 (USP7)

Xuebin Wang [1], Ning Liu [2], Nuan Li [3], Shaoyong Lu [2,3,*] and Zongtao Chai [4,5,*]

1 Department of Pharmacy, Shanghai Children's Hospital, School of Medicine, Shanghai Jiao Tong University, Shanghai 200062, China; binxuewang@sjtu.edu.cn
2 Key Laboratory of Protection, Development and Utilization of Medicinal Resources in Liupanshan Area, Ministry of Education, Peptide & Protein Drug Research Center, School of Pharmacy, Ningxia Medical University, Yinchuan 750004, China; 20160126@nxmu.edu.cn
3 Medicinal Chemistry and Bioinformatics Center, Shanghai Jiao Tong University School of Medicine, Shanghai 200025, China; linuan123@sjtu.edu.cn
4 Department of Hepatic Surgery, Shanghai Geriatric Medical Center, Shanghai 201104, China
5 Department of Liver Surgery and Transplantation, Liver Cancer Institute and Zhongshan Hospital, Fudan University, Shanghai 200032, China
* Correspondence: lushaoyong@sjtu.edu.cn (S.L.); chai.zongtao@zs-hospital.sh.cn (Z.C.)

Abstract: Ubiquitin-specific protease 7 (USP7), a deubiquitinase enzyme responsible for removing ubiquitin (Ub) from target proteins, plays a crucial role in oncogenic pathways and has been implicated in various human diseases. X-ray crystallography has revealed distinct conformations of USP7, including apo (ligand-free), allosteric inhibitor-, and Ub-bound states. However, the dynamic mechanisms underlying the allosteric inhibition of USP7 remain unclear. This study investigates the effect of allosteric inhibitor binding on the dynamics of USP7 through multiple replica molecular dynamics simulations. Our results demonstrate that Ub binding stabilizes the USP7 conformation, while allosteric inhibitor binding increases flexibility and variability in the fingers and palm domains of USP7. Furthermore, our analysis of USP7 local regions reveals that allosteric inhibitor binding not only restrains the dynamics of the C-terminal Ub binding site, thereby impeding the accessibility of Ub to USP7, but also disrupts the proper alignment of the catalytic triad (Cys223-His464-Asp481) in USP7. Additionally, community network analysis indicates that intra-domain communications within the fingers domain in USP7 are significantly enhanced upon allosteric inhibitor binding. This study reveals that the binding of an allosteric inhibitor induces a dynamic shift in enzyme's conformational equilibrium, effectively disrupting its catalytic activity through allosteric modulation.

Keywords: molecular dynamics simulations; conformational dynamics; allosteric inhibition; protein–protein interactions; allosteric communication

1. Introduction

The ubiquitin-proteasome system (UPS), a highly conserved eukaryotic protein quality control mechanism, governs intracellular proteostasis through targeted substrate degradation [1]. This system operates via a coordinated enzymatic cascade involving ubiquitin-activating enzymes (E1), ubiquitin-conjugating enzymes (E2), ubiquitin ligases (E3), deubiquitinases (DUBs), ubiquitin (Ub), and the 26S proteasome complex [2]. Central to UPS functionality is ubiquitination—a dynamic enzymatic process wherein Ub molecules are covalently linked to specific protein substrates. This post-translational modification is sequentially mediated by the E1-E2-E3 enzymatic triad, ultimately marking substrates

https://doi.org/10.3390/biom15060749

for proteasomal recognition and degradation. Conversely, deubiquitination, catalyzed by DUBs, removes Ub from target proteins, reversing the ubiquitination process. These processes regulate protein stability, function, and localization [1].

There are ~100 DUBs evolutionarily grouped into two major classes based on their enzymatic mechanism: cysteine proteases and metalloproteases. Among these, ubiquitin-specific protease 7 (USP7), a cysteine protease extensively documented in studies [3–8], exemplifies structural complexity through its multi-domain architecture. This 1102-residue enzyme contains four functionally specialized regions: (i) an N-terminal disordered segment (residues 1–50), featuring a polyglutamine (Poly-Q) repeat, (ii) a TRAF-homology domain (residues 53–205) mediating protein interactions, (iii) a central catalytic core (residues 208–560) responsible for Ub recognition and isopeptidase activity, and (iv) a C-terminal quintuple UBL domain array (residues 564–1102) modulating substrate specificity [9–11]. Architecturally, the catalytic domain adopts a tripartite papain-like fold—comprising fingers, palm, and thumb subdomains (Figure 1A)—a structural blueprint conserved across USP family members [12,13]. Mechanistic crystallographic analyses reveal that Ub binding within the fingers–palm interface induces conformational restructuring of the catalytic triad (Cys223/His464/Asp481), transitioning these residues into catalytically competent orientations essential for efficient Ub chain hydrolysis (Figure 1B). Crucially, this domain serves dual roles: direct Ub engagement and spatial coordination of substrate deubiquitination events.

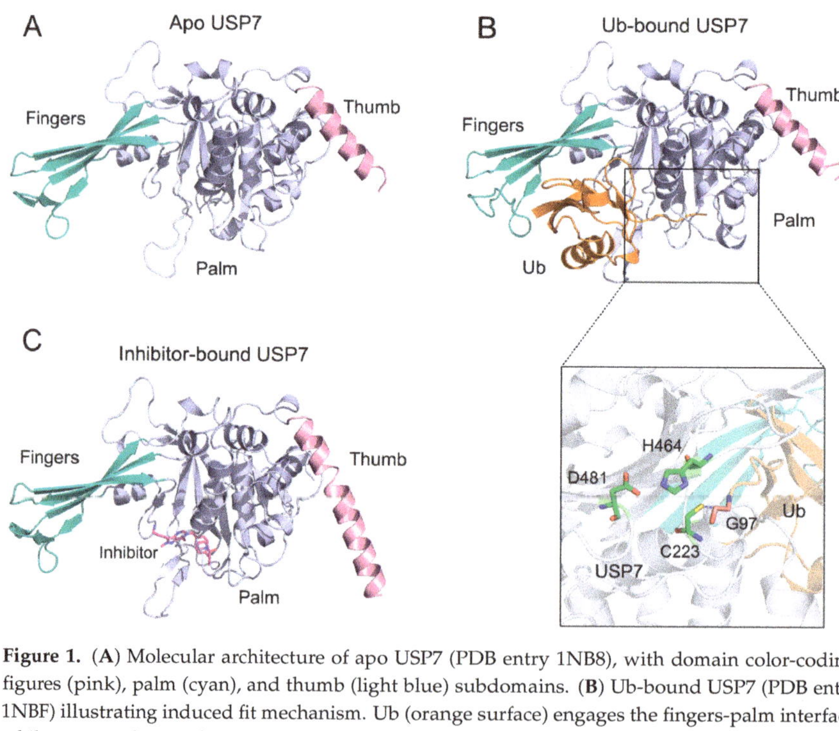

Figure 1. (**A**) Molecular architecture of apo USP7 (PDB entry 1NB8), with domain color-coding: figures (pink), palm (cyan), and thumb (light blue) subdomains. (**B**) Ub-bound USP7 (PDB entry 1NBF) illustrating induced fit mechanism. Ub (orange surface) engages the fingers-palm interface, while structural magnification reveals catalytic triad reconfiguration (Cys223/His464/Asp481) and thioester bond formation between USP7 Cys223 sulfur atom and Ub Gly97 carboxyl oxygen. (**C**) Allosteric inhibition complex (PDB entry 5N9T) showing compound **5** (stick model) binding distal to the catalytic cleft. Secondary structure elements maintain the standard coloring scheme across panels.

Recently, a small-molecule inhibitor (compound 4) was reported to exhibit high affinity and specificity for USP7 (IC_{50} = 6 ± 2 nM) both in vitro and in human cells [14]. The X-ray co-crystal structure of the catalytic domain of USP7 in complex with compound 5, a close analog of compound 4, reveals occupancy of a non-canonical allosteric pocket within the palm subdomain (Figure 1C), distinct from Ub-binding regions. Despite minimal global structural perturbations compared to apo USP7 (Cα RMSD 0.38 Å), the inhibitor-bound state exhibits marked divergence from Ub-complexed conformations (Cα RMSD 0.79 Å), suggesting ligand-specific domain reconfigurations. Moreover, allosteric inhibitor binding may misalign the catalytic triad, leading to a catalytically non-productive conformation. While these existing structural snapshots (apo, Ub-bound, inhibitor-bound) provide useful information about Ub and inhibitor interaction with the enzyme, the underlying dynamic mechanism regarding the allosteric inhibition of USP7 catalytic domain remains to be fully elucidated because the overall conformational dynamics of USP7 catalytic domain are rarely captured by structural methods.

All-atom molecular dynamics (MD) simulations offer atomically detailed information on protein conformational dynamics, sampling multiple timescales from femtoseconds to microseconds to reach biologically relevant timescales (microseconds and beyond), such as signal transduction, enzyme catalysis, allosteric transitions, and protein-ligand or protein-protein binding [15–17]. Thus, MD simulations can complement experiments by providing atomic-level detail necessary for monitoring the progression of conformational changes in proteins [18–24]. To address the mechanism of allosteric inhibition of USP7 catalytic domain, we employed multi-replica MD simulations to sample three functional states (apo, Ub-bound, inhibitor-bound). An integrated conformational analysis method, including dynamic cross-correlation matrix analysis and network analysis, was used to reveal state-specific dynamic signatures. The simulation-derived conformational landscapes provide critical insights for allosteric inhibitor design.

2. Materials and Methods

2.1. System Preparation

The starting coordinates were obtained from the Protein Data Bank (www.rcsb.org) for three systems: apo USP7 (PDB entry 1NB8) [25], USP7 in complex with Ub (PDB entry 1NBF) [25], and USP7 bound to compound 5 (PDB entry 5N9T) [14]. Structural analysis revealed that in the USP7-Ub complex, the catalytic cysteine residue (Cys223) of USP7 establishes a covalent linkage with the C-terminal glycine (Gly97) of Ub. For modeling purposes, the (*R*)-trifluoromethyl moiety in compound 5 was computationally modified to an (*R*)-methyl group to represent compound 4. Any incomplete side chain coordinates were reconstructed using UCSF Chimera's modeling tools [26].

2.2. MD Simulations

All MD simulations were executed using the Amber (Version: 18) 18 software package [27]. Hydrogen atoms were added to three simulated systems (apo USP7, USP7-Ub complex, and USP7-compound 4 complex) using the *tleap* module. Force field assignments included the Amber ff14SB for protein residues [28] and the General Amber Force Field (GAFF) for the small-molecule inhibitor (compound 4) [29]. The covalent bond between USP7 Cys223 and the C-terminal carboxyl group of Ub's Gly97 was explicitly defined in *tleap*. Compound 4 underwent geometry optimization using the semi-empirical AM1 method, followed by bond-charge correction (BCC) charge assignment via the *Antechamber* module [30]. Protonation states of amino acid residues were determined at physiological pH using PROPKA (Version: 3.0) [31]. Each system was solvated in a truncated octahedron periodic box filled with TIP3P water [32], maintaining a minimum 10 Å water layer around the protein surface, and neutralized with Na^+ counterions.

Energy minimization involved two phases: 20,000 cycles with positional restraints on protein atoms and 50,000 cycles of unrestrained minimization. Systems were then gradually heated from 0 K to 300 K over 100 ps under constant volume (NVT ensemble), followed by 200 ps equilibration. Production simulations were conducted in the NPT ensemble (300 K, 1 atm) using three independent 1000 ns trajectories initialized with random velocities. Electrostatic interactions were computed using the Particle Mesh Ewald (PME) method [33], and hydrogen-containing bonds were constrained via the SHAKE algorithm [34]. A 2 fs integration time step was employed.

All simulations utilized the *pmemd.cuda* GPU-accelerated module in Amber 18, with temperature and pressure regulation achieved through the Langevin thermostat/barostat (1.0 ps coupling time) [35]. This protocol ensured reproducible trajectories for subsequent structural and dynamical analyses.

2.3. DCCM Analysis

To characterize residue-level dynamical coupling, cross-correlation coefficients (C_{ij}) between Cα atoms were computed using trajectory-averaged calculations. The normalized covariance metric was defined as [36–39]:

$$C(i,j) = \frac{\langle \Delta r_i \cdot \Delta r_j \rangle}{\sqrt{< \Delta r_i >^2 \cdot < \Delta r_j >^2}}$$

where Δri (Δrj) denotes the displacement vector of residue *i* (*j*) relative to its equilibrium position, and angle brackets represent averaging across concatenated replica trajectories.

2.4. Community Network Analysis

The residue interactions network of USP7 was generated from the DCCM through NetworkView [40,41], a Tcl plugin integrated with VMD [42]. In this representation, each residue's Cα atom served as a network node, with edges established between node pairs maintaining spatial proximity (≤4.5 Å) for over 75% of the simulation duration [43,44]. Edge weights were determined by the logarithmic transformation:

$$d_{i,j} = -log(\mid C_{i,j} \mid)$$

where Cij values originated from the Cα-based cross-correlation analysis.

3. Results and Discussion

3.1. Allosteric Modulation Amplifies USP7 Structural Plasticity

To comprehensively investigate the conformational sampling of the USP7 catalytic domain (residues 208–560, hereafter referred to as USP7) in its apo, inhibitor-bound (compound 4), and Ub-bound states, we implemented MD simulations across three functional states. Each simulation comprised three replicates, with each replicate running for 1 microsecond (μs), resulting in a total simulation time of 3 μs per system. To reveal the convergence and structural stability of USP7 across different states through multiple replicate MD simulations, the Cα RMSD was monitored. MD simulations of all systems generally achieved conformational equilibrium after approximately 300 ns (Figure 2). Consequently, the equilibrium trajectories from 300 to 1000 ns were combined from the three replicates, yielding a single consolidated MD trajectory of 2100 ns for post-processing analyses.

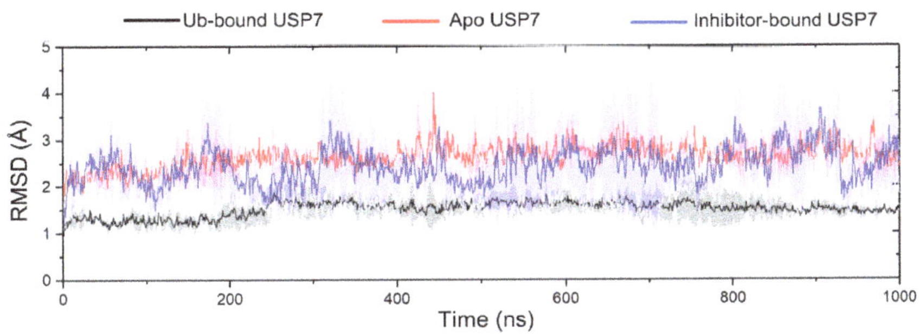

Figure 2. The root-mean-square deviation (RMSD) of Cα atoms for USP7 in the Ub-bound (black line), apo (red line), and inhibitor-bound (blue line) states along the 1000 ns MD simulations. The shaded region denotes variability (standard deviation) across three independent MD replicates.

The RMSD values for USP7 in the Ub-bound, apo, and inhibitor-bound states were 1.51 ± 0.23, 2.64 ± 0.25, and 2.45 ± 0.66 Å for the USP7, respectively. These results indicated that Ub binding significantly contributes to the stabilization of conformational plasticity of USP7, whereas the inhibitor-bound enzyme retains considerable flexibility, similar to the apo form.

Quantitative residue-wise dynamics were characterized through Cα-root mean square fluctuation (Cα-RMSF) profiling. While it is indeed commonly observed that inhibitor-bound protein complexes exhibit reduced conformational flexibility compared to apo forms, our MD simulations revealed a striking exception in USP7. The RMSF profiling demonstrated preserved global conformational sampling capacity in USP7 between apo and inhibitor-engaged states (Figure 3), with the most remarkable fluctuations observed in the fingers domain (residues 333–348 and 370–400). Comparing to apo USP7, inhibitor binding notably increases the structural flexibility of the fingers domain. The comparable RMSF values between apo USP7 and inhibitor-bound USP7 suggest a unique allosteric regulation mechanism: the catalytic domain maintains intrinsic flexibility even when occupied by inhibitors. These findings challenge the paradigm of ligand-induced rigidity and offer new insights for developing allosteric inhibitors targeting flexible enzymes. Conversely, Ub binding markedly stabilizes this region due to direct interactions between Ub and the fingers domain of USP7.

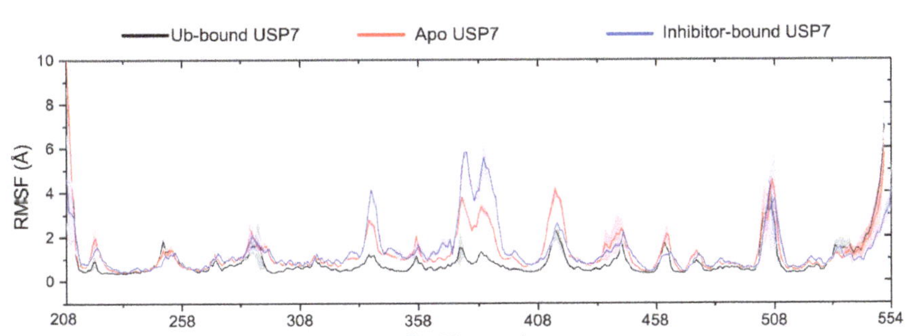

Figure 3. Root mean square fluctuation (RMSF) profiles of Cα atoms are shown for Ub-bound (black), apo (red), and inhibitor-bound (blue) USP7. Shaded regions reflect statistical variability across triplicate MD simulations (standard deviation).

We next compared the NMR data of the USP7 catalytic domain in both its unbound state and in complex with Ub. Ub interaction elicited pronounced chemical shift perturbations localized to the palm subdomain of USP7, as documented in prior work [9], indicating a stable palm-Ub binding interface upon Ub binding. Consistent with these experimental observations, our RMSF analysis of the apo USP7 catalytic domain versus the USP7 catalytic domain-Ub complex revealed that, in the complex system, the USP7 catalytic domain exhibited reduced RMSF.

3.2. Allosteric Inhibitor Binding Enhances the Internal Motions of USP7

To elucidate the influence of allosteric inhibitor interactions on USP7 conformational dynamics, Cα-based dynamic cross-correlation matrices (DCCM) were generated using the CPPTRAJ module in Amber 18 [45–47]. The color-coded modes are utilized to display the degree of correlated or anti-correlated motions between residues. Orange represents the correlated motions ($C_{ij} > 0$) between specific residues, while magenta indicates the anti-correlated motions ($C_{ij} < 0$) between residues. Weak pairwise couplings ($|C_{ij}| < 0.4$) were filtered as white space in DCCM visualizations, reflecting statistically insignificant inter-residue motion coordination.

As shown in Figure 4, allosteric inhibitor binding significantly alters the movement pattern of USP7. For Ub-bound USP7 (Figure 4A), there is minimal intradomain motion, indicating that Ub binding strongly stabilizes the conformational flexibility of USP7. For apo USP7 (Figure 4B), significant anti-correlated motions are observed between residues 320 and 440, which belong to the fingers domain. In contrast, allosteric inhibitor binding strengthens both correlated and anti-correlated motions within residues 320–440 (fingers domain). In addition, allosteric inhibitor binding slightly increases anti-correlated motions between residues 260 and 270 (palm domain) and residues 320 and 440 (fingers domain), as well as correlated motions between residues 505 and 515 (palm domain) and residues 320 and440 (fingers domain). Notably, the changes in RMSF values in the inhibitor-bound USP7, particularly in the fingers domain (residues 333–348 and 370–400) and palm domain (residues 505–515), align with the DCCM findings.

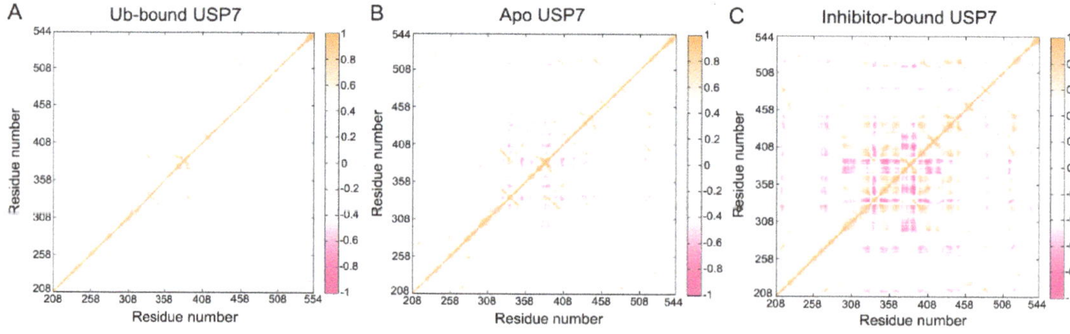

Figure 4. Dynamic coordination landscapes derived from Cα atom trajectories demonstrate conformational coupling differences between: (**A**) Ub-bound, (**B**) apo, and (**C**) inhibitor-bound USP7 states. A bidirectional color scheme ($-1 \leq C_{ij} \leq 1$) quantifies residue pair correlations, with white masking coefficients $|C_{ij}| < 0.4$ following statistical validation through trajectory subsampling.

3.3. Allosteric Inhibitor Binding Restricts Ub Accessibility to USP7

Structural analysis of the USP7-Ub interface identifies three critical recognition elements: α1/α4 helical domains and the β-hairpin BL2 loop (residues 456–464), which cooperatively stabilize the Ub C-terminus. To quantify allosteric perturbations induced by inhibitor binding, representative structural complexes of USP7 in different states were obtained using cluster analysis of MD trajectories [48]. Figure 5 illustrates the backbone superimposition of USP7 across apo, inhibitor-bound, and Ub-bound states. Notably, the conformations of helix α4 and the BL2 loop are similar between these ligand-free and inhibitor-complexed conformations, but differ significantly from those in the Ub-bound state. Specially, both the ligand-free and inhibitor-bound states exhibit a "down" conformation of the BL2 loop and an "outward" conformation of helix α4, while the Ub-bound state shows an "up" conformation of the BL2 loop and an "inward" conformation of helix α4. Consequently, the pronounced conformational changes in the BL2 loop and helix α4 upon allosteric inhibitor binding may lead to a contraction of the C-terminal Ub binding site, thereby restraining the accessibility of Ub to the USP7.

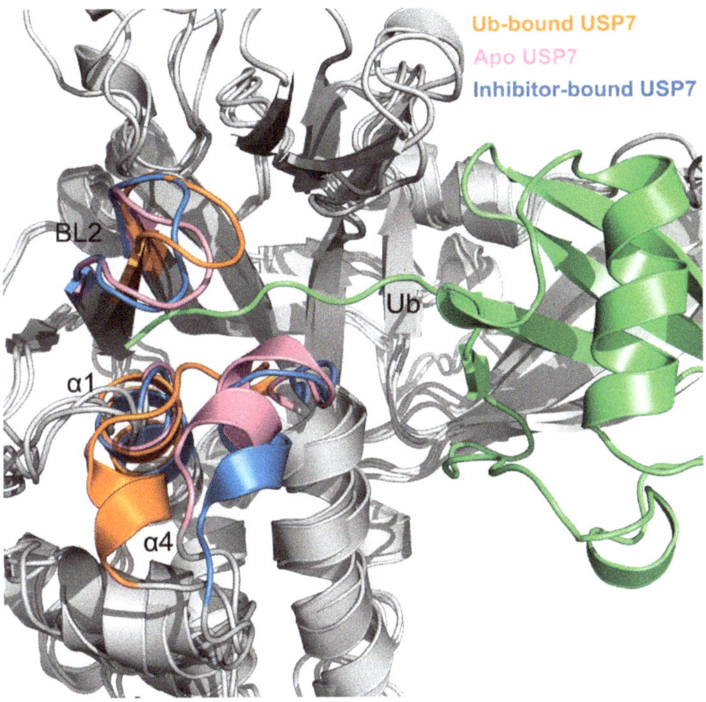

Figure 5. Superimposition of the USP7 backbone reveals distinct conformational shifts in key structural regions, specially the BL2 loop and helices α1 and α4 across its Ub-bound (orange), apo (pink), and inhibitor-bound (blue) states. The Ub is depicted in green.

To test the hypothesis that allosteric inhibitor binding disturbs the Ub binding site in USP7, two distances defining key features of the C terminal Ub binding site were calculated. Two key interatomic distances were analyzed to quantify conformational changes: the Cα-Cα distance between Tyr224 (located in helix α1) and His461 (within the BL2 loop), and the Cα-Cα distance between Phe291 (positioned in helix α4) and His461 (BL2 loop). These metrics were selected to characterize the overall conformational dynamics of the C-terminal Ub binding site. As shown in Figure 6A, in the Ub-bound USP7, the distance between Tyr224 and His461 remains stable along the simulations, with the value of 13.47 ± 1.35 Å. In the inhibitor-bound USP7, the distance stabilizes at 9.17 ± 0.33 Å throughout the simulations. In contrast, in the apo USP7, the distance fluctuates between values observed in the inhibitor-bound and Ub-bound states, averaging 10.93 Å. Similarly, as depicted in Figure 6B, the distance between Phe291 and His461 remains stable at 13.35 ± 1.68 Å in the Ub-bound USP7 and at 8.66 ± 0.56 Å in the inhibitor-bound USP7. However, in the apo USP7, the distance fluctuates between these two values, averaging 10.72 Å.

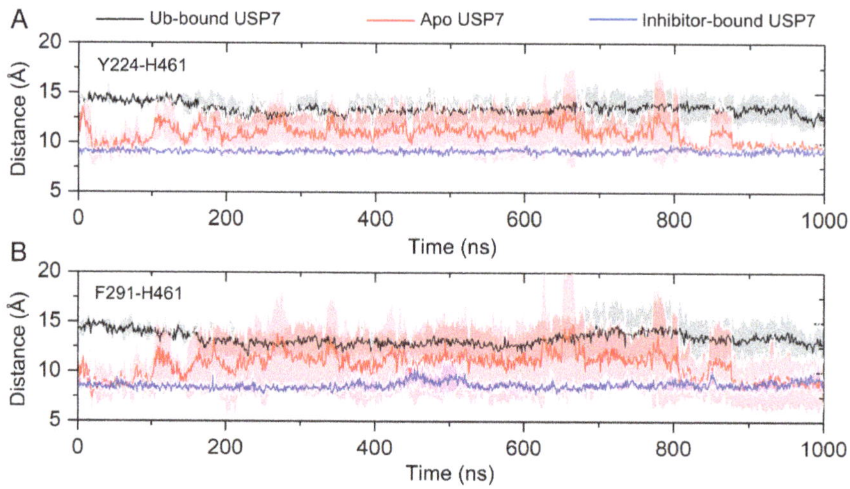

Figure 6. (**A**) The Cα-Cα distance between Tyr224 (helix α1) and His461 (BL2 loop) for the Ub-bound (black line), apo (red line), and inhibitor-bound (blue line) USP7 along the 1000 ns MD simulations. (**B**) The Cα-Cα distance between Phe291 (helix α4) and His461 (BL2 loop) for the Ub-bound (black line), apo (red line), and inhibitor-bound (blue line) USP7 along the 1000 ns MD simulations. The shaded region denotes variability (standard deviation) across three independent MD replicates.

To probe how allosteric inhibitors modulate USP7's functional dynamics, we statistically analyzed the conformational sampling of two geometric parameters associated with the C-terminal Ub-binding domain. Figure 7A visualizes the statistical distribution of distances between Tyr224 and His461 Cα atoms across simulations. In the Ub-bound USP7, the distribution exhibits two peaks at 11.5 and 14.5 Å. In the inhibitor-bound USP7, the distribution has a predominant peak at 9.1 Å. In the apo USP7, the distribution shows two peaks, with a major peak at 9.7 Å and a minor peak at 14.0 Å. Figure 7B presents the statistical distribution for distances between Phe291 and His461 Cα atoms. In the Ub-bound USP7, the distribution has two peaks at 11.1 and 13.8 Å. In the inhibitor-bound USP7, the distribution has a predominant peak at 8.7 Å. In the apo USP7, the distribution exhibits three peaks at 7.2, 10.8, and 15.1 Å. Taken together, these results suggest that allosteric inhibitor binding leads to a closed C-terminal Ub binding site, thereby impeding Ub's spatial entry to USP7's catalytic interface. Conversely, MD trajectories of apo USP7 demonstrate conformational plasticity in the C-terminal domain, facilitating Ub binding.

3.4. Allosteric Inhibitor Binding Disrupts the Catalytic Triad Arrangement

To resolve conformational perturbations in USP7's catalytic triad (Cys223, His464, Asp481) induced by allosteric regulation, we performed structural alignment of three distinct complexes: USP7-Ub holoenzyme, inhibitor-bound USP7, and apo USP7 (Figure 8). Structural analysis of the USP7-Ub complex reveals precise catalytic triad coordination essential for enzymatic competence. The nucleophilic thiol of Cys223 engages Ub's C-terminal Gly97 through thioester bond formation, while His464 maintains catalytic geometry by simultaneously coordinating the Cys223 sulfhydryl group and Asp481 carboxylate moiety. However, both the inhibitor-bound and apo USP7 exhibit a misaligned catalytic triad, with significant conformational changes in catalytic Cys223 and His464, adopting novel conformations not observed in the Ub-bound USP7. These data indicated that allosteric inhibitor binding may disrupt the proper alignment of the catalytic triad.

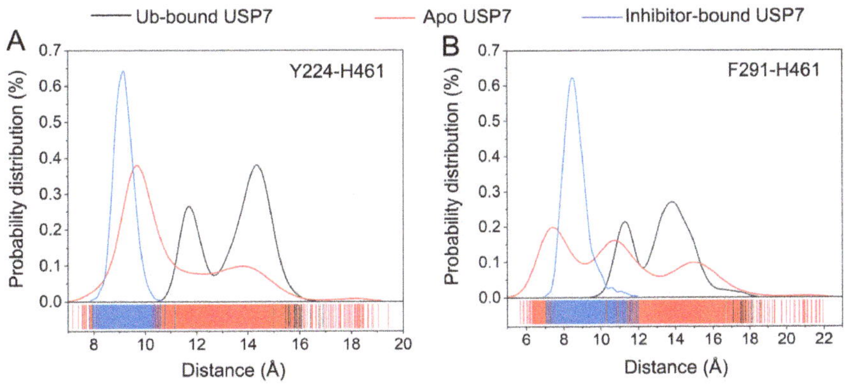

Figure 7. (**A**) Statistical distribution of distances between Tyr224 and His461 Cα atoms for the Ub-bound (black line), apo (red line), and inhibitor-bound (blue line) USP7. (**B**) Statistical distribution of distances between Phe291 and His461 Cα atoms for the Ub-bound (black line), apo (red line), and inhibitor-bound (blue line) USP7.

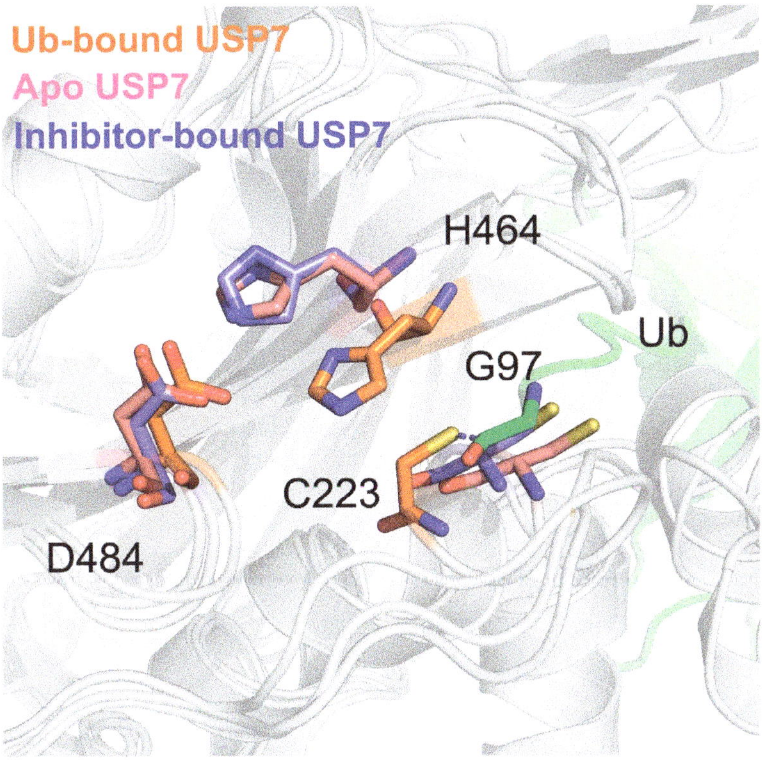

Figure 8. The overlap of USP7's catalytic triad (Cys223, His464, Asp481) in the Ub-bound (orange), apo (pink), and inhibitor-bound (blue) states. The Gly97 at the C-terminal Ub is showing in green.

To test the hypothesis that allosteric inhibitor binding leads to a misaligned catalytic triad, the inter-residue distances among Cys223, His464, and Asp481 were calculated. As illustrated in Figure 9A, the Cα-Cα distance between Cys223 and His464 maintains stability throughout the simulations, with values of 6.09 ± 0.17 Å for Ub-bound USP7 and 8.02 ± 0.32 Å for inhibitor-bound USP7. In contrast, the distance fluctuates in apo USP7 during the 0–350 ns simulation, averaging 7.08 Å and reaching up to 7.7 Å in subsequent MD simulations. For the inter-residue distance between the two Cα atoms of Cys223 and Asp481 (Figure 9B), stability is observed across all three simulated systems, with respective values of 7.93 ± 0.25 Å for Ub-bound USP7, 11.62 ± 0.36 Å for inhibitor-bound USP7, and 11.20 ± 0.55 Å for apo USP7. The inter-residue distance between the two Cα atoms of His464 and Asp481 shows no appreciable difference among the three simulated systems (Figure 9C), with values of 9.59 ± 0.31 Å for Ub-bound USP7, 8.80 ± 0.37 Å for inhibitor-bound USP7, and 8.90 ± 0.41 Å for apo USP7.

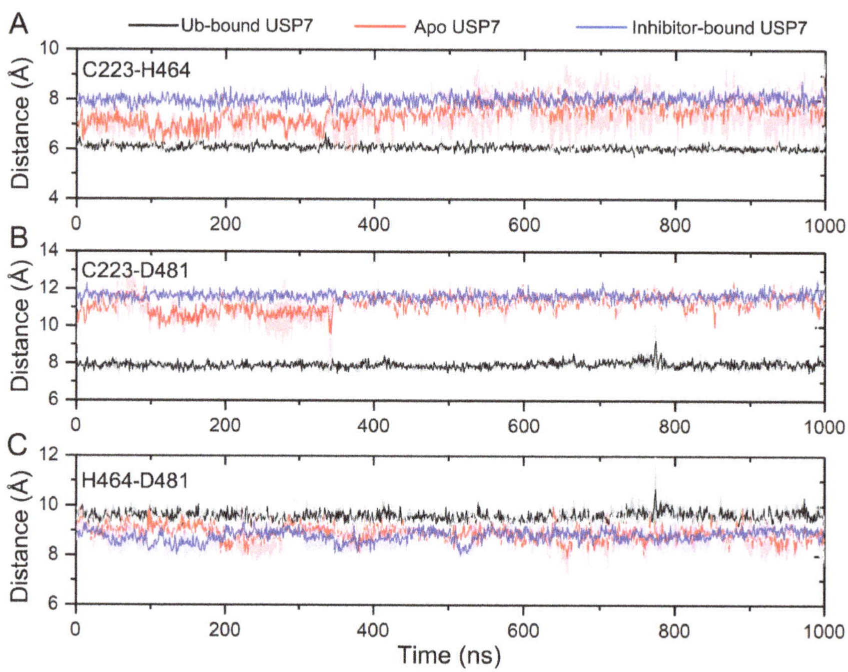

Figure 9. The Cα-Cα distances between Cys223 and His464 (**A**), between Cys223 and Asp481 (**B**), and between His464 and Asp481 (**C**) for the Ub-bound (black line), apo (red line), and inhibitor-bound (blue line) USP7 along the 1000 ns MD simulations. The shaded region denotes variability (standard deviation) across three independent MD replicates.

Quantitative assessment of catalytic triad dynamics was performed through conformational sampling analysis. Figure 10A presents the statistical distribution for the Cα-Cα distance between Cys223 and His464. The distribution displays a predominant peak for all three simulated systems, with a peak at 6.1 Å for Ub-bound USP7, 8.1 Å for inhibitor-bound USP7, and 7.4 Å for apo USP7. Similarly, the probability distribution for the distance between the two Cα atoms of Cys223 and Asp481 also exhibits a predominant peak for all three simulated systems (Figure 10B), with a peak at 7.9 Å for Ub-bound USP7, 11.6 Å for inhibitor-bound USP7, and 11.1 Å for apo USP7. For the probability distribution for the Cα-Cα distance between H464 and Asp481 (Figure 10C), a single predominant peak is observed at 9.6 Å for Ub-bound USP7, 8.8 Å for inhibitor-bound USP7, and 8.9 Å for apo USP7. Cumulatively, the calculated inter-residue distances and their probability distributions suggested that allosteric inhibitor binding disrupts the proper alignment of the catalytic triad.

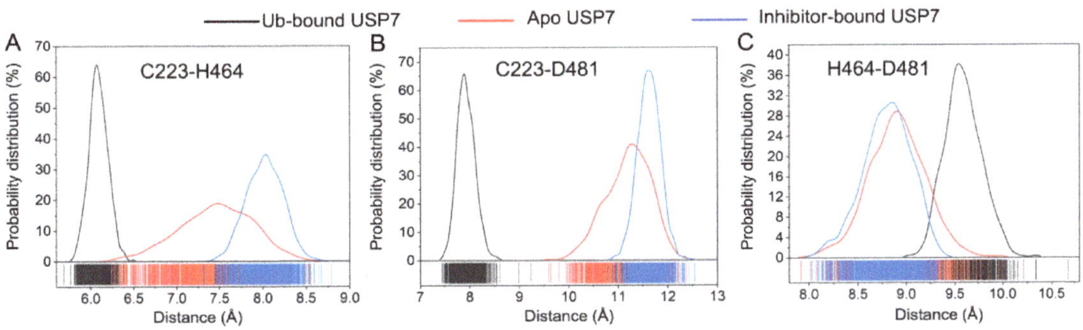

Figure 10. Statistical distributions of the Cα-Cα distance between Cys223 and His464 (**A**), between Cys223 and Asp481(**B**), and between His464 and Asp481 (**C**) for the Ub-bound (black line), apo (red line), and inhibitor-bound (blue line) USP7.

3.5. Topological Reorganization of Allosteric Networks

To map ligand-induced perturbations in USP7's signal propagation pathways, we conducted community network analysis (CNA) to evaluate dynamic networks in three different states. CNA is a robust method for exploring conformational changes in proteins by decoupling correlated motions between residues [49–52]. If a cluster of residues have the same movements, they are classified into the same community. Two communities are connected by an edge whose thickness means the intensity of the community coupling between them [53,54]. Figure 11 illustrates the 3D representations of community networks of apo, Ub-bound, and inhibitor-bound USP7. In general, the community networks in apo (Figure 11A) and inhibitor-bound (Figure 11B) USP7 are similar, while those in Ub-bound USP7 exhibit alterations (Figure 11C).

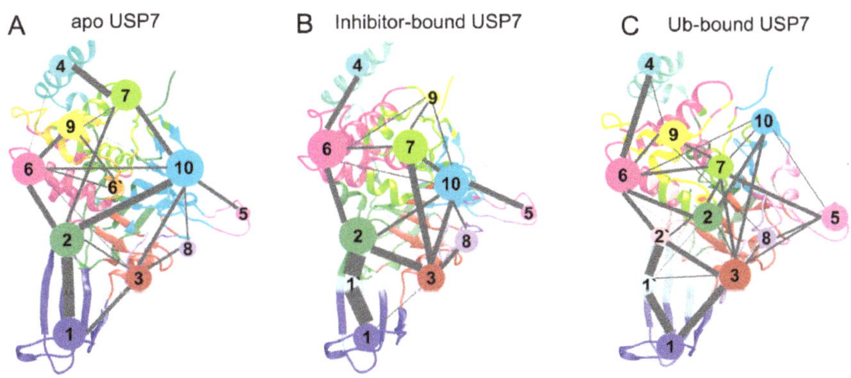

Figure 11. Community networks of USP7. States include: (**A**) apo, (**B**) inhibitor-bound, and (**C**) Ub-bound configurations. Edge thickness encodes inter-community contact persistence.

In both the apo and inhibitor-bound USP7, all residues were classified into 11 communities, whereas Ub-bound USP7 had 12 communities. Specially, in apo USP7, communities 1, 2, and 6 encompass the fingers domain, representing the binding site for Ub. Upon inhibitor binding, community 1 in the apo state splits into two communities (1 and 1′). Furthermore, in inhibitor-bound USP7, communities 1 and 2 from the apo state divide into four communities (1, 1′, 2 and 2′). Compared to apo USP7, allosteric inhibitor binding strengthens the coupling between communities 1 and 2 via community 1′. Conversely, in Ub-bound USP7, communication among communities 1, 1′, 2, and 2 weakens upon Ub binding compared to the communication between communities 1 and 2 in apo USP7 among communities 1, 1′, and 2 in inhibitor-bound USP7. These results indicate that allosteric inhibitor binding enhances intra-domain coupling with the fingers domain of USP7, suggesting increased flexibility consistent with RMSD and RMSF results. This enhanced flexibility may hinder the binding of Ub to the catalytic site, thus inactivating USP7.

4. Conclusions

Given the multifaced nature of USP7 in disease, it is becoming an increasingly attractive target for cancer treatment. Accurately elucidating the allosteric inhibition mechanism of USP7 is crucial for gaining deeper insights into its biological function. In this study, all-atom MD simulations totaling 3 μs of simulation time, comprising three independent 1 μs simulations, were carried out on apo, inhibitor-, and Ub-bound USP7 to enhance conformational sampling and explore the allosteric mechanism of USP7 by the inhibitor.

Through analyses, such as RMSF analysis, cross-correlation, network mapping, and detailed evaluations of USP7-inhibitor and USP7-Ub interactions, we gained valuable insights into the allosteric inhibition mechanism mediated by the inhibitor. The results suggest that allosteric inhibitor binding increases the conformational flexibility of USP7 and markedly impacts the intradomain movements of the fingers domain (residues 320–440) and palm domain (residues 260–270 and residues 505–515). Structurally, the fingers domain constitutes the binding site for Ub; therefore, alterations in flexibility, motion patterns, and dynamic behavior of USP7 contribute to its allosteric inhibition. Analysis of the C-terminal Ub binding site and the arrangement of the catalytic triad indicates that allosteric inhibitor binding leads to closure of the C-terminal Ub binding site, thereby restraining Ub accessibility to USP7. This also disrupts the proper alignment of catalytic triad (Cys223-His464-Asp481) within USP7. Additionally, network community analysis suggests that allosteric inhibitor binding strengthens intra-domain communications within the fingers domain of USP7, potentially hindering Ub binding to USP7.

While the present investigation centers on the USP7 catalytic domain, subsequent research will employ integrative structural modeling approaches to develop a full-length USP7-allosteric inhibitor complex. This future work will leverage the catalytic domain-inhibitor architecture established in our current study to elucidate the structural basis of allosteric regulation across the entire enzyme.

Author Contributions: S.L. and Z.C. conceived and supervised the study. X.W., N.L. (Ning Liu) and N.L. (Nuan Li) developed the methodology and performed formal analyses. X.W. conducted validation experiments and data curation. N.L. (Ning Liu) and N.L. (Nuan Li) contributed to investigation and data interpretation. The original draft was prepared by X.W., with critical review and editing by S.L. and Z.C. Visualization and resource provision were handled by X.W. and S.L., respectively. S.L. and Z.C. jointly administered the project. Z.C. acquired funding. All authors reviewed and approved the final manuscript.

Funding: This work was supported through research grants from the National Natural Science Foundation of China (Grant No. 82172846).

Institutional Review Board Statement: Not applicable.

Informed Consent Statement: Not applicable.

Data Availability Statement: Data are contained within the article.

Conflicts of Interest: The authors declare no conflicts of interest.

References

1. Chen, Y.; Xue, H.; Jin, J. Applications of Protein Ubiquitylation and Deubiquitylation in Drug Discovery. *J. Biol. Chem.* **2024**, *300*, 107264. [CrossRef] [PubMed]
2. Daviet, L.; Colland, F. Targeting Ubiquitin Specific Proteases for Drug Discovery. *Biochimie* **2008**, *90*, 270–283. [CrossRef] [PubMed]
3. Nininahazwe, L.; Liu, B.; He, C.; Zhang, H.; Chen, Z.S. The Emerging Nature of Ubiquitin-Specific Protease 7 (USP7): A New Target in Cancer Therapy. *Drug Discov. Today* **2021**, *26*, 490–502. [CrossRef]
4. Rawat, R.; Starczynowski, D.T.; Ntziachristos, P. Nuclear Deubiquitination in the Spotlight: The Multifaceted Nature of USP7 Biology in Disease. *Curr. Opin. Cell Biol.* **2019**, *58*, 85–94. [CrossRef]
5. Sun, L.-L.; Zhao, N.-L.; Sun, J.; Yuna, H.-F.; Wang, Y.-F.; Hou, C.-Y.; Lv, P.; Zhang, H.-H.; Yang, G.; Zhang, N.-N.; et al. Inhibition of USP7 enhances CD8+ T cell activity in liver cancer by suppressing PRDM1-mediated FGL1 upregulation. *Acta Pharmacol. Sin.* **2024**, *45*, 1686–1700. [CrossRef]
6. Di Lello, P.; Pastor, R.; Murray, J.M.; Blake, R.A.; Cohen, F.; Crawford, T.D.; Drobnick, J.; Drummond, J.; Kategaya, L.; Kleinheinz, T.; et al. Discovery of Small-Molecule Inhibitors of Ubiquitin Specific Protease 7 (USP7) Using Integrated NMR and in Silico Techniques. *J. Med. Chem.* **2017**, *60*, 10056–10070. [CrossRef]
7. Leger, P.R.; Hu, D.X.; Biannic, B.; Bui, M.; Han, X.; Karbarz, E.; Maung, J.; Okano, A.; Osipov, M.; Shibuya, G.M.; et al. Discovery of Potent, Selective, and Orally Bioavailable Inhibitors of USP7 with in Vivo Antitumor Activity. *J. Med. Chem.* **2020**, *63*, 5398–5420. [CrossRef]
8. Oliveira, R.I.; Guedes, R.A.; Salvador, J.A.R. Highlights in USP7 Inhibitors for Cancer Treatment. *Front. Chem.* **2022**, *10*, 1005727. [CrossRef]
9. Rougé, L.; Bainbridge, T.W.; Kwok, M.; Tong, R.; Di Lello, P.; Wertz, I.E.; Maurer, T.; Ernst, J.A.; Murray, J. Molecular Understanding of USP7 Substrate Recognition and C-Terminal Activation. *Structure* **2016**, *24*, 1335–1345. [CrossRef]
10. Cheng, J.; Li, Z.; Gong, R.; Fang, J.; Yang, Y.; Sun, C.; Yang, H.; Xu, Y. Molecular Mechanism for the Substrate Recognition of USP7. *Protein Cell* **2015**, *6*, 849–852. [CrossRef]
11. Zhou, J.; Wang, J.; Chen, C.; Yuan, H.; Wen, X.; Sun, H. USP7: Target Validation and Drug Discovery for Cancer Therapy. *Med. Chem.* **2018**, *14*, 3–18. [CrossRef] [PubMed]
12. Pozhidaeva, A.; Bezsonova, I. USP7: Structure, Substrate Specificity, and Inhibition. *DNA Repair* **2019**, *76*, 30–39. [CrossRef] [PubMed]
13. Özen, A.; Rougé, L.; Bashore, C.; Hearn, B.R.; Skelton, N.J.; Dueber, E.C. Selectively Modulating Conformational States of USP7 Catalytic Domain for Activation. *Structure* **2018**, *26*, 72–84.e7. [CrossRef]
14. Gavory, G.; O'Dowd, C.R.; Helm, M.D.; Flasz, J.; Arkoudis, E.; Dossang, A.; Hughes, C.; Cassidy, E.; McClelland, K.; Odrzywol, E.; et al. Discovery and Characterization of Highly Potent and Selective Allosteric USP7 Inhibitors. *Nat. Chem. Biol.* **2018**, *14*, 118–125. [CrossRef]
15. Liu, S.; Zhou, X.; Li, M.; Zhao, W.; Zhou, S.; Cheng, K.; Xu, Q.; Chen, C.; Wen, X.; Sun, H.; et al. Discovery of Ubiquitin-Specific Protease 7 (USP7) Inhibitors with Novel Scaffold Structures by Virtual Screening, Molecular Dynamics Simulation, and Biological Evaluation. *J. Chem. Inf. Model.* **2020**, *60*, 3255–3264. [CrossRef]
16. Zhao, L.-H.; Yuan, Q.-N.; Dai, A.-T.; He, X.-H.; Chen, C.-W.; Zhang, C.; Xu, Y.-W.; Zhou, Y.; Wang, M.W.; Yang, D.-H.; et al. Molecular Recognition of Two Endogenous Hormones by the Human Parathyroid Hormone Receptor-1. *Acta Pharmacol. Sin.* **2023**, *44*, 1227–1237. [CrossRef]
17. Velázquez-Libera, J.L.; Caballero, J.; Alzate-Morales, J.; Ruiz-Pernía, J.J.; Tuñón, I. Understanding the Interactions of Ubiquitin-Specific Protease 7 with Its Substrates through Molecular Dynamics Simulations: Insights into the Role of Its C-Terminal Domains in Substrate Recognition. *J. Chem. Inf. Model.* **2024**, *64*, 4134–4148. [CrossRef]
18. Ni, D.; Wei, J.; He, X.; Rehman, A.U.; Li, X.; Qiu, Y.; Pu, J.; Lu, S.; Zhang, J. Discovery of Cryptic Allosteric Sites Using Reversed Allosteric Communication by a Combined Computational and Experimental Strategy. *Chem. Sci.* **2021**, *12*, 464–476. [CrossRef]
19. Lu, S.; He, X.; Yang, Z.; Chai, Z.; Zhou, S.; Wang, J.; Rehman, A.U.; Ni, D.; Pu, J.; Sun, J.; et al. Activation Pathway of a G Protein-Coupled Receptor Uncovers Conformational Intermediates as Targets for Allosteric Drug Design. *Nat. Commun.* **2021**, *12*, 4721. [CrossRef]
20. Chai, X.; Hu, X.-P.; Wang, X.-Y.; Wang, H.-T.; Pang, J.-P.; Zhou, W.-F.; Liao, J.-N.; Shan, L.-H.; Xu, X.-H.; Xu, L.; et al. Computational Guided Discovery of Novel Non-steroidal AR-GR dual antagonists demonstrating potency against antiandrogen resistance. *Acta Pharmacol. Sin.* **2023**, *44*, 1500–1518. [CrossRef]
21. Chen, J.; Zeng, Q.; Wang, W.; Sun, H.; Hu, G. Decoding the Identification Mechanism of an SAM-III Riboswitch on Ligands through Multiple Independent Gaussian-Accelerated Molecular Dynamics Simulations. *J. Chem. Inf. Model.* **2022**, *62*, 6118–6132. [CrossRef] [PubMed]
22. Liu, H.-L.; Zhong, H.-Y.; Zhang, Y.-X.; Xue, H.-R.; Zhang, Z.-S.; Fu, K.-Q.; Cao, X.-D.; Xiong, X.-C.; Guo, D. Structural basis of tolvaptan binding to the vasopressin V2 receptor. *Acta Pharmacol. Sin.* **2024**, *45*, 2441–2449. [CrossRef] [PubMed]

23. Chen, J.; Wang, X.; Pang, L.; Zhang, J.Z.H.; Zhu, T. Effect of Mutations on Binding of Ligands to Guanine Riboswitch Probed by Free Energy Perturbation and Molecular Dynamics Simulations. *Nucleic Acids Res.* **2019**, *47*, 6618–6631. [CrossRef]
24. Wei, J.; Chen, F.; Lu, X.; Fan, J.; Li, M.; Huang, J.; Liu, N.; Zhang, J.; Chai, Z.; Lu, S. In Silico Identification and Experimental Validation of Long-Range Allosteric Inhibition of Staphylococcus Aureus Cas9 Catalytic Activity by An Anti-CRISPR Protein AcrIIA14. *Int. J. Biol. Macromol.* **2025**, *310*, 143324. [CrossRef] [PubMed]
25. Hu, M.; Li, P.; Li, M.; Li, W.; Yao, T.; Wu, J.-W.; Gu, W.; Cohen, R.E.; Shi, Y. Crystal Structure of a UBP-Family Deubiquitinating Enzyme in Isolation and in Complex with Ubiquitin Aldehyde. *Cell* **2002**, *111*, 1041–1054. [CrossRef]
26. Pettersen, E.F.; Goddard, T.D.; Huang, C.C.; Couch, G.S.; Greenblatt, D.M.; Meng, E.C.; Ferrin, T.E. UCSF Chimera—A Visualization System for Exploratory Research and Analysis. *J. Comput. Chem.* **2004**, *25*, 1605–1612. [CrossRef]
27. Salomon-Ferrer, R.; Case, D.A.; Walker, R.C. An Overview of the Amber Biomolecular Simulation Package. Wiley Interdiscip. *Rev. Comput. Mol. Sci.* **2013**, *3*, 198–210. [CrossRef]
28. Maier, J.A.; Martinez, C.; Kasavajhala, K.; Wickstrom, L.; Hauser, K.E.; Simmerling, C. Ff14SB: Improving the Accuracy of Protein Side Chain and Backbone Parameters from Ff99SB. *J. Chem. Theory Comput.* **2015**, *11*, 3696–3713. [CrossRef]
29. Wang, J.; Wolf, R.M.; Caldwell, J.W.; Kollman, P.A.; Case, D.A. Development and Testing of a General Amber Force Field. *J. Comput. Chem.* **2004**, *25*, 1157–1174. [CrossRef]
30. Wang, L.; Schauperl, M.; Mobley, D.L.; Bayly, C.; Gilson, M.K. A Fast, Convenient, Polarizable Electrostatic Model for Molecular Dynamics. *J. Chem. Theory Comput.* **2024**, *20*, 1293–1305. [CrossRef]
31. Olsson, M.H.M.; SØndergaard, C.R.; Rostkowski, M.; Jensen, J.H. PROPKA3: Consistent Treatment of Internal and Surface Residues in Empirical pKa Predictions. *J. Chem. Theory Comput.* **2011**, *7*, 525–537. [CrossRef] [PubMed]
32. Jorgensen, W.L.; Chandrasekhar, J.; Madura, J.D.; Impey, R.W.; Klein, M.L. Comparison of Simple Potential Functions for Simulating Liquid Water. *J. Chem. Phys.* **1983**, *79*, 926. [CrossRef]
33. Darden, T.; York, D.; Pedersen, L. Particle Mesh Ewald: An N.Long(N)Method for Ewald Sums in Large Systems. *J. Chem. Phys.* **1993**, *98*, 10089–10092. [CrossRef]
34. Ryckaert, J.-P.; Ciccotti, G.; Berendsen, H.J.C. Numerical Integration of the Cartesian Equations of Motion of a System with Constraints: Molecular Dynamics of n-Alkanes. *J. Comput. Phys.* **1977**, *23*, 327–341. [CrossRef]
35. Wu, X.; Brooks, B.R. Self-Guided Langevin Dynamics Simulation Method. *Chem. Phys. Lett.* **2003**, *381*, 512–518. [CrossRef]
36. Zhang, H.; Ni, D.; Fan, J.; Li, M.; Zhang, J.; Hua, C.; Nussinov, R.; Lu, S. Markov State Models and Molecular Dynamics Simulations Reveal the Conformational Transition of the Intrinsically Disordered Hypervariable Region of K-Ras4B to the Ordered Conformation. *J. Chem. Inf. Model.* **2022**, *62*, 4222–4231. [CrossRef]
37. Huang, J.-P.; Yun, S.-T.; Zhao, J.X.; Wang, X.T.; Wang, X.-C.; Guo, X.Y.; San, D.-M.; Zhou, Y.-X. The Two-step Strategy for Enhancing the Specific Activity and Thermostability of Alginate Lyase AlyG2 with Mechanism for Improved Thermostability. *Int. J. Biol. Macromol.* **2024**, *273*, 132685. [CrossRef]
38. Zhang, Y.; Dou, W.; Zhao, Z.; Li, G.; Li, C.; Chen, X.; Mou, L. Stereo-Selectivity of Enantiomeric Inhibitors to Ubiquitin-Specific Protease 7 (USP7) Dissected by Molecular Docking, Molecular Dynamics Simulations, and Binding Free Energy Calculations. *Mol. Divers.* **2025**, *29*, 1725–1735. [CrossRef]
39. Liu, J.; Jin, Q.; Xia, J.; Liu, L.; Liu, Y.; Jiang, L. Computational Insights into the Resistance of the Gatekeeper Mutation V564F of FGFR2 Against Pemigatinib. *J. Comput. Biophys. Chem.* **2024**, *23*, 891–900. [CrossRef]
40. Eargle, J.; Luthey-Schulten, Z. Network View: 3D Display and Analysis of Protein·RNA Interaction Networks. *Bioinformatics* **2012**, *28*, 3000–3001. [CrossRef]
41. Pan, F.; Wang, C.-N.; Yu, Z.-H.; Wu, Z.-R.; Wang, Z.; Lou, S.; Li, W.-H.; Liu, G.-X.; Li, T.; Zhao, Y.-Z.; et al. NADPHnet: A Novel Strategy to Predict Compounds for Regulation of NADPH Metabolism via Network-Based Methods. *Acta Pharmacol. Sin.* **2024**, *45*, 2199–2211. [CrossRef] [PubMed]
42. Humphrey, W.; Dalke, A.; Schulten, K. VMD: Visual Molecular Dynamics. *J. Mol. Graph.* **1996**, *14*, 33–38. [CrossRef] [PubMed]
43. Miao, Y.; Caliman, A.D.; McCammon, J.A. Allosteric Effects of Sodium Ion Binding on Activation of the M3 Muscarinic G-Protein-Coupled Receptor. *Biophys. J.* **2015**, *108*, 1796–1806. [CrossRef] [PubMed]
44. Lei, J.; Qi, R.; Tang, Y.; Wang, W.; Wei, G.; Nussinov, R.; Ma, B. Conformational Stability and Dynamic of the Cancer-associated Isoform Δ133p53β Are Modulated by p53 Peptides and p53-specific DNA. *FASEB J.* **2019**, *33*, 4225–4235. [CrossRef]
45. Roe, D.R.; Cheatham, T.E. PTRAJ and CPPTRAJ: Software for Processing and Analysis of Molecular Dynamics Trajectory Data. *J. Chem. Theory Comput.* **2013**, *9*, 3084–3095. [CrossRef]
46. Xu, G.; Wu, B.; Liang, W.; Zhang, J.; Xu, S.; Du, T.; Li, M.; Zheng, G.; Zhou, B.; Wang, Z.; et al. Computational Revealing Infigratinib Resistance to the FGFR2 N549H and N549K Mutations. *J. Exp. Nanosci.* **2025**, *20*, 2442305. [CrossRef]
47. Zheng, G.; Liu, W.; Kang, Y.; Xu, B.; Qiu, X.; Du, T.; Xu, S.; Chen, R.; Cheng, H.; Cai, C. Probing the Role of Zinc Ion in Metallo-β-Lactamase Inhibitor Binding by Using Multiple Molecular Dynamics Simulations. *Results Chem.* **2025**, *15*, 102171. [CrossRef]

48. Shao, J.; Tanner, S.W.; Thompson, N.; Cheatham, T.E. Clustering Molecular Dynamics Trajectories: 1. Characterizing the Performance of Different Clustering Algorithms. *J. Chem. Theory Comput.* **2007**, *3*, 2312–2334. [CrossRef]
49. Zhang, M.; Ao, J.; Liu, N.; Chen, T.; Lu, S. Exploring the Constitutive Activation Mechanism of the Class A Orphan GPR20. *Acta Pharmacol. Sin.* **2025**, *46*, 500–511. [CrossRef]
50. Li, M.; Lan, X.; Shi, X.; Zhu, C.; Lu, X.; Pu, J.; Lu, S.; Zhang, J. Delineating the Stepwise Millisecond Allosteric Activation Mechanism of the Class C GPCR Dimer mGlu5. *Nat. Commun.* **2024**, *15*, 7519. [CrossRef]
51. Basith, S.; Manavalan, B.; Lee, G. Unveiling Local and Global Conformational Changes and Allosteric Communications in SOD1 Systems Using Molecular Dynamics Simulation and Network Analyses. Comput. *Biol. Med.* **2024**, *168*, 107688. [CrossRef]
52. Zhang, Y.; Guo, J.; Liu, Y.; Qu, Y.; Li, Y.Q.; Mu, Y.; Li, W. An Allosteric Mechanism for Potent Inhibition of SARS-CoV-2 Main Proteinase. *Int. J. Biol. Macromol.* **2024**, *265*, 130644. [CrossRef]
53. Li, X.; Hou, C.; Yang, M.; Luo, B.; Mao, N.; Chen, K.; Chen, Z.; Bai, Y. The Effect of Phosphorylation on the Conformational Dynamics and Allostery of the Association of Death-associated Protein Kinase with Calmodulin. *J. Biolmol. Struct. Dyn.* **2024**, 1–9. [CrossRef]
54. Wang, Y.; Qiao, X.; Zhu, R.; Zhou, L.; Zhang, Q.; Lu, S.; Chai, Z. Computational Elucidation of a Monobody Targeting the Phosphatase Domain of SHP2. *Biomolecules* **2025**, *15*, 217. [CrossRef]

biomolecules

MDPI

Article

Computational Modeling Study of the Molecular Basis of dNTP Selectivity in Human Terminal Deoxynucleotidyltransferase

Egor O. Ukladov [1,2,†], Timofey E. Tyugashev [1,†] and Nikita A. Kuznetsov [1,2,*]

1 Institute of Chemical Biology and Fundamental Medicine, Siberian Branch of Russian Academy of Sciences, Novosibirsk 630090, Russia; e.ukladov@g.nsu.ru (E.O.U.); tyugashev@niboch.nsc.ru (T.E.T.)
2 Department of Natural Sciences, Novosibirsk State University, Novosibirsk 630090, Russia
* Correspondence: nikita.kuznetsov@niboch.nsc.ru; Tel.: +7-(383)-363-5175
† These authors contributed equally in this work.

Abstract: Human terminal deoxynucleotidyl transferase (TdT) can catalyze template-independent DNA synthesis during the V(D)J recombination and DNA repair through nonhomologous end joining. The capacity for template-independent random addition of nucleotides to single-stranded DNA makes this polymerase useful in various molecular biological applications involving sequential stepwise synthesis of oligonucleotides using modified dNTP. Nonetheless, a serious limitation to the applications of this enzyme is strong selectivity of human TdT toward dNTPs in the order dGTP > dTTP ≈ dATP > dCTP. This study involved molecular dynamics to simulate a potential impact of amino acid substitutions on the enzyme's selectivity toward dNTPs. It was found that the formation of stable hydrogen bonds between a nitrogenous base and amino acid residues at positions 395 and 456 is crucial for the preferences for dNTPs. A set of single-substitution and double-substitution mutants at these positions was analyzed by molecular dynamics simulations. The data revealed two TdT mutants—containing either substitution D395N or substitutions D395N+E456N—that possess substantially equalized selectivity toward various dNTPs as compared to the wild-type enzyme. These results will enable rational design of TdT-like enzymes with equalized dNTP selectivity for biotechnological applications.

Keywords: terminal deoxynucleotidyl transferase; TdT; enzyme; polymerase; dNTP; molecular dynamics; molecular modeling; rational design; DNA synthesis

1. Introduction

Terminal deoxynucleotidyl transferase (TdT) is responsible for template-independent random addition of nucleotides to single-stranded DNA during the V(D)J recombination and DNA repair via nonhomologous end joining [1,2]. TdT belongs to the X family of DNA polymerases, alongside polymerases X, β, λ, and μ. Its distinctive feature is the ability to catalyze template-independent DNA synthesis in the 5′→3′ direction; this property is attributed to the length and structure of TdT loop 1 acting as a "complementary base" by interacting with a dNTP and the DNA primer [3]. TdT is capable of catalyzing reactions with various divalent metal ions, such as Mg^{2+}, Mn^{2+}, and Co^{2+} [4].

TdT is currently widely utilized in biotechnological and molecular biological applications. Its first use was the rapid amplification of cDNA ends (RACE) method demonstrated in the 1980s [5]. In this technique, TdT is used to add a homopolymeric tail to the 3′ end of cDNA, thereby facilitating primer hybridization. Another method involving this enzyme is the TdT-dUTP nick end-labeling (TUNEL) assay [6], which allows us to attach dUTP to 3′ ends of DNA at sites of strand breaks. Subsequently, these sites are visualized with antibodies against dUTP. This approach helps us to detect DNA fragmentation during apoptosis. Nonetheless, one of the most promising applications of TdT is thought to be chemical–enzymatic synthesis of DNA oligonucleotides. The synthesis of a desired DNA sequence from 3′-protected dNTPs has already been demonstrated, e.g., from 3′-O-(2-nitrobenzyl)-2′-deoxyribonucleoside triphosphates [7] or benzoyl/pivaloyl derivatives [8]. In 2018, DNA synthesis using conjugates of TdT and dNTPs was reported, where the enzyme can attach a nucleotide to a primer followed by the detachment of the TdT molecule [9]. Besides, TdT has found applications in more exotic areas such as the development of DNA origami and nanostructures, DNA biosensors, DNA-based information storage, and aptamer synthesis [10].

Nevertheless, a major limitation of all the aforementioned methods is the pronounced selectivity of TdT toward various dNTPs. For human TdT, in the presence of Mg^{2+} ions, the selectivity can be ranked as dGTP > dTTP ≈ dATP > dCTP [4]. By means of an analysis of X-ray crystallographic structures of TdT from *Mus musculus* [3], alignment of TdT orthologs [11], and molecular dynamics (MD) simulations of pre- and post-catalytic complexes of wild-type human TdT [4], amino acid residues Asp395 and Glu456 have been found to be potentially capable of stable interaction with the nitrogenous base of a nucleotide being inserted. Therefore, these residues may play an important role in the dNTP selectivity. Unfortunately, residue Asp395 is poorly resolved in crystal structures mimicking pre- or post-catalytic complexes between TdT and single-stranded DNA primers. In complexes of TdT with double-stranded DNA, Asp395 is engaged in interactions either with the 5′-OH group of a 3′-terminal nucleotide of a template strand (PDB ID 5D46) or with the 3′-flanking guanine base of the template strand (PDB ID 5D49) [12]. During alanine scanning, it was found that substitution of D395A leads to a slight reduction in polymerase activity, and dGTP incorporation occurred more slowly as compared to the three other nucleotides [13]. All data obtained up to now indicate the involvement of Asp395 in the substrate selectivity of TdT. The other residue, Glu456, is less well studied; however, it is known that positional homologs of Glu456 in the PolX family of enzymes interact with the base of an incoming dNTP [14], but the type of the residue at this position varies widely across the entire PolX family [11]. Therefore, the objective of this study was to determine the role of amino acid residues of the active site and loop 1 in the catalytic activity of TdT by in silico mutagenesis and to assess the impact of relevant mutations through MD simulations.

2. Results and Discussion

In our previous report [4], through experimental analysis of the purified wild-type enzyme and its mutants, it was found that substitutions of Asp395 and Glu456 in human TdT significantly influence the enzymatic activity. Notably, an analysis of MD trajectories of the wild-type-protein–DNA–dNTP–Mg^{2+} complex in that work revealed stable hydrogen bonds between dGTP (the most readily incorporable dNTP) and these two residues (Figure 1). Therefore, to assess the influence of substitutions of Asp395 and Glu456 on the selection of dNTPs and enzymatic activity, in the present study, we performed detailed in silico analysis.

Aspartic and glutamic acids, owing to their side-chain functional groups, can act only as acceptors of hydrogen bonds. Therefore, it was hypothesized that replacing them with asparagine or glutamine (which are similar in size but capable of acting as acceptors and donors of hydrogen bonds) would lead to changes in the enzyme's selectivity. In addition, positional substitution D395E was found in approximately a quarter of all analyzed vertebrate species [11]. Unique substitutions have also been found: D395N in the amphibian *Ambystoma mexicanum* and D395K+E456Q in the fish *Cottoperca gobio*, respectively. Therefore, a set of single-substitution and double-substitution mutants of human TdT was designed for in silico analysis: D395E, D395N, D395Q, E456N, E456Q, D395N+E456N, D395Q+E456Q, D395Q+E456N, and D395K+E456Q.

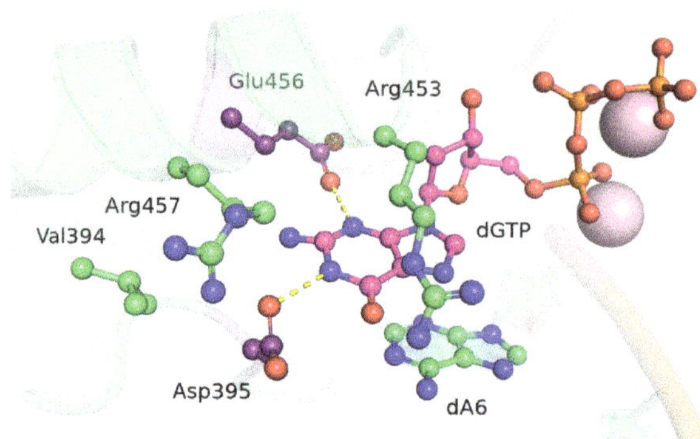

Figure 1. Structure of the active site in the precatalytic complex of TdT with dGTP. Residues forming the most stable hydrogen bonds (yellow dashed lines) with the nitrogenous base of dGTP (pink) are highlighted in violet.

2.1. Substitution of Asp395 with Negatively Charged Amino Acid Residue Glu

Substitution D395E—found in approximately one-quarter of vertebrate species [11]—did not yield results distinguishable from the wild type. In the precatalytic complex of TdT, the Glu395 residue was found to form stable hydrogen bonds with the adenine amino group (Figure 2a) and the N1 atom and exocyclic amino group of guanine (Figure 2b), while not interacting with pyrimidine bases (Figure 2c,d). This outcome was expected because pyrimidines are less stabilized by stacking interactions, with thymine primarily acting as a hydrogen bond acceptor through atoms O_2 and O_4.

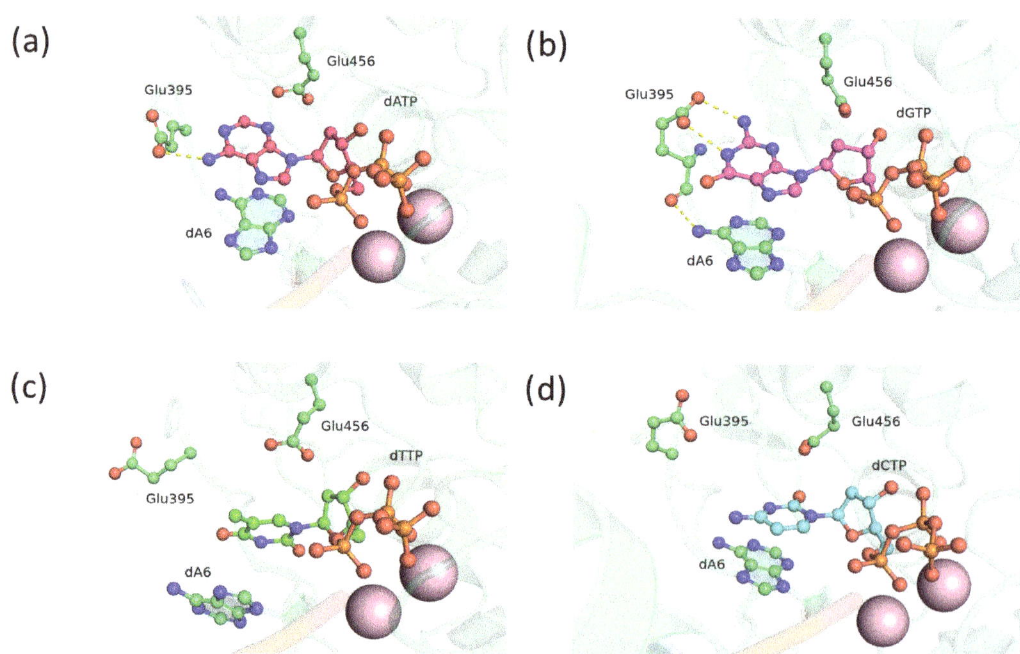

Figure 2. Precatalytic complexes of TdT D395E with (**a**) dATP, (**b**) dGTP, (**c**) dTTP, or (**d**) dCTP. Side chains of amino acid residues 395 and 456 and residues forming hydrogen bonds with them are visualized. Hydrogen bonds are depicted with yellow dashed lines.

2.2. Substitution of Asp395 with Uncharged Amino Acid Residues Asn and Gln

Next, we focused on asparagine and glutamine substitutions at position 395. First, substitutions of Asp395 with Asn and Gln were modeled and analyzed. In the precatalytic complex of TdT D395Q, residue Gln395 forms bonds with atoms N1 and O6 of guanine (Figure 3b), N3 of thymine (Figure 3c), or the N3 atom and amino group of cytosine (Figure 3d). Additionally, stable flipped positionings of Gln395 exist, where the residue does not have any hydrogen bonds with the nitrogenous base of adenosine (Figure 3a). Nevertheless, the D395Q mutant, as anticipated, is capable of entering into hydrogen bonds with nitrogenous bases as an acceptor and donor.

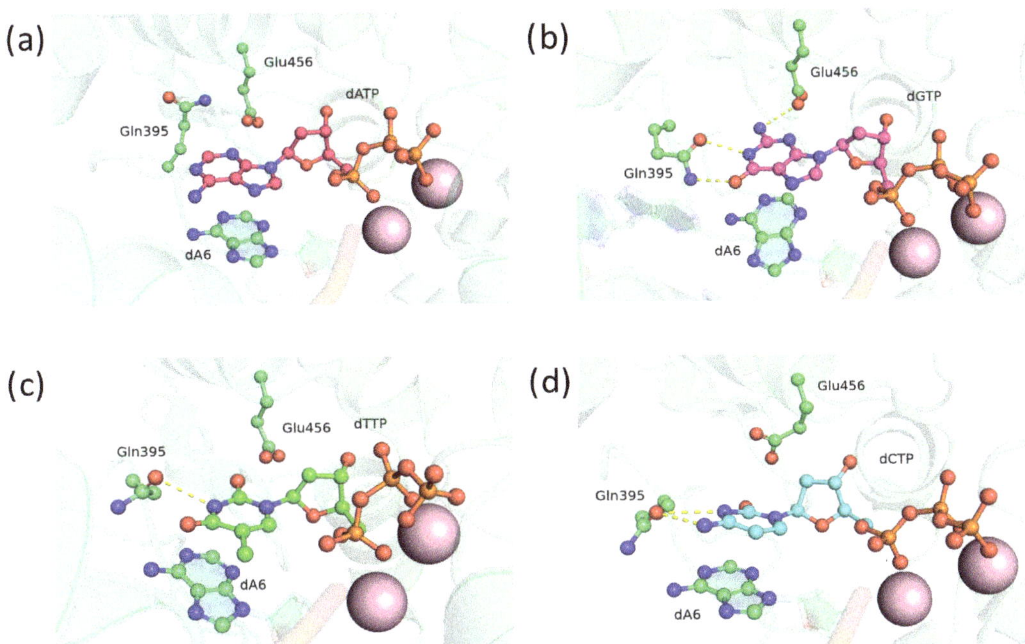

Figure 3. Precatalytic complexes of TdT D395Q with (**a**) dATP, (**b**) dGTP, (**c**) dTTP, or (**d**) dCTP. Side chains of amino acid residues 395 and 456 and residues forming hydrogen bonds with them are visualized. Hydrogen bonds are indicated by yellow dashed lines.

In the precatalytic complex of TdT D395N, the Asn395 residue has bonds with the N1 and amino group of adenine (Figure 4a), the N1 atom and amino group of guanine (Figure 4b), N3 and O4 of thymine (Figure 4c), or with the amino group of cytosine (Figure 4d). Apparently, due to the smaller size of the asparagine side chain, it is less prone to flipping and breaking hydrogen bonds with the nitrogenous bases as compared to glutamine. Besides, asparagine forms hydrogen bonds with the 3′-terminal adenine through the oxygen of the peptide bond and the amide group, and this arrangement may further stabilize the relative positioning of the nucleotide being embedded and the 3′-terminal nucleotide. Substitution of D395N gave slightly better results than substitution of D395Q, probably because of the difference in the size of their side chains.

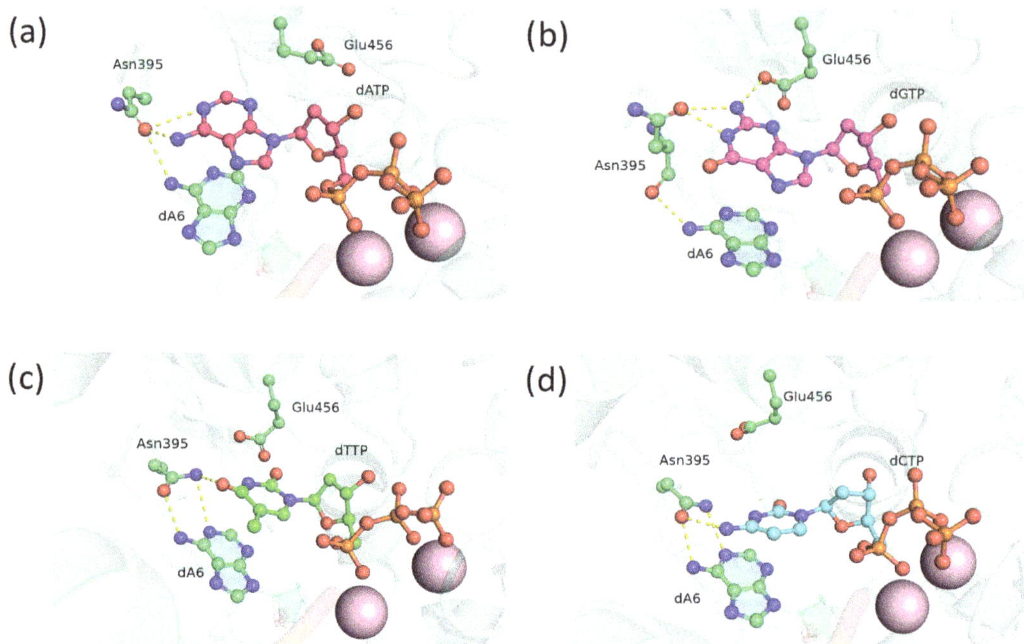

Figure 4. Precatalytic complexes of TdT D395N with (**a**) dATP, (**b**) dGTP, (**c**) dTTP, or (**d**) dCTP. Side chains of amino acid residues 395 and 456 and residues forming hydrogen bonds with them are visualized. Hydrogen bonds are indicated by yellow dashed lines.

2.3. Glu456 Substitution with Uncharged Amino Acid Residues Gln and Asn

Next, an in silico mutational analysis of glutamic acid at position 456 was conducted. After replacement of Glu456 with Gln in the precatalytic complex, residue Gln456 appears to be too bulky and does not rotate toward the nitrogenous bases, resulting in minimal hydrogen bonding with the bases (Figure 5a,c,d), except for guanine, which is the most easily incorporable (Figure 5b). The original Asp395 formed hydrogen bonds with adenine, guanine, or cytosine as acceptors (Figure 5a,b,d) while also interacting with the 3′-terminal adenine (Figure 5b,c).

For asparagine mutant E456N, in the precatalytic complex, residue Asn456 has bonds with the N_3 atom of adenine (Figure 6a), the amino group of guanine (Figure 6b), O_2 of thymine (Figure 6c), or O_2 of cytosine (Figure 6d). Unlike glutamine, asparagine, owing to its smaller size, is capable of easily rotating its side chain and entering into stable hydrogen bonds with all types of DNA nitrogenous bases. For position 456, the difference in the side-chain size proved to be more critical. This effect may be attributed to the fact that residue 456 is located in a stable α-helix, while residue 395 is in unstructured loop 1. Nevertheless, for both positions 395 and 456, substitution with asparagine is more conducive to equalized dNTP selectivity.

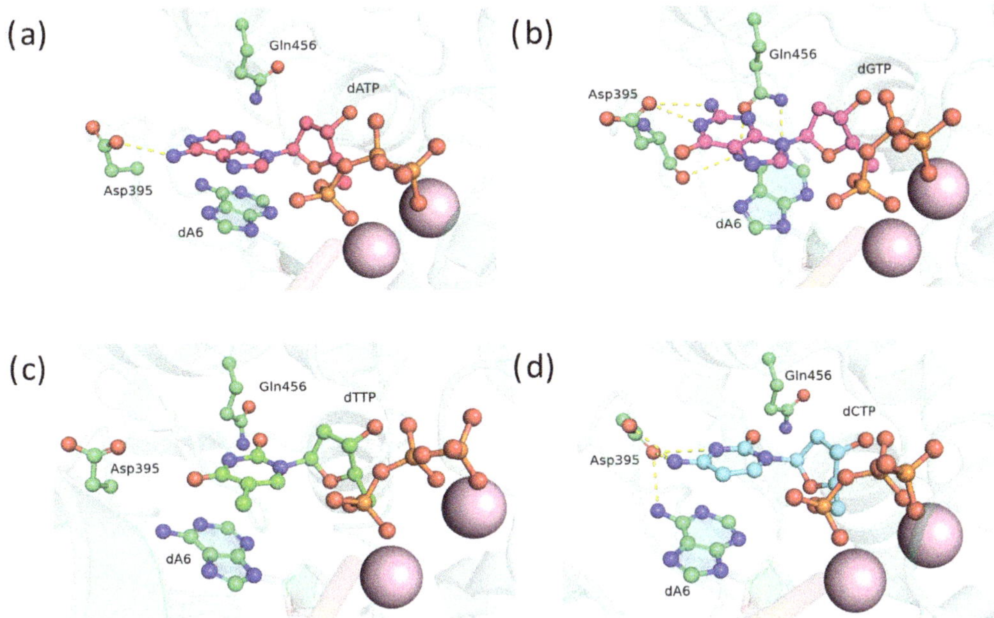

Figure 5. Precatalytic complexes of TdT E456Q with (**a**) dATP, (**b**) dGTP, (**c**) dTTP, or (**d**) dCTP. Side chains of amino acid residues 395 and 456 and the residues forming hydrogen bonds with them are visualized. Hydrogen bonds are shown as yellow dashed lines.

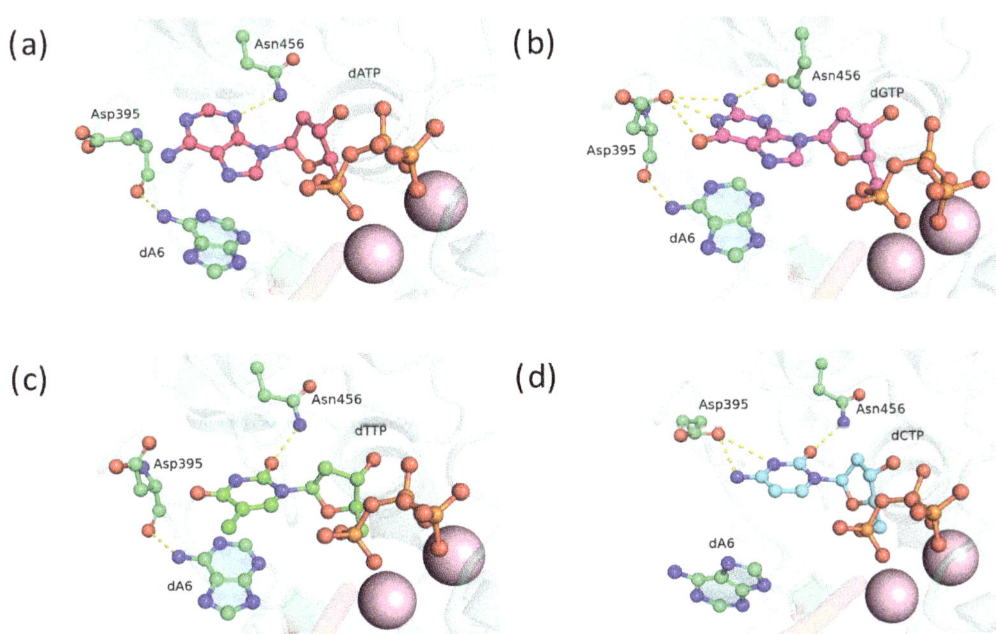

Figure 6. Precatalytic complexes of TdT E456N with (**a**) dATP, (**b**) dGTP, (**c**) dTTP, or (**d**) dCTP. Side chains of amino acid residues 395 and 456 and the residues engaged in hydrogen bonds with them are visualized. Hydrogen bonds are shown as yellow dashed lines.

2.4. The Double Substitution of Asp395 and Glu456

Because of the data on the effects of stand-alone substitutions of Asp395 and Glu456 with glutamines or asparagines, several double substitutions became interesting: D395K+E456Q, D395Q+E456Q, D395Q+E456N, and D395N+E456N. We hypothesized that the combined effect of two substitutions might be greater than the effect of single ones.

First, mutant D395K+E456Q was analyzed (the substitutions are analogous to those in the TdT of the fish *C. gobio*; Figure 7). Unlike the wild type, where Glu395 forms hydrogen bonds with the nitrogenous base of an incoming nucleotide, the side chain of Lys395 is turned out, thus engaging in stable interactions with Glu180 and Glu185 (Figure 7). Such reorientation should potentially stabilize the positioning of loop 1 (which is responsible for template-independent properties of the enzyme) by means of an additional ionic bridge with the protein backbone. Nevertheless, due to the oxygen atom in the peptide bond, in some MD simulations, Lys395 started to occasionally form hydrogen bonds with the amino group of the adenosine that is located at the 3′ end of the DNA primer (Figure 7a–c).

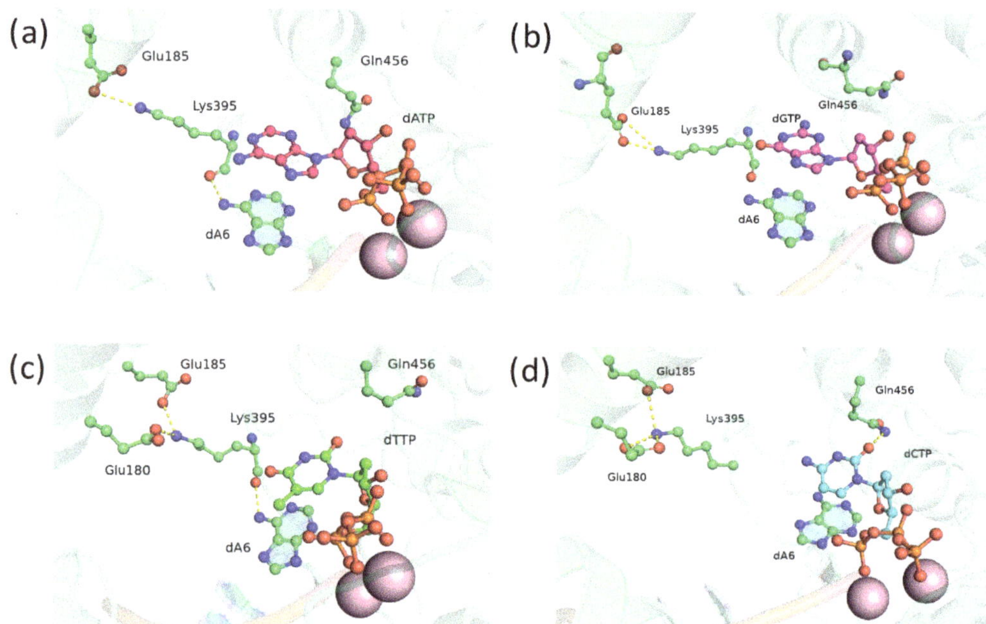

Figure 7. Precatalytic complexes of TdT D395K+E456Q with (**a**) dATP, (**b**) dGTP, (**c**) dTTP, or (**d**) dCTP. Side chains of amino acid residues 395 and 456 and residues forming hydrogen bonds with them are visualized. Hydrogen bonds are presented as yellow dashed lines.

Residue Gln456 appeared to be excessively bulky and rarely tilted toward a dNTP. Nevertheless, in some cases, hydrogen bond formation between the side chain of Gln456 and the base of a dNTP was observed (Figure 7b,d). Moreover, Gln456 acted as an acceptor with dGTP (Figure 7b) and as a donor with dCTP (Figure 7d).

After that, we analyzed the potential ability of a positively charged amino acid residue at position 395 to stabilize the positioning of loop 1. For this purpose, we modeled the complex of the D395R mutant with dGTP, which is the most easily attachable (Figure 8). Indeed, similarly to lysine, residue Arg395 entered into a hydrogen bond with the amino group of the 3′-terminal adenosine via the oxygen of the peptide bond. Additionally, Arg395 engaged in a greater number of interactions with residues Glu180 and Glu185. Therefore, the arginine residue at position 395 may stabilize loop 1 more effectively than lysine can.

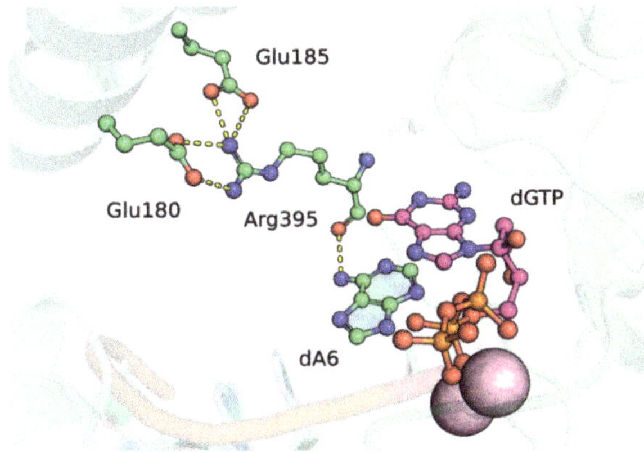

Figure 8. The precatalytic complex of TdT D395R with dGTP. Residues forming hydrogen bonds are visualized. Hydrogen bonds are indicated by yellow dashed lines.

Thus, it can be concluded that contacts between loop 1 and the rest of the protein may be important for maintaining the catalytically competent state of the complex of TdT with dNTP and DNA. On the other hand, besides the correct positioning of loop 1 for catalysis, stabilization of the incoming dNTP by hydrogen bonds between it and the protein is necessary, as evidenced by the lower activity of D395K+E456Q compared to the wild type [11].

In the precatalytic complex of double mutant TdT D395Q+E456Q, residue Gln395 formed stable hydrogen bonds with the N1 atom and amino group of adenine (Figure 9a), O_6 and N_1 of guanine (Figure 9b), the N_3 atom of thymine (Figure 9c), or the amino group of cytosine (Figure 9d). Furthermore, Gln395 was capable of entering into a hydrogen bond with the amino group of the 3′-terminal adenine through the oxygen of the peptide bond (Figure 9b,d). Despite its size, residue Gln456 had hydrogen bonds with the amino group of guanine (Figure 9b) or the O_2 atom of thymine (Figure 9c). Apparently, the presence of two glutamine residues at positions 395 and 456 increased the chances of both residues' forming hydrogen bonds with deoxynucleotide analogs, thereby contributing to a stabler positioning of the nucleotide in the active site of TdT.

Of all nine mutations potentially altering substrate selectivity of human terminal deoxynucleotidyl transferase, the most interesting result was obtained with the double asparagine substitution in mutant D395N+E456N (Figure 10). The asparagine residue was small enough to freely turn toward the dNTP at the 456th position, while not engaging in stable interactions with the protein backbone at the 395th position. Additionally, the amide group in the asparagine side chain can act not only as a hydrogen bond acceptor (Figure 10b) but also as a donor (Figure 10a,d) owing to the nitrogen atom.

Residue Asn395 formed stable hydrogen bonds with all types of nitrogenous bases of dNTP: with the adenine amino group (Figure 10a), the N1 atom of guanine (Figure 10b), N_3 and O_4 of thymine (Figure 10c), or the N_3 atom and amino group of cytosine (Figure 10d). Moreover, Asn395 is able to form two hydrogen bonds with nitrogenous bases (Figure 10b,d). In addition, Asn395 frequently interacts with the 3′-terminal adenine of the DNA primer (Figure 10a,c,d).

Residue Asn456 almost always turned toward dNTP and formed a stable hydrogen bond with it as an acceptor in case of dGTP and dCTP (Figure 10b,d), or as a donor with dATP and dTTP (Figure 10a,c). Asn456 entered into bonds with N_3 of adenine (Figure 10a), the amino group of guanine (Figure 10b), or the O_2 atom of thymine or cytosine (Figure 10c,d). The formation of two hydrogen bonds with the nitrogenous base was not observed for Asn456.

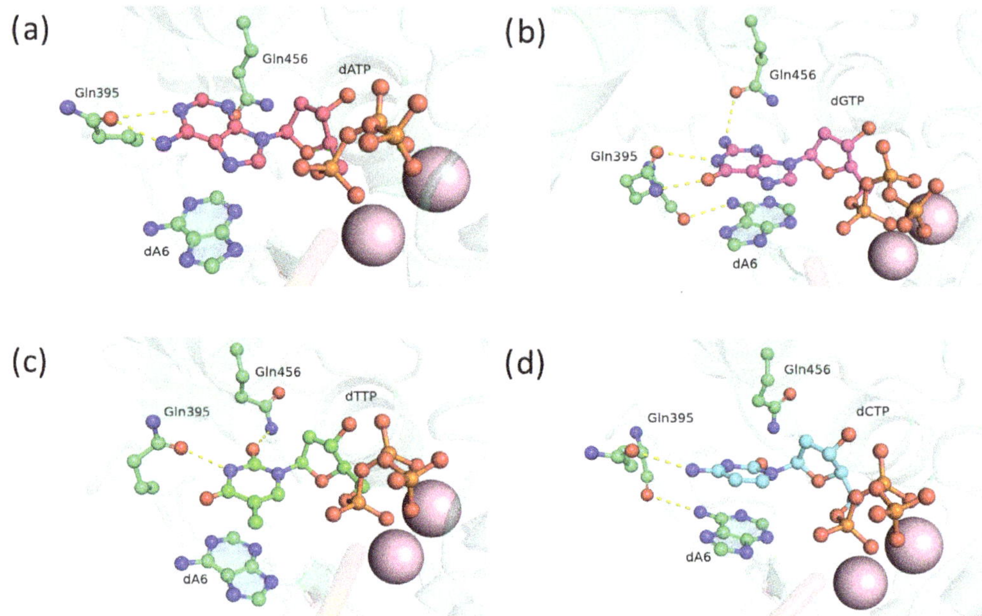

Figure 9. Precatalytic complexes of TdT D395Q+E456Q with (**a**) dATP, (**b**) dGTP, (**c**) dTTP, or (**d**) dCTP. Side chains of amino acid residues 395 and 456 and the residues forming hydrogen bonds with them are visualized. Hydrogen bonds are shown as yellow dashed lines.

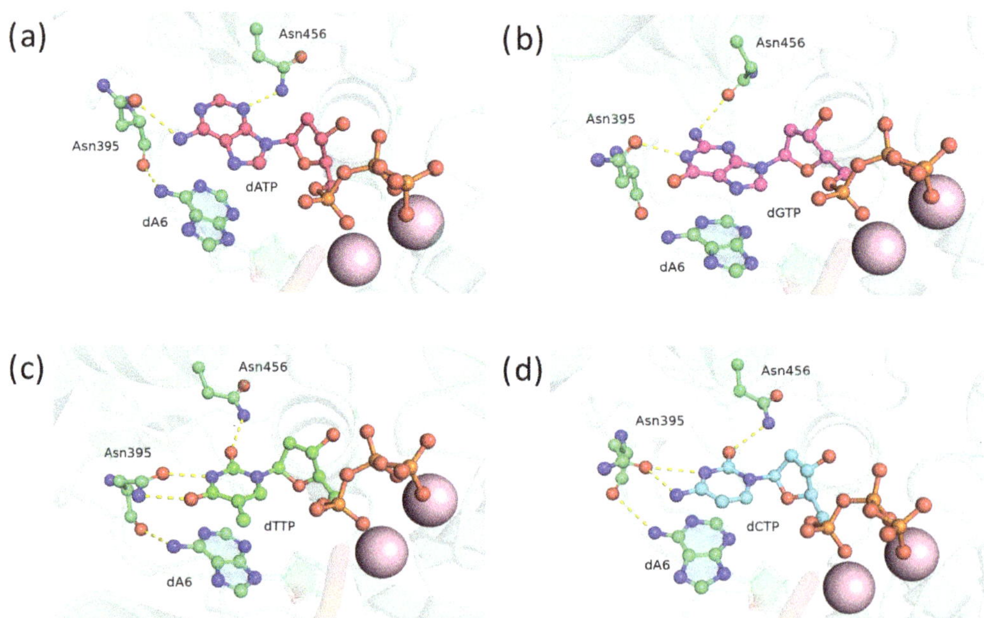

Figure 10. Precatalytic complexes of TdT D395N+E456N with (**a**) dATP, (**b**) dGTP, (**c**) dTTP, or (**d**) dCTP. Side chains of amino acid residues 395 and 456 and residues forming hydrogen bonds with them are visualized. Hydrogen bonds are indicated by yellow dashed lines.

On the basis of the findings about paired glutamine and asparagine substitutions, we hypothesized that residue 395 should be replaced by Gln to form stabler bonds with pyrimidine nucleotides, while residue 456 should be replaced by shorter Asn to more likely rotate toward a dNTP. Accordingly, mutant D395Q+E456N was simulated and analyzed.

Residue Gln395 formed bonds with the amino group of adenine (Figure 11a), N1 and O6 atoms of guanine (Figure 11b), N3 of thymine (Figure 11c), or the amino group of cytosine (Figure 11d). Similarly to the single mutant, Gln395 is capable of forming a bond with the amino group of the 3′-terminal adenine (Figure 11b). Residue Asn456 entered into hydrogen bonds with the amino group of guanine (Figure 11b) or the O_2 atom of thymine or cytosine (Figure 11c,d). Thus, we demonstrated that the proposed amino acid substitutions in principle can form hydrogen bonds with the nitrogenous base of dNTPs.

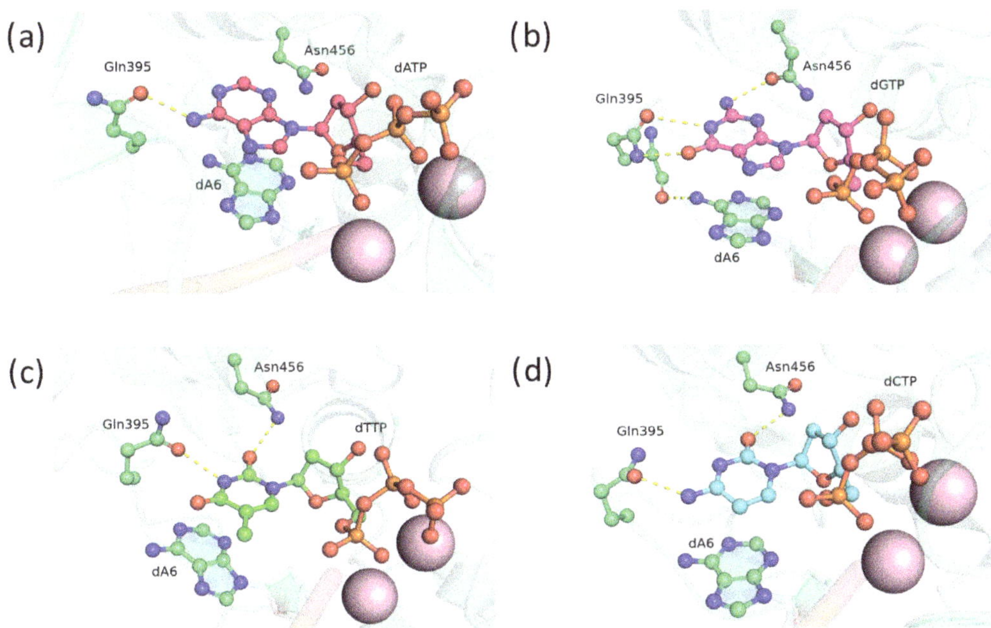

Figure 11. Precatalytic complexes of TdT D395Q+E456N with (**a**) dATP, (**b**) dGTP, (**c**) dTTP, or (**d**) dCTP. Side chains of amino acid residues 395 and 456 and residues forming hydrogen bonds with them are visualized. Hydrogen bonds are indicated by yellow dashed lines.

2.5. Average Number of Hydrogen Bonds between the Protein and the Nucleobase of an Incoming dNTP

Previously [4], using the MMPBSA approach for free-energy estimates of pre- and post-catalytic complexes formed by the wild-type enzyme, it was found that due to the highly charged nature of both ligands and protein-active sites, free-energy estimates have significant fluctuations and thereby are a very inconclusive parameter for characterizing the stability of the modeled complexes. At the same time, the number of hydrogen bonds between the active site residues and the nucleobase as well as H-bond lifetime parameter, which together provided the average number of H-bonds within the simulation trajectory, revealed correlations with the experimental results. Thereby, this parameter was selected as the in silico properties predictor within the present study of mutant forms of the enzyme.

To assess the quantitative contribution of the proposed substitutions, the average number of hydrogen bonds between the protein and the base of an embedded or 3′-terminal nucleotide was evaluated for pre- and post-catalytic complexes, respectively (Figure 12). First of all, one can clearly see the huge excess (more than 5-fold) of hydrogen bonds formed by the wild-type enzyme with dGTP as compared to the other nucleotides. Nevertheless, other substantial differences in the H-bond network were documented for some mutants in their precatalytic complexes (Figure 12a). For instance, for single mutant D395N interacting with dATP, the number of hydrogen bonds was 0.76 compared to 0.29 for the wild type, and with dCTP, the number of hydrogen bonds was 1.28 compared to 0.26 for the wild type. Mutant D395Q also manifested increased dCTP selectivity: the number of bonds with dCTP was 0.73 compared to 0.26 for the wild type, but no significant differences were found for complexes with other dNTPs. On average, E456N formed 1.13 hydrogen bonds with dATP, and 1.12 with dCTP. From this finding, it can be concluded that single asparagine substitutions D395N and E456N yielded better results as compared to glutamine substitutions D456Q and E456Q.

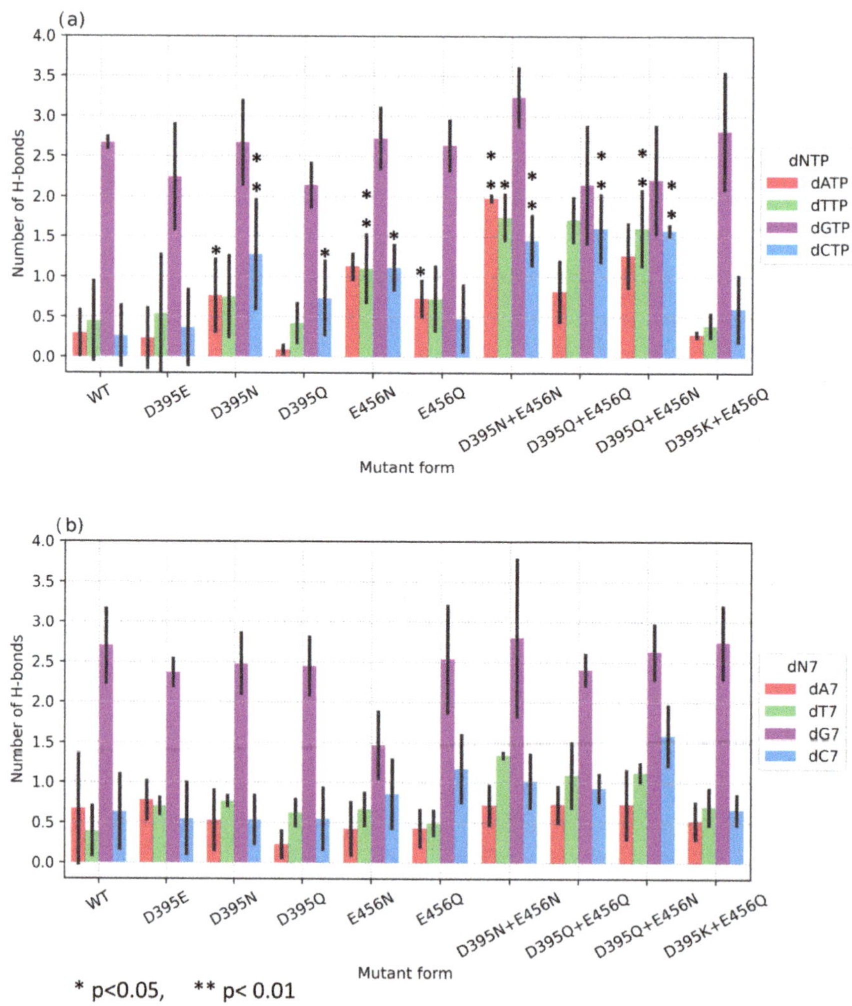

Figure 12. Average numbers of hydrogen bonds between dNTP or 3′-terminal nucleotide and amino acid residues of the protein for (**a**) a precatalytic complex and (**b**) a post-catalytic complex. Mutant enzymes are plotted on the X-axis. The average number of hydrogen bonds for three or more independent replicates of each complex is plotted on the Y-axis. Error bars represent the standard deviation among replicates.

Much more pronounced equalization of dNTP selectivity was achieved with all double mutants, which showed an increase in the number of H-bonds with dATP, dCTP, and dTTP, but not dGTP (Figure 12a). Statistically significant differences were found for most mutants, except for natural variant D395K+E456Q, when compared with the wild-type enzyme. For instance, for D395Q+E456Q, the average number of hydrogen bonds with dCTP was 1.61. Double substitution D395Q+E456N caused an increase in the average number of hydrogen bonds with dATP up to 1.27, and for dCTP up to 1.58. Mutant D395N+E456N showed the most statistically significant differences for all three deoxynucleotides by achieving 1.98 H-bonds with dATP, 1.45 with dCTP, and 1.74 with dTTP.

Notably, for post-catalytic complexes, no significant differences were found, indicating a lower tendency of residues 395 and 456 to form hydrogen bonds with an already embedded nucleotide. The decrease in stability of the positioning of the 3′-terminal nucleotide in the active site is necessary for the enzyme to transition to the next catalytic action and to avoid substrate inhibition.

Additionally, it could be noted that enzyme efficacy depends not only on the amino acid residues conformation and contacts, but also conformation of the 2′-deoxyribose moiety of 3′-terminal nucleotide in the primer. The crystal structures of catalytic intermediates suggest that complex formation happens with 3′-terminal primer 2′-deoxyribose in C2′-endo conformation and a shift to C3′-endo conformation occurs prior to the catalytic step. The simulations revealed that the 3′-terminal nucleotide 2′-deoxyribose samples both C3′-endo and C2′-endo, as well as other states, like O4′-endo. However, current molecular dynamics force fields are known to disfavor C3′-endo conformation in DNA [15,16], even in simulations of protein–DNA complexes where contact with the protein enforces the C3′-endo conformation [17]. While our simulations reveal some effects on the conformation of the 2′-deoxyribose moiety of 3′-terminal nucleotide, explicitly considering them would be a dubious effort due to aforementioned force-field deficiencies.

2.6. *Average Number of Hydrogen Bonds between the Protein and the Nucleobase of a 3′-Terminal Nucleotide*

Another necessary condition for the catalytic reaction to proceed is a proper mutual arrangement of planes of nitrogenous bases of an incoming dNTP and 3′-terminal nucleotide. Therefore, hydrogen bonds with the 3′-terminal nucleotide (adenosine in the simulated models) were calculated for pre- and post-catalytic complexes (Figure 13).

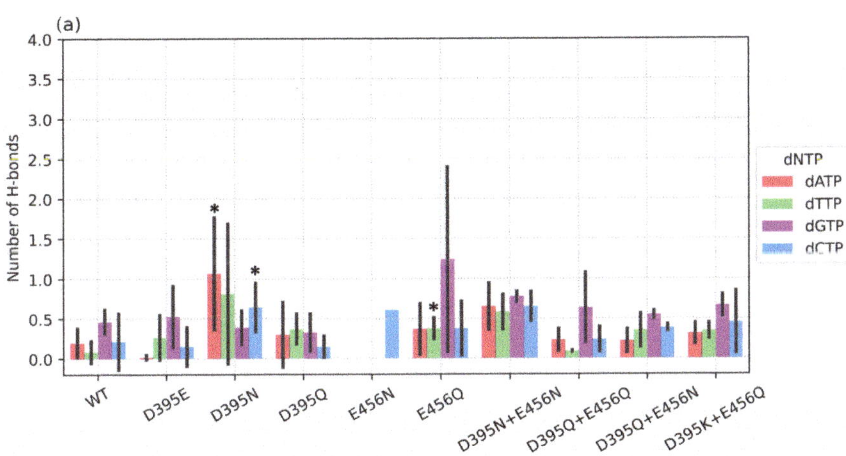

Figure 13. *Cont.*

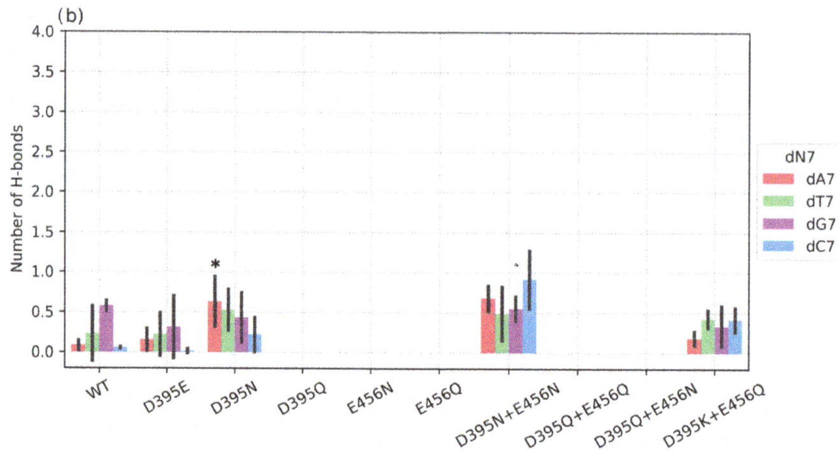

Figure 13. The average number of hydrogen bonds between the 3′-terminal nucleotide and amino acid residues in a pre- (**a**) or post-catalytic (**b**) complex.

It should be noted that total number of hydrogen bonds with a 3′-terminal nucleotide was much lower compared with the incoming dNTP. Nevertheless, statistically significant increases in the number of hydrogen bonds for the precatalytic complex were registered for D395N with dATP and dCTP (Figure 13a) and for D395N+E456N with dTTP (Figure 13a). From this result, it can be concluded that substitution D395N leads to an increase both in the number of hydrogen bonds with the base of an embedded dNTP and in the number of hydrogen bonds with the 3′-terminal adenine. Moreover, in post-catalytic complexes of D395N and of D395N+E456N, an increased level of H-bonding was found, too (Figure 13b), indicating a greater role of amino acid residue 395—compared to position 456—in the formation of hydrogen bonds with both dNTP and DNA termini.

3. Materials and Methods

3.1. The Protein Model

Pre- and post-catalytic complexes of human TdT were examined by homology modeling with Modeller [18] by means of X-ray structures of mouse TdT complexes representing pre- and post-catalytic states [3] as well as time-resolved X-ray structures of a catalytic complex of human TdT [19].

The initial structures of pre- and post-catalytic complexes were constructed for the wild-type enzyme and for corresponding mutants (D395E, D395Q, D395N, E456Q, E456N, D395Q+E456Q, D395Q+E456N, D395N+E456N, and D395K+E456Q) in complex with either hexameric DNA (5′-GGAAGA-3′) and four different dNTPs or heptameric DNA (5′-GGAAGAN-3′) and various 3′-terminal nucleotides, respectively. Additionally, an elongated complex was modeled for pyrimidine nucleotides: mutant D396N+E456N interacting with either dTTP and 5′-GGAAGT-3′ or dCTP and 5′-GGAAGC-3′.

3.2. Modeling Parameters

MD simulations were carried out using the GROMACS software package [20] with AMBER force-field parameters version 14SB for the protein [21] and OL15 for DNA [22]. The TIP3P water model was employed with respective ion parameters [23,24].

Parameterization of dNTPs was performed similarly [25], utilizing the RED server for charge optimization [26–28]. Magnesium ions in the enzyme's active site were described via a distributed charge model [29]. Protonation states of protein amino acid residues were determined from data contained within the literature and stored on the H++ server [30].

Energy minimization was performed until a maximum force threshold of 500 kJ/(mol·nm) was reached, with a total of 50,000 steps and a step size of emstep = 0.01. After energy minimization, two subsequent equilibrations were conducted in NVT and NPT ensembles, respectively. The parameters for the NVT ensemble were a total time of 1 ns (dt = 0.002 ps, n-steps = 500,000), a V-rescale thermostat with equilibration and velocity generation temperatures set to 300 K, and constraints on hydrogen bond vibrations applied by the LINCS method.

For the NPT ensemble, the parameters were as follows: a total time of 1 ns (dt = 0.002 ps, n-steps = 500,000), the V-rescale thermostat [31], the Parrinello–Rahman barostat [32], with constraints on hydrogen bond vibrations applied by the LINCS method.

The MD simulations were conducted with the same parameters as in the NPT ensemble, except with a duration of 100 ns (dt = 0.002 ps, n-steps = 50,000,000). At least three replicates were obtained for each model.

3.3. Data Processing

Data extraction from MD trajectories was performed with the help of standard functions of the GROMACS software. Compilation of summary tables was carried out via bash scripting commands. Python 3.10 programming language with numpy, pandas, matplotlib.pyplot, seaborn, and scipy.stats modules was utilized for graph plotting and statistical analysis. The Shapiro–Wilk test was employed to determine the statistical significance.

Representative frames from molecular trajectories were selected using clustering of frames based on the nitrogenous base of a nucleotide being inserted or the 3′-terminal nucleotide for pre- and post-catalytic complexes, respectively, as well as amino acid residues of the active site. The threshold for the root mean square deviation used for clustering was set to 0.13 nm.

4. Conclusions

A detailed analysis of the molecular interactions that ensure substrate selectivity of human terminal deoxynucleotidyl transferase was carried out in this study. MD models of pre- and post-catalytic complexes were obtained for the wild-type enzyme and a set of single-substitution and double-substitution mutants. MD simulations of mutants featuring substitution D395E, D395Q, D395N, E456Q, or E456N or double substitution D395Q+E456Q, D395Q+E456N, or D395N+E456N enabled us to deeply analyze the interactions of this enzyme with each dNTP and to find key amino acid residues participating in the dNTP selection. It was found that stabilization of the nucleobase of an incoming dNTP by an H-bond network may be crucial for dNTP selectivity. Among all tested mutants of TdT, the formation of an additional hydrogen bond in the precatalytic complex between amino acid residues and dATP, dTTP, or dCTP was observed for the mutants carrying substitution D395N or E456N. Moreover, two additional hydrogen bonds were found in cases of double mutants D395Q+E456Q, D395Q+E456N, and D395N+E456N. It was shown that TdT mutant D395N+E456N forms the most stable hydrogen bonds between the introduced asparagine residues and each dNTP.

Finally, it can be concluded that we successfully analyzed the factors that alter the dNTP selectivity of human terminal deoxynucleotidyl transferase, and these results will facilitate the rational design of TdT-like enzymes with equalized dNTP selectivity for biotechnological applications.

Author Contributions: E.O.U. and T.E.T. conceived and designed the experiments; E.O.U. and T.E.T. conducted the experiments; E.O.U., T.E.T. and N.A.K. analyzed the data; E.O.U., T.E.T. and N.A.K. writing—original draft preparation; E.O.U., T.E.T. and N.A.K. writing—review and editing; N.A.K. supervision; N.A.K. project administration; N.A.K. funding acquisition. All authors have read and agreed to the published version of the manuscript.

Funding: This work was supported by the Ministry of Science and Higher Education of the Russian Federation, agreement No. 075-15-2022-263.

Institutional Review Board Statement: Not applicable.

Informed Consent Statement: Not applicable.

Data Availability Statement: Data are available from N.A.K. and T.E.T. upon request. Tel.: +7-(383)-363-5174, e-mail: nikita.kuznetsov@niboch.nsc.ru (N.A.K.); tyugashev@niboch.nsc.ru (T.E.T.).

Acknowledgments: Computational resources were provided by the Computational Center of Novosibirsk State University.

Conflicts of Interest: The authors declare no conflicts of interest.

References

1. Mahajan, K.N.; Gangi-Peterson, L.; Sorscher, D.H.; Wang, J.; Gathy, K.N.; Mahajan, N.P.; Reeves, W.H.; Mitchell, B.S. Association of Terminal Deoxynucleotidyl Transferase with Ku. *Proc. Natl. Acad. Sci. USA* **1999**, *96*, 13926–13931. [CrossRef] [PubMed]
2. Mahajan, K.N.; Nick McElhinny, S.A.; Mitchell, B.S.; Ramsden, D.A. Association of DNA Polymerase μ (Pol μ) with Ku and Ligase IV: Role for Pol μ in End-Joining Double-Strand Break Repair. *Mol. Cell Biol.* **2002**, *22*, 5194–5202. [CrossRef] [PubMed]
3. Gouge, J.; Rosario, S.; Romain, F.; Beguin, P.; Delarue, M. Structures of Intermediates along the Catalytic Cycle of Terminal Deoxynucleotidyltransferase: Dynamical Aspects of the Two-Metal Ion Mechanism. *J. Mol. Biol.* **2013**, *425*, 4334–4352. [CrossRef] [PubMed]
4. Kuznetsova, A.A.; Tyugashev, T.E.; Alekseeva, I.V.; Timofeyeva, N.A.; Fedorova, O.S.; Kuznetsov, N.A. Insight into the Mechanism of DNA Synthesis by Human Terminal Deoxynucleotidyltransferase. *Life Sci. Alliance* **2022**, *5*, e202201428. [CrossRef] [PubMed]
5. Hrdlickova, R.; Toloue, M.; Tian, B. RNA-Seq Methods for Transcriptome Analysis. *Wiley Interdiscip. Rev. RNA* **2017**, *8*, e1364. [CrossRef] [PubMed]
6. Kyrylkova, K.; Kyryachenko, S.; Leid, M.; Kioussi, C. Detection of Apoptosis by TUNEL Assay. *Methods Mol. Biol.* **2012**, *887*, 41–47. [CrossRef] [PubMed]
7. Mathews, A.S.; Yang, H.; Montemagno, C. Photo-Cleavable Nucleotides for Primer Free Enzyme Mediated DNA Synthesis. *Org. Biomol. Chem.* **2016**, *14*, 8278–8288. [CrossRef] [PubMed]
8. Flamme, M.; Katkevica, D.; Pajuste, K.; Katkevics, M.; Sabat, N.; Hanlon, S.; Marzuoli, I.; Püntener, K.; Sladojevich, F.; Hollenstein, M. Benzoyl and Pivaloyl as Efficient Protecting Groups for Controlled Enzymatic Synthesis of DNA and XNA Oligonucleotides. *Asian J. Org. Chem.* **2022**, *11*, e202200384. [CrossRef]
9. Palluk, S.; Arlow, D.H.; De Rond, T.; Barthel, S.; Kang, J.S.; Bector, R.; Baghdassarian, H.M.; Truong, A.N.; Kim, P.W.; Singh, A.K.; et al. De Novo DNA Synthesis Using Polymerase-Nucleotide Conjugates. *Nat. Biotechnol.* **2018**, *36*, 645–650. [CrossRef]
10. Ashley, J.; Potts, I.G.; Olorunniji, F.J. Applications of Terminal Deoxynucleotidyl Transferase Enzyme in Biotechnology. *ChemBioChem* **2023**, *24*, e202200510. [CrossRef]
11. Kuznetsova, A.A.; Senchurova, S.I.; Gavrilova, A.A.; Tyugashev, T.E.; Mikushina, E.S.; Kuznetsov, N.A. Substrate Specificity Diversity of Human Terminal Deoxynucleotidyltransferase May Be a Naturally Programmed Feature Facilitating Its Biological Function. *Int. J. Mol. Sci.* **2024**, *25*, 879. [CrossRef] [PubMed]
12. Loc'h, J.; Rosario, S.; Delarue, M. Structural Basis for a New Templated Activity by Terminal Deoxynucleotidyl Transferase: Implications for V(D)J Recombination. *Structure* **2016**, *24*, 1452–1463. [CrossRef] [PubMed]
13. Romain, F.; Barbosa, I.; Gouge, J.; Rougeon, F.; Delarue, M. Conferring a Template-Dependent Polymerase Activity to Terminal Deoxynucleotidyltransferase by Mutations in the Loop1 Region. *Nucleic Acids Res.* **2009**, *37*, 4642–4656. [CrossRef] [PubMed]
14. Bienstock, R.J.; Beard, W.A.; Wilson, S.H. Phylogenetic Analysis and Evolutionary Origins of DNA Polymerase X-Family Members. *DNA Repair* **2014**, *22*, 77–88. [CrossRef] [PubMed]
15. Zgarbová, M.; Jurečka, P.; Šponer, J.; Otyepka, M. A- to B-DNA Transition in AMBER Force Fields and Its Coupling to Sugar Pucker. *J. Chem. Theory Comput.* **2018**, *14*, 319–328. [CrossRef] [PubMed]
16. Strelnikov, I.A.; Kovaleva, N.A.; Klinov, A.P.; Zubova, E.A. C-B-A Test of DNA Force Fields. *ACS Omega* **2023**, *8*, 10253–10265. [CrossRef] [PubMed]
17. Jurečka, P.; Zgarbová, M.; Černý, F.; Salomon, J. Multistate B- to A- Transition in Protein-DNA Binding–How Well Is It Described by Current AMBER Force Fields? *J. Biomol. Struct. Dyn.* **2024**, 1–11. [CrossRef] [PubMed]
18. Webb, B.; Sali, A. Protein Structure Modeling with MODELLER. *Methods Mol. Biol.* **2014**, *1137*, 1–15. [CrossRef] [PubMed]
19. Jensen, M.A.; Davis, R.W. Template-Independent Enzymatic Oligonucleotide Synthesis (TiEOS): Its History, Prospects, and Challenges. *Biochemistry* **2018**, *57*, 1821. [CrossRef]
20. Abraham, M.J.; Murtola, T.; Schulz, R.; Páll, S.; Smith, J.C.; Hess, B.; Lindah, E. GROMACS: High Performance Molecular Simulations through Multi-Level Parallelism from Laptops to Supercomputers. *SoftwareX* **2015**, *1–2*, 19–25. [CrossRef]
21. Maier, J.A.; Martinez, C.; Kasavajhala, K.; Wickstrom, L.; Hauser, K.E.; Simmerling, C. Ff14SB: Improving the Accuracy of Protein Side Chain and Backbone Parameters from Ff99SB. *J. Chem. Theory Comput.* **2015**, *11*, 3696–3713. [CrossRef] [PubMed]
22. Galindo-Murillo, R.; Robertson, J.C.; Zgarbová, M.; Šponer, J.; Otyepka, M.; Jurečka, P.; Cheatham, T.E. Assessing the Current State of Amber Force Field Modifications for DNA. *J. Chem. Theory Comput.* **2016**, *12*, 4114–4127. [CrossRef] [PubMed]
23. Jorgensen, W.L.; Chandrasekhar, J.; Madura, J.D.; Impey, R.W.; Klein, M.L. Comparison of Simple Potential Functions for Simulating Liquid Water. *J. Chem. Phys.* **1983**, *79*, 926–935. [CrossRef]

24. Joung, I.S.; Cheatham, T.E. Determination of Alkali and Halide Monovalent Ion Parameters for Use in Explicitly Solvated Biomolecular Simulations. *J. Phys. Chem. B* **2008**, *112*, 9020–9041. [CrossRef] [PubMed]
25. Meagher, K.L.; Redman, L.T.; Carlson, H.A. Development of Polyphosphate Parameters for Use with the AMBER Force Field. *J. Comput. Chem.* **2003**, *24*, 1016–1025. [CrossRef] [PubMed]
26. Vanquelef, E.; Simon, S.; Marquant, G.; Garcia, E.; Klimerak, G.; Delepine, J.C.; Cieplak, P.; Dupradeau, F.Y. R.E.D. Server: A Web Service for Deriving RESP and ESP Charges and Building Force Field Libraries for New Molecules and Molecular Fragments. *Nucleic Acids Res.* **2011**, *39*, W511–W517. [CrossRef]
27. Dupradeau, F.Y.; Pigache, A.; Zaffran, T.; Savineau, C.; Lelong, R.; Grivel, N.; Lelong, D.; Rosanski, W.; Cieplak, P. The R.E.D. Tools: Advances in RESP and ESP Charge Derivation and Force Field Library Building. *Phys. Chem. Chem. Phys.* **2010**, *12*, 7821–7839. [CrossRef] [PubMed]
28. Bayly, C.I.; Cieplak, P.; Cornell, W.D.; Kollman, P.A. A Well-Behaved Electrostatic Potential Based Method Using Charge Restraints for Deriving Atomic Charges: The RESP Model. *J. Phys. Chem.* **1993**, *97*, 10269–10280. [CrossRef]
29. Jiang, Y.; Zhang, H.; Tan, T. Rational Design of Methodology-Independent Metal Parameters Using a Nonbonded Dummy Model. *J. Chem. Theory Comput.* **2016**, *12*, 3250–3260. [CrossRef]
30. Anandakrishnan, R.; Aguilar, B.; Onufriev, A.V. H++ 3.0: Automating PK Prediction and the Preparation of Biomolecular Structures for Atomistic Molecular Modeling and Simulations. *Nucleic Acids Res.* **2012**, *40*, W537–W541. [CrossRef]
31. Bussi, G.; Donadio, D.; Parrinello, M. Canonical Sampling through Velocity Rescaling. *J. Chem. Phys.* **2007**, *126*, 014101. [CrossRef] [PubMed]
32. Parrinello, M.; Rahman, A. Polymorphic Transitions in Single Crystals: A New Molecular Dynamics Method. *J. Appl. Phys.* **1981**, *52*, 7182–7190. [CrossRef]

biomolecules

MDPI

Article

Computational Elucidation of a Monobody Targeting the Phosphatase Domain of SHP2

Yang Wang [1,2,†], Xin Qiao [1,†], Ruidi Zhu [1], Linxuan Zhou [1], Quan Zhang [1], Shaoyong Lu [1,2,*] and Zongtao Chai [3,4,*]

1 Medicinal Chemistry and Bioinformatics Center, Shanghai Jiao Tong University School of Medicine, Shanghai 200025, China; spursanity@sjtu.edu.cn (Y.W.)
2 Key Laboratory of Protection, Development and Utilization of Medicinal Resources in Liupanshan Area, Ministry of Education, Peptide & Protein Drug Research Center, School of Pharmacy, Ningxia Medical University, Yinchuan 750004, China
3 Department of Hepatic Surgery, Shanghai Geriatric Medical Center, Shanghai 201104, China
4 Department of Liver Surgery and Transplantation, Liver Cancer Institute and Zhongshan Hospital, Fudan University, Shanghai 200032, China
* Correspondence: lushaoyong@sjtu.edu.cn (S.L.); chai.zongtao@zs-hospital.sh.cn (Z.C.)
† These authors contributed equally to this work.

Abstract: Src homology 2 (SH2) domain-containing phosphatase 2 (SHP2) is a key regulator in cellular signaling pathways because its dysregulation has been implicated in various pathological conditions, including cancers and developmental disorders. Despite its importance, the molecular basis of SHP2's regulatory mechanism remains poorly understood, hindering the development of effective targeted therapies. In this study, we utilized the high-specificity monobody Mb11 to investigate its interaction with the SHP2 phosphatase domain (PTP) using multiple replica molecular dynamics simulations. Our analyses elucidate the precise mechanisms through which Mb11 achieves selective recognition and stabilization of the SHP2-PTP domain, identifying key residues and interaction networks essential for its high binding specificity and regulatory dynamics. Furthermore, the study highlights the pivotal role of residue C459 in preserving the structural integrity and functional coherence of the complex, acting as a central node within the interaction network and underpinning its stability and efficiency. These findings have significantly advanced the understanding of the mechanisms underlying SHP2's involvement in disease-related signaling and pathology while simultaneously paving the way for the rational design of targeted inhibitors, offering significant implications for therapeutic strategies in SHP2-associated diseases and contributing to the broader scope of precision medicine.

Keywords: Src homology 2 domain-containing phosphatase 2; Mb11; molecular dynamics simulation; targeted drug design; network analysis

1. Introduction

Src homology 2 (SH2) domain-containing phosphatase 2 (SHP2) is a non-receptor protein tyrosine phosphatase that plays an essential role in regulating signaling pathways involved in metabolism, proliferation, differentiation, migration, transcription, and oncogenesis [1–3]. Structurally, SHP2 comprises two tandem SH2 domains, a central phosphatase catalytic (PTP) domain, and a C-terminal region enriched with multiple functional phosphorylation sites [4] (Figure 1A). The N-terminal SH2 (N-SH2) and C-terminal SH2 (C-SH2) domains cooperate to mediate interactions with phosphotyrosine-containing ligands or adaptor proteins. Under basal conditions, the N-SH2 domain forms a non-covalent interaction with the PTP domain, maintaining SHP2 in an autoinhibited conformation [4,5]. Upon stimulation by external signals such as growth factors or cytokines, bidentate phosphotyrosine ligands bind to the SH2 domains, inducing a conformational change that releases the N-SH2 domain from the PTP active site. This structural rearrangement exposes the catalytic center of the PTP domain, enabling enzymatic activation [6,7] (Figure 1B).

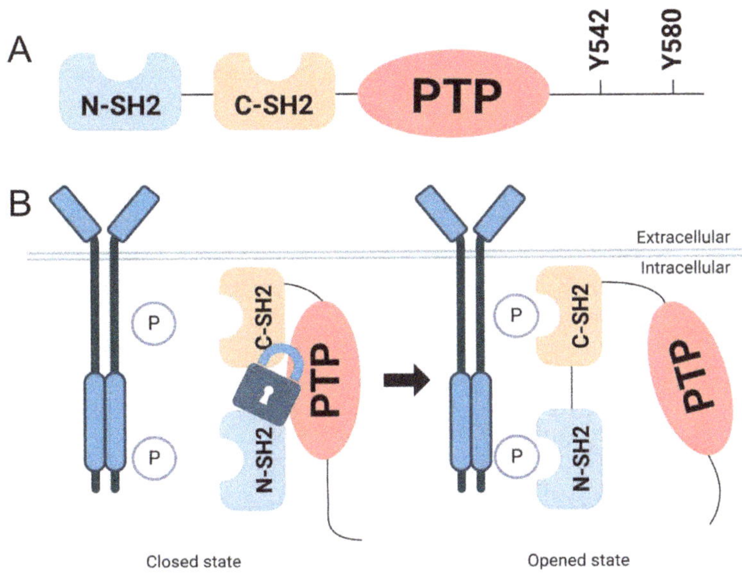

Figure 1. SHP2 protein structure and its activation mechanism. (**A**) SHP2 protein contains two SH2 domains, N-SH2 (blue) and C-SH2 (orange), followed by a central PTP domain (red). The C-terminal region contains phosphorylation sites, including Y542 and Y580, which play critical roles in SHP2's regulatory functions. (**B**) The left part (closed state) depicts SHP2 in its autoinhibited conformation. In this state, the N-SH2 domain interacts with the PTP domain, blocking the active site and preventing catalytic activity. This interaction functions as a 'lock', maintaining SHP2 in a dormant state. The right part (opened state) illustrates SHP2 activation upon stimulation by external signals, such as growth factors or cytokines. Phosphotyrosine-containing ligands bind to the SH2 domains, inducing a conformational change that releases the N-SH2 domain from the PTP active site. This structural rearrangement exposes the catalytic center of the PTP domain, enabling enzymatic activity.

The SHP2-PTP domain is critically involved in oncogenesis [8–10]. In breast and gastric cancers, SHP2 enhances RAS/ERK signaling by dephosphorylating inhibitory regulatory sites on adaptor proteins such as Gab1, thereby promoting cellular proliferation and resistance to targeted therapies [11–14]. In non-small cell lung cancer (NSCLC), PTP activity mediates resistance to EGFR inhibitors, such as osimertinib, by modulating the PI3K/AKT signaling pathway [15,16]. Additionally, activating mutations in SHP2, such as E76K, disrupt the autoinhibitory interaction between the N-SH2 and PTP domains in acute myeloid leukemia (AML), resulting in constitutive PTP activation and accelerated leukemic cell proliferation [9,17]. Moreover, the SHP2-PTP domain also plays a pivotal role in immune evasion [18,19]. Within the PD-1 signaling pathway, SHP2 suppresses T-cell activation by dephosphorylating critical components of the T-cell receptor (TCR) complex, including CD28 and ZAP70, thereby impairing the host's anti-tumor immune response [20–22]. In solid tumors such as melanoma and hepatocellular carcinoma (HCC), SHP2 activity positively correlates with PD-L1 expression, further underscoring its central role in modulating the tumor immune microenvironment [23,24].

Beyond cancers, the SHP2-PTP domain is implicated in the pathogenesis of non-malignant diseases. Gain-of-function mutations, such as D61G and E76K in Noonan syndrome, lead to the constitutive activation of the RAS signaling pathway, disrupting tightly regulated developmental processes [25]. In contrast, loss-of-function mutations associated with LEOPARD syndrome impair SHP2 phosphatase activity, resulting in dysregulated signaling and developmental abnormalities [26].

Despite substantial progress in elucidating the role of the SHP2-PTP domain, the molecular mechanisms remain poorly understood. Recently, a monobody, Mb (SHP2PTP_11, hereafter referred to as Mb11), was identified as a highly specific binder of the SHP2-PTP domain [27], which may serve as a valuable tool for elucidating its regulatory mechanisms. Mb11, as demonstrated by Biolayer Interferometry (BLI), exhibits exceptional binding affinity for SHP2-PTP, with a dissociation constant (K_D) of 2.7 nM. Notably, Mb11 displayed significantly reduced binding affinity to the SHP2-PTP(C459S) mutant, with a K_D value of 120 nM. This striking difference offers a critical opportunity to investigate the molecular determinants underlying Mb11's interaction with SHP2-PTP.

Molecular dynamics (MD) simulations have become an indispensable tool for elucidating dynamic conformational changes and protein–protein interaction mechanisms at atomic resolution [28,29]. In this study, large-scale MD simulations were conducted to systematically explore the interaction mechanism between Mb11 and the PTP domain, with a specific focus on how Mb11 influences the conformation and dynamic properties of the PTP active site. The research aims to identify the key amino acid residues involved in Mb11-PTP interactions and to elucidate the primary forces stabilizing the Mb11-SHP2-PTP complex. These findings will provide critical molecular-level insights into Mb11's mechanism of action and expand the theoretical framework for understanding SHP2 signaling regulation. Moreover, the study will provide a mechanistic basis for the rational design of potent SHP2 inhibitors and facilitate a deeper understanding of their regulatory pathways.

2. Results

2.1. Mb11 Targets the Deep Catalytic Pocket of the PTP Domain

Protein–protein docking analysis identified three potential binding modes for the Mb11-SHP2-PTP complex [30,31], labeled sequentially as Modes 1–3 (Figure 2A–C), which were subsequently used to construct models of the C459S mutant complex. Molecular Mechanics/Generalized Born Surface Area (MM/GBSA) calculations were performed to evaluate the Gibbs binding free energy ($\Delta G_{binding}$) of Modes 1–3 and their corresponding C459S mutants (Tables S1–S3) [32], ultimately identifying the most stable binding conformation (Figure 2C). The $\Delta G_{binding}$ of wild-type Mb11-SHP2-PTP was -48.43 ± 11.97 kcal/mol, whereas the $\Delta G_{binding}$ of C459S mutant was -37.11 ± 10.62 kcal/mol (Table S3). The mutation of the key catalytic site residue significantly weakens the binding affinity of Mb11.

Structural analysis of the final binding conformation revealed that Mb11 is directly positioned within the catalytic site of the PTP domain (Figure 2D), further supporting its role as a competitive inhibitor that blocks substrate access to the active site. Additionally, the binding epitope of Mb11 on SHP2-PTP shares certain similarities with the interface between the PTP and N-SH2 (Figure 2E,F) domains in the full-length SHP2 structure. However, Mb11's binding is relatively more focused on the deep catalytic pocket of the PTP domain. Despite targeting similar surface regions, Mb11 and N-SH2 exhibit significant differences in the types of amino acids involved in binding (Figure 2G,H). Specifically, Mb11 does not mimic the PTP binding interface of N-SH2 but instead employs a distinct recognition pattern to interact with the catalytic surface. This unique binding mode may serve as the basis for Mb11's specific interaction with SHP2-PTP, which merits further investigation.

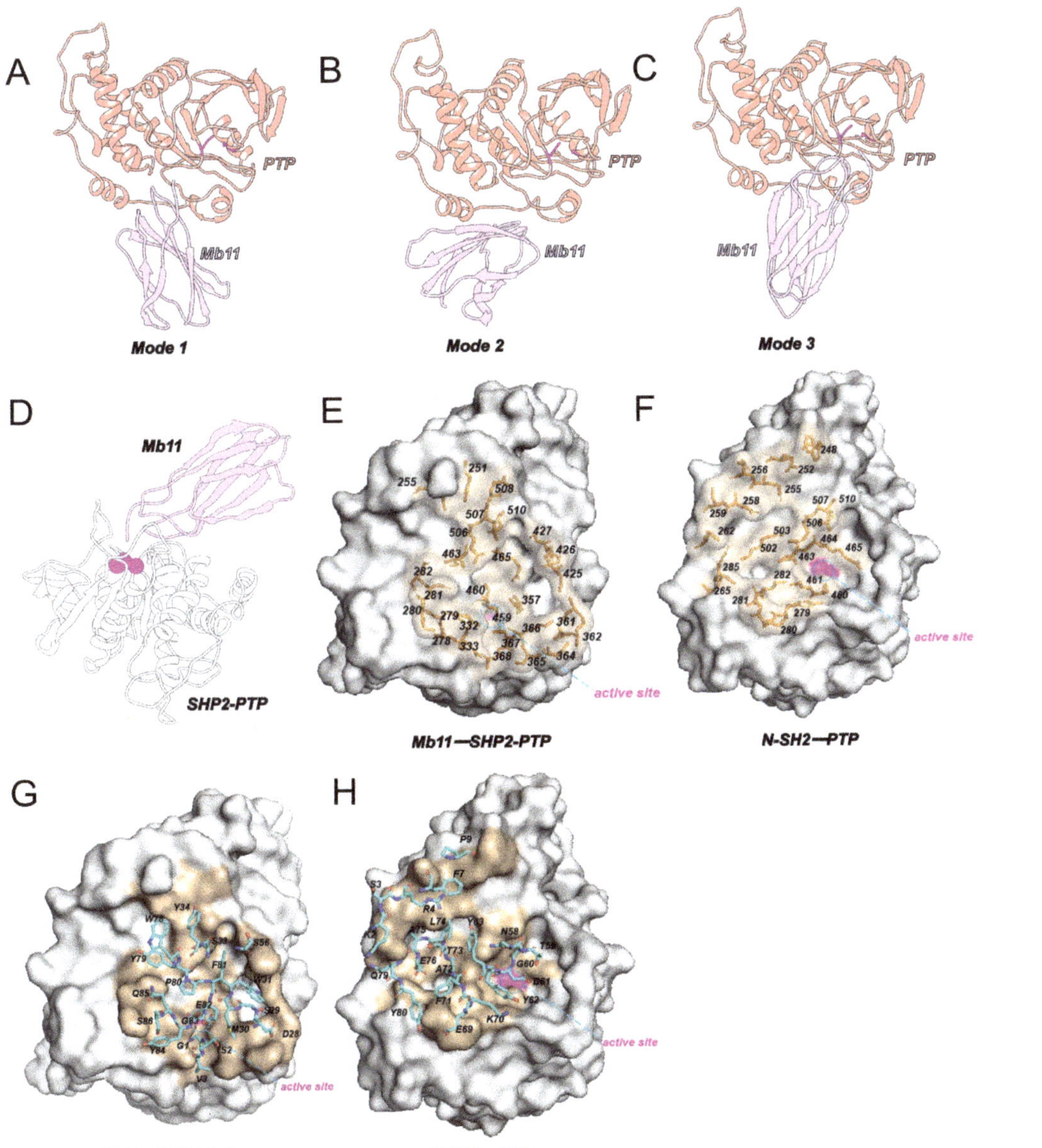

Figure 2. Static structural analysis of Mb11-SHP2-PTP. (**A–C**) Protein docking performed using the ClusPro2 server identified three potential binding modes of Mb11 with SHP2-PTP. Mb11 is shown in purple, SHP2-PTP in red, and the disulfide bond between C367 and C459 is highlighted in magenta. (**D**) The Mb11-SHP2-PTP complex is depicted, with Mb11 represented in purple and SHP2-PTP in white to enhance visual clarity. The critical residue, C459, is highlighted as a magenta sphere. (**E**) Detailed view of the Mb11-SHP2-PTP interaction interface, emphasizing critical residues within the defined epitope. PTP is depicted as a surface model, with the epitope region shown in wheat and rendered at 50% transparency. Residues identified as being within 4 Å of Mb11 heavy atoms are represented as orange sticks and labeled. The active site is outlined with a dashed circle, while the surface of C459 is highlighted in magenta. (**F**) Zoomed-in representation of the N-SH2-PTP interaction within the autoinhibited SHP2 structure, showcasing key PTP residues located within 4 Å of N-SH2 heavy atoms. Labeling conventions align with those used in (**E**). (**G**) Magnified view of the

Mb11-SHP2-PTP interface, focusing on residues comprising the Mb11 paratope. PTP is shown as an opaque wheat-colored surface, with paratope residues (defined as those within 4 Å of PTP heavy atoms) displayed as sticks and labeled. The active site is marked with a dashed circle, and the C459 surface is highlighted in magenta. (**H**) Close-up depiction of the N-SH2-PTP interface within the autoinhibited SHP2 structure, concentrating on N-SH2 residues within 4 Å of PTP heavy atoms. Labeling methods follow those in (**G**). (**E–H**) are consistently oriented to facilitate comparative structural analysis.

2.2. Mb11 Significantly Enhances the Stability of SHP2-PTP

Large-scale MD simulations were employed to investigate the differential binding of Mb11 to wild-type and C459S mutant SHP2-PTP. To ensure clarity and consistency, the residue sequence and secondary structure of Mb11-SHP2-PTP are provided in Figure S1, which serve as the reference standard for subsequent discussions of residue-level interactions and structural analyses.

To assess the overall stability of the simulated systems, the root-mean-square deviations (RMSDs) of Cα atoms relative to the initial structure were calculated for each system (Figure S2A). RMSD analysis revealed that all systems gradually reached equilibrium after approximately 300 ns. Subsequent analyses focused on the trajectories within the equilibrium state. Further calculations showed that the RMSD value for the wild-type Mb11-SHP2-PTP complex was 3.53 $\pm$ 0.16 Å, whereas that of the C459S mutant was 5.69 $\pm$ 0.59 Å. This indicates that the wild-type complex exhibited lower dynamic fluctuation and a more stable binding mode, preliminarily supporting the high affinity of Mb11 for SHP2-PTP.

Next, the root-mean-square fluctuations (RMSFs) of Cα atoms relative to their initial positions were calculated for each residue (Figure S2B). The RMSF analysis revealed a generally similar distribution pattern across systems, with the wild-type Mb11-SHP2-PTP complex exhibiting consistently lower RMSF values. To further explore the relationship between residue-level flexibility and binding stability, residues involved in the Mb11-SHP2-PTP interface, as identified from the static structure (Figure 2E,G), were highlighted on the RMSF plot (blue). These interface residues in the wild-type complex displayed notably lower RMSF values, underscoring their role in stabilizing protein–ligand interactions.

To further elucidate the distinct conformational dynamics triggered by the C459S mutation, dynamic cross-correlation matrices (DCCMs) were calculated for the Cα atoms (Figure 3A,B). In the DCCM plots, the diagonal regions represent the motion of residues relative to themselves, while off-diagonal regions indicate the correlated motions between different residues. Based on the color-coding scheme, red signifies positively correlated motions between residues, while green denotes anti-correlated motions. To facilitate interpretation, key regions were highlighted using colored boxes in the matrix plots. The purple box represents the internal correlated motions within Mb11, the red and green boxes correspond to the intra-molecular correlated motions within wild-type and C459S mutant SHP2-PTP, respectively, and the blue box indicates inter-molecular correlated motions between Mb11 and SHP2-PTP. The results revealed that both intra- and inter-molecular correlated motions in the wild-type Mb11-SHP2-PTP complex were significantly weaker than those in the C459S mutant. This diminished correlation suggests that the wild-type complex exhibits more coordinated and stable overall dynamics, further supporting its superior binding dynamics.

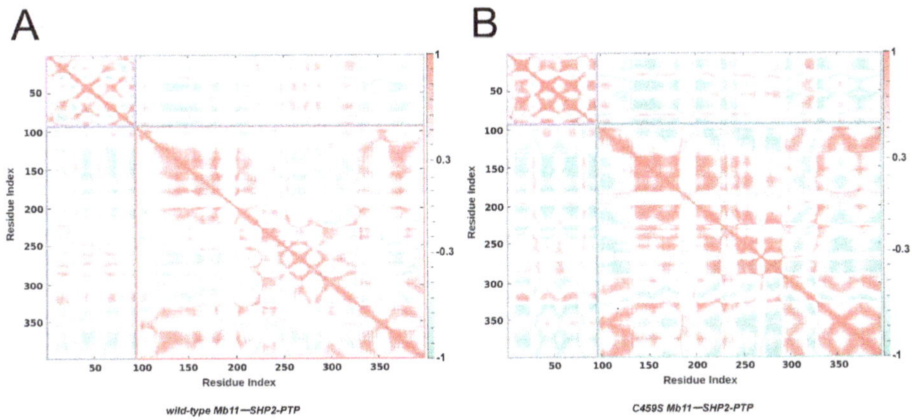

Figure 3. Comparative analysis of dynamic behavior in wild-type and mutant Mb11-SHP2-PTP complexes. (**A**) DCCMs for wild-type Mb11-SHP2-PTP. (**B**) DCCMs for C459S mutant complex. The purple box denotes intra-Mb11 correlated motions, the red and green boxes represent intra-SHP2-PTP correlations in the wild-type and mutant systems, respectively, and the blue box indicates inter-molecular correlations between Mb11 and SHP2-PTP. The color scale is displayed on the right. Interactions with absolute correlation coefficients below 0.3 are shown in white to improve clarity.

2.3. Wild-Type Mb11-SHP2-PTP Exhibits a More Compact Conformation

To understand the conformations during the simulations, principal component analysis (PCA) was performed on the wild-type and C459S mutant Mb11-SHP2-PTP complexes (Figure 4A,B). The two systems were projected onto a shared PCA space to ensure direct comparability of their conformational landscapes. By extracting the two most significant principal components (PC1 and PC2), the high-dimensional conformational space was reduced to a two-dimensional (2D) distribution map for each system. The results indicate that the wild-type Mb11-SHP2-PTP complex exhibits significant conformational convergence, with its distribution concentrated within a narrow range along PC1, forming compact cluster structures. This distribution pattern suggests that the wild-type complex has lower conformational fluctuations in the conformational space, tending towards a more stable conformation and a tighter energy state. In contrast, the conformational distribution of the C459S mutant complex shows a marked expansion in coverage along PC1, displaying a highly dispersed distribution pattern. Notably, M3′ and M4′ deviate significantly from the primary clusters and are located at the distal regions of the PC1 and PC2 space, highlighting the heterogeneity of the conformational space. This dispersed distribution pattern clearly indicates that the C459S mutant complex occupies a higher free-energy state within the conformational space.

Next, we investigated the PCA eigenvalue distributions of the two systems to further elucidate their differences in conformational dynamic properties (Figure S3A,B). The cumulative variances of PC1 and PC2 for the wild-type complex are relatively low (40.1% and 27.1%), with an eigenvalue decay curve displaying a gradual slope, indicative of more evenly distributed conformational motions across multiple PCs. This distribution reflects the greater conformational stability and the restricted conformational space of the wild-type complex. In contrast, the C459S mutant exhibits significantly higher variances for PC1 and PC2 (73.0% and 13.5%), with a steep eigenvalue decay after PC1. This sharp decline indicates that the mutant's motions are predominantly concentrated in a single dominant principal direction, suggesting larger amplitude motions, enhanced conformational flexibility, and broader sampling of the conformational space. These findings further substantiate that the mutant explores a more extensive conformational landscape, while the wild-type remains confined to a narrower free energy basin, indicative of superior stability.

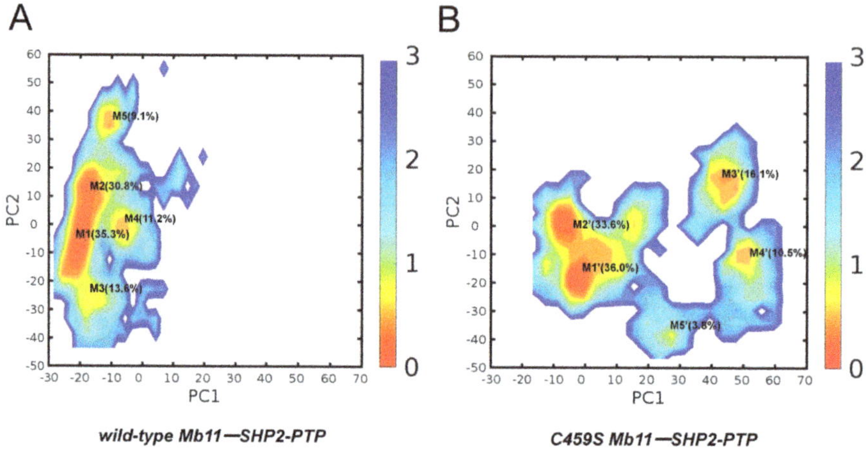

Figure 4. PCA and free energy landscapes of wild-type and C459S mutant Mb11-SHP2-PTP complexes. (**A**,**B**) depict the free energy landscapes of the wild-type and C459S mutant systems, respectively, based on PC1 and PC2. Free energy values are expressed in kcal/mol, with the corresponding color scale shown on the right. K-means clustering of the PCA projections identifies the five most dominant conformational clusters for each system, with their respective occupancy percentages annotated near the cluster centroids.

Figure S4A,B illustrates the five representative conformations of each system complemented by RMSD values obtained from structural alignments (Figure S4C,D), which, from a new perspective, validate the differences in conformational flexibility and stability between both systems. Specifically, the C459S mutant displays markedly higher RMSD values and greater structural divergence among its representative conformations. In contrast, the wild-type Mb11-SHP2-PTP complex exhibits lower RMSD values and more similar representative conformations, indicating a more compact and stable conformational state. These findings reinforce the distinct advantage of the wild-type complex in terms of conformational stability.

To further investigate the conformational properties of wild-type and C459S mutant Mb11-SHP2-PTP complexes, the Defined Secondary Structure of Proteins (DSSP) method was used to analyze dynamic secondary structure changes throughout the simulation (Figure S5A,B). Comparative results indicated that in the wild-type Mb11-SHP2-PTP system, α-helices and β-sheets exhibited a higher proportion and maintained stronger temporal continuity. This suggested a more compact and stable overall conformation. In contrast, the C459S mutant showed an increase in bend, turn, and disordered regions (denoted as 'none' in Figure S5). These unstable structures were more widely distributed throughout the mutant system and persisted for extended periods, reflecting looser and more dynamic conformational characteristics. These trends corroborated the PCA findings, further confirming the distinct conformational stability and dynamics of both systems.

2.4. Robust Inter-Protein Interactions Drive an Observed 'Approaching' Trend

To quantitatively evaluate the similarity of dynamic behavior between the two complexes, we performed a Root Mean Square Inner Product (RMSIP) analysis on the first 10 PCs of the two systems. The analysis revealed an RMSIP value of 0.33, which is significantly lower than the ideal value of 1.0, indicating substantial differences in the overall dynamic behavior within the principal subspaces of the two systems. Further analysis of the overlap matrix showed that the overall degree of overlap between the PCs of the two systems was generally low. As illustrated in Figure S6, most overlap values are close to 0, indicating that the two systems exhibit markedly different dynamic behavior along their major motion directions. The limited similarity between the PCs suggests that the mutation has profoundly impacted the global dynamic modes of the system, significantly altering the characteristics of its dynamical subspace.

To graphically illustrate the dominant motions within different regions of Mb11-SHP2-PTP, PC1 was mapped onto the initial structure for each system (Figure 5A,B). The analysis revealed an 'approaching' motion at the interface of wild-type Mb11-SHP2-PTP, particularly between loop M1 and P2, as well as loop M2 and P1 (specific residue composition shown in Table S4), indicating enhanced interaction and structural convergence. In contrast, the C459S mutant system exhibited a weak 'departing' motion, characterized by reduced inter-regional interactions and an increased spatial separation between these loops.

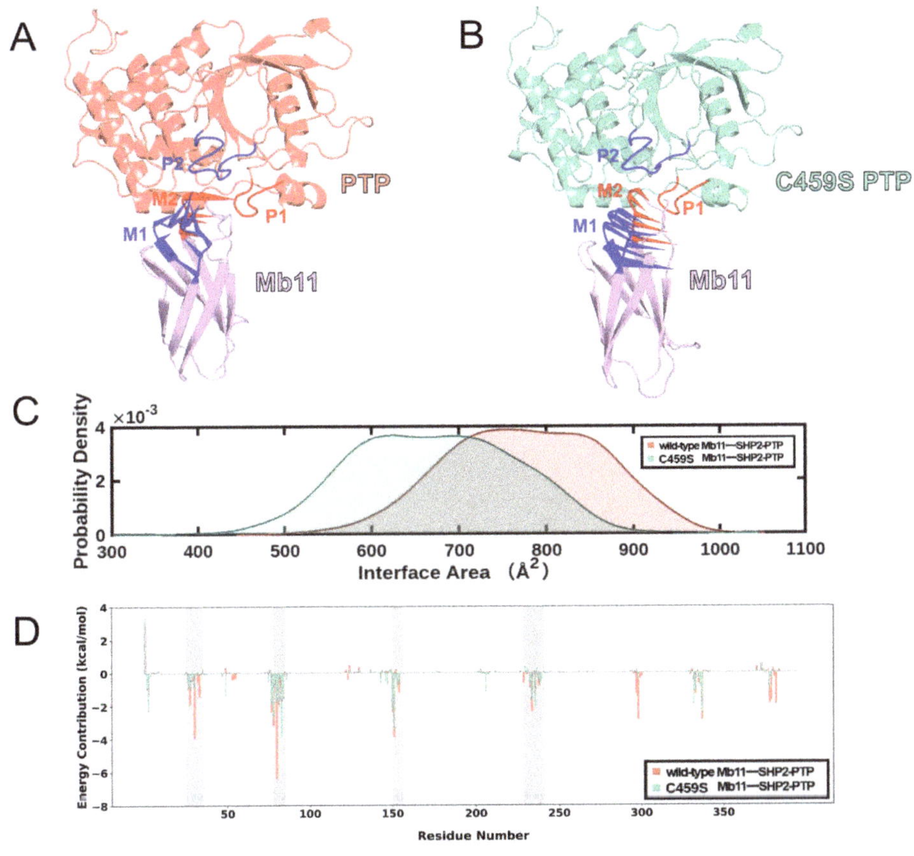

Figure 5. Inter-protein interactions drive the 'approaching' trend in Mb11-SHP2-PTP systems. (**A**,**B**) Porcupine plots illustrate the primary motion along PC1 in the wild-type and C459S mutant Mb11-SHP2-PTP systems. Mb11 is shown in purple, the wild-type SHP2-PTP in red, and the C459S mutant SHP2-PTP in green. Loops M1 and P2 are marked in deep blue, and Loops M2 and P1 are marked in vivid red, highlighting the distinct motion patterns between the two systems. (**C**) Probability distribution of the interface area in the wild-type (red) and C459S mutant (green) systems during simulations. (**D**) Per-residue decomposition of binding free energy in the wild-type (red) and C459S mutant (green) systems. Key residue regions, including Loops M1, M2, P1, and P2, are highlighted with a semi-transparent gray background.

Building on the observed 'approaching' dynamics, we hypothesized that the protein–protein interface (PPI) in the wild-type Mb11-SHP2-PTP complex would be larger. As illustrated in Figure 5C, the wild-type complex displayed a prominent peak at an interface area of approximately 750 $Å^2$, indicating a more uniform and tightly packed PPI. In contrast, the C459S mutant exhibited a bimodal distribution, with two dominant peaks around 625 $Å^2$ and 700 $Å^2$, reflecting a less homogeneous and potentially less stable interface.

Generally, larger interface areas are linked with stronger binding interactions due to lower protein–protein interaction energies. To further quantify this relationship, we employed the Molecular Mechanics Poisson–Boltzmann Surface Area (MM/PBSA) method to calculate the $\Delta G_{binding}$ between Mb11 and SHP2-PTP (Table 1) [32]. The results showed that the wild-type complex exhibited a more favorable $\Delta G_{binding}$ of -57.79 ± 13.04 kcal/mol, compared to -49.09 ± 11.70 kcal/mol for the C459S mutant. The stronger binding free energy in the wild-type complex is consistent with its larger interface area and highlights the critical role of stable protein–protein interactions in maintaining structural integrity and functional specificity.

Table 1. Binding free energies (kcal/mol) of wild-type and C459S mutant Mb11-SHP2-PTP systems computed by the MM-PBSA method.

Energy Component (kcal/mol)	Wild-Type Mb11-SHP2-PTP	C459S Mb11-SHP2-PTP
ΔG_{gas}	-482.92 ± 67.09	-498.85 ± 95.77
ΔG_{solv}	425.13 ± 59.68	449.76 ± 88.69
$\Delta G_{binding}$	-57.79 ± 13.04	-49.09 ± 11.70

Next, we decomposed $\Delta G_{binding}$ on a per-residue mode (Figure 5D). In the wild-type Mb11-SHP2-PTP system, the intermolecular interaction network was strengthened either by enhancing favorable negative Gibbs free energy contributions or mitigating unfavorable positive contributions. Key interaction regions, including Loops M1, M2, P1, and P2, are highlighted (gray). Residues within these regions exhibited significantly negative energy contributions in the wild-type system, underscoring their pivotal role in supporting enhanced conformational stability and dynamic behavior.

2.5. Deciphering the Critical Interacting Residues That Drive the Observed Differences

To preliminarily explore the molecular interaction mechanisms underlying the 'approaching' dynamics, we defined a 'stable contact' as a residue–residue interaction persisting for at least 70% of the total simulation time. The analysis revealed that in the wild-type Mb11-SHP2-PTP system, extensive stable hydrophobic interactions serve as the cornerstone of the overall interaction network (Figure 6A,B). Moreover, five key stable polar interactions were identified during the simulations (D27-K237, D28-R235, D28-K239, F81-Y152, and E82-R151). These residue pairs demonstrated not only close spatial proximity but also complementary electrostatic properties (Figure 6C), which explains the observed 'approaching' trend.

To comprehensively evaluate the influence of key interacting residues on the stability of the Mb11-SHP2-PTP complex, we performed a systematic analysis of salt bridges and hydrogen bonds across both systems (Table 2). The findings revealed that a small subset of highly persistent salt bridges and hydrogen bonds, with occupancies exceeding 50% in the wild-type system, constituted the core of the interfacial interaction network, whereas these interactions were markedly attenuated in the C459S mutant system.

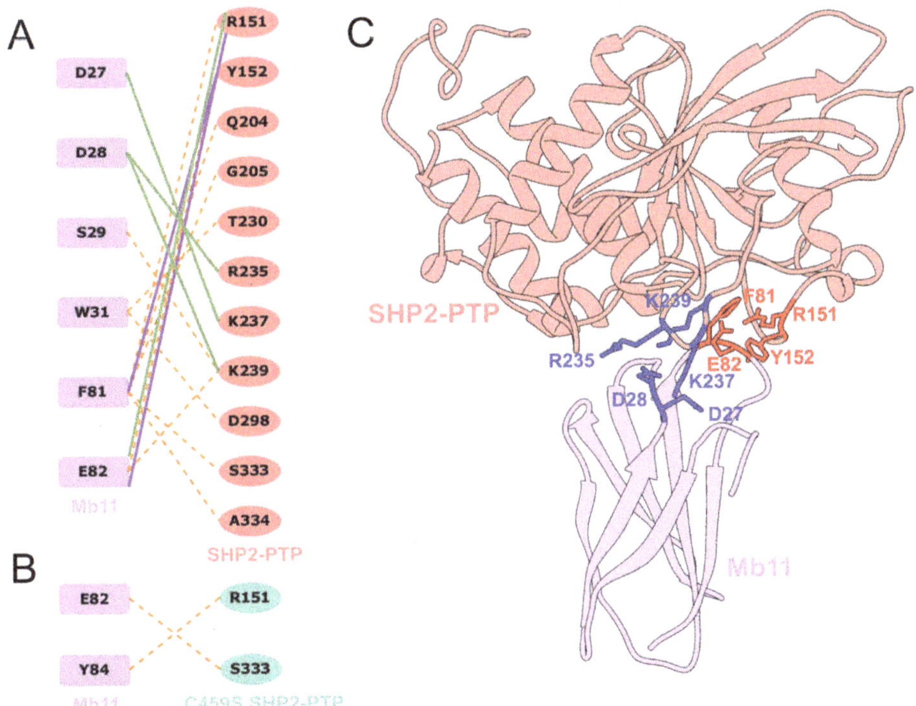

Figure 6. Key residue interactions between Mb11 and SHP2-PTP in wild-type and C459S mutant systems. Schematic representation of residue–residue interactions in the wild-type (**A**) and C459S mutant (**B**) Mb11-SHP2-PTP system. Purple blocks represent Mb11 residues, while red ovals and green ovals represent wild-type and C459S mutant SHP2-PTP residues, respectively. Dashed orange lines represent hydrophobic contacts, solid green lines represent salt bridges, and solid purple lines represent hydrogen bonds. (**C**) Three-dimensional structure showing the interaction interface between Mb11 (purple) and SHP2-PTP (red). Five key stable polar interactions are highlighted, with deep blue and vivid red residues illustrating the approaching trend between M1 and P2, and M2 and P1, respectively.

Table 2. Systematic analysis of salt bridges and hydrogen bonds in both systems.

Residue Pair	Interaction Type	Wild-Type Mb11-SHP2-PTP	C459S Mutant Mb11-SHP2-PTP
F81-Y152	Hydrogen bond	63.20%	2.57%
E82-R151	Hydrogen bond	34.00%	1.13%
D27-K237	Salt bridge	54.00%	24.33%
D28-R235	Salt bridge	46.10%	19.80%
D28-K239	Salt bridge	36.00%	9.33%
E82-R151	Salt bridge	77.80%	3.40%

In the wild-type complex, the salt bridge E82-R151 exhibited exceptional stability, with an occupancy of 77.8%, underscoring its pivotal role as a structural anchor driven by strong electrostatic attraction. Similarly, the salt bridge D27-K237 maintained an occupancy of 54.0%, while the hydrogen bond F81-Y152 demonstrated an even higher occupancy of 63.2%, collectively reinforcing the electrostatic framework. In stark contrast, these core interactions were significantly disrupted in the C459S mutant system. The occupancy of the salt bridge E82-R151 plummeted to 3.4%, and the hydrogen bond F81-Y152 was reduced to a mere 2.57%, highlighting the mutation's profound destabilizing effect on the primary interaction network. Beyond these core interactions, secondary contributions from the salt bridges D28-R235 and D28-K239, as well as the hydrogen bond E82-R151 in the wild-type system, provided additional stabilizing support. Thus, we have elucidated the molecular mechanisms underpinning the stability of the wild-type Mb11-SHP2-PTP complex.

2.6. The Interaction Network Unveils the Driving Forces Behind the Binding Specificity

Variation in polar interaction occupancy contributes to Mb11's binding selectivity; however, the fundamental mechanisms driving this difference remain elusive. Therefore, Community Network Analysis (CNA) was performed using the Girvan–Newman algorithm [33]. Variational coupling between communities was quantitatively assessed across all simulation trajectories. Residues within 4.5 Å of each other for at least 75% of the simulation time were grouped into the same community, which were considered cooperative functional units within the overall protein structure. The community composition and their interconnections are visualized in Figure 7A,B.

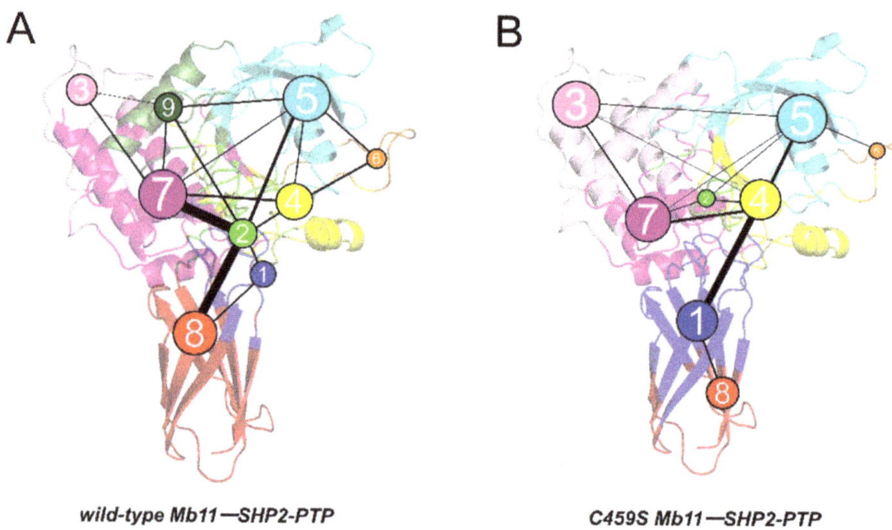

Figure 7. Community networks overlaid on the 2D structure of the wild-type (**A**) and C459S mutant (**B**) Mb11-SHP2-PTP systems. Each circle denotes a distinct community, with its size reflecting the number of residues it contains. The thickness of the lines connecting the circles represents the strength of intercommunity interactions.

Preliminary analysis reveals that Communities 1 and 8 primarily comprise Mb11 residues, while Communities 3–7 and 9 are largely composed of SHP2-PTP residues. However, in the wild-type complex, Community 2 occupies the protein interface and predominantly includes residues involved in key stabilizing interactions, serving as a central hub for interfacial connectivity. Community 2 forms strong connections with multiple other communities, particularly Communities 7 and 8, maintaining the integration and functional coordination of the interfacial network. In the mutant system, the composition and position of Community 2 change significantly. Community 2 shifts deeper into the SHP2-PTP structure, becoming entirely composed of SHP2-PTP residues, with most of its components reassigned to Communities 1 and 4. Moreover, the C459S mutation disrupts the hub-like functionality of Community 2, markedly weakening or even eliminating its connections with other communities. These changes underscore the critical differences in community organization between the systems, where the highly coordinated network in the wild-type provides a robust foundation for interfacial stability and binding efficiency.

Further analysis reveals that the C459S mutation induces significant changes in binding patterns. In the wild-type complex, C459 is located within Community 2, acting as a pivotal node in the electrostatic interaction and hydrogen bond network at the interface. This positioning reinforces the hub functionality of Community 2, allowing it to coordinate interactions across multiple communities, thereby enhancing interfacial coordination and network integrity. This network contributes to the superior dynamic stability and binding efficiency of the wild-type complex. In the mutant system, however, C459 is reassigned to Community 4, resulting in the loss of Community 2's functionality. Post-mutation, the coordination of the interfacial electrostatic and hydrogen bond network is significantly weakened, with intercommunity connections relying solely on limited interactions between Communities 1 and 4. This leads to reduced residue integration at the interface and fragmentation of the protein–protein interaction network. These changes highlight the destabilizing impact of the C459S mutation on interfacial network stability and complex functionality.

In summary, C459 serves as a core hub within the interfacial electrostatic and hydrogen bond network, integrating key residues and coordinating interactions between communities to maintain the structural stability and dynamic coordination of the complex. The C459S mutation disrupts this critical hub, resulting in a significant reduction in interfacial network coordination and structural integrity and ultimately impairing the binding efficiency and functional specificity of the complex.

3. Discussion

SHP2 plays an indispensable role in cellular signal regulation, with aberrant activity of its PTP domain closely associated with various cancers and developmental disorders. However, the molecular mechanisms remain largely unexplored. This knowledge gap not only limits our understanding of the biological functions of SHP2 but also significantly hinders the development of efficient targeted therapeutics. To address this issue, this study employed the high-specificity monobody Mb11 in complex with the SHP2-PTP domain and its C459S mutant as the research model. Using large-scale MD simulations, we systematically investigated the binding mechanisms and dynamic behavior of Mb11 with SHP2-PTP.

Preliminary analyses revealed that the Mb11-SHP2-PTP complex exhibited greater overall conformational stability and reduced molecular flexibility compared to the C459S mutant. DCCM analysis demonstrated a tighter protein–protein interface interaction pattern in the wild-type complex. PCA analysis further supported this conclusion, showing a more concentrated low-energy state distribution and a dynamic 'approaching' trend at the interface of the wild-type complex. MM/PBSA calculations indicated that the more stable interface interactions in the Mb11-SHP2-PTP complex were the driving factors behind this 'approaching' trend. Key residues responsible for the robust interface interactions were determined through a systematic analysis of salt bridges and hydrogen bonds. Finally, the visualized community network highlighted the changes in network topology before and after mutation, further revealing the critical role of the C459 residue as a central hub in the interface network.

This study addressed the key scientific question of Mb11's selective targeting of SHP2-PTP and systematically elucidated the molecular mechanisms underlying its binding specificity. Mb11 achieves high selectivity by precisely targeting the deep catalytic pocket of SHP2-PTP, a process dependent on multiple highly stable and functionally cooperative polar interactions at the interface, including the salt bridges D27-K237 and E82-R151, as well as the hydrogen bond F81-Y152. These interactions not only enhance intermolecular electrostatic attraction but also stabilize interface binding through the integration of a hydrophobic network. More importantly, C459 acts as a core hub in the interface network, playing an irreplaceable role in coordinating dynamic interactions between interface residues, maintaining network topology, and ensuring interfacial cooperativity. The integrity of its structure directly determines the high binding affinity and functional specificity of the complex, while its mutation disrupts these interaction networks, leading to a significant loss of interface coordination and binding efficiency. This study not only uncovers the molecular basis of Mb11-SHP2-PTP binding but also highlights the decisive role of key residues in interface functionality, providing valuable theoretical guidance for the rational design of SHP2-targeted therapeutics.

Future research should focus on optimizing the structural properties of the monobody to enhance its specificity for SHP2-PTP, particularly in pathological conditions associated with SHP2 dysfunction. Additionally, it is necessary to investigate the dynamic behavior and functional mechanisms of Mb11 in diverse physiological and pathological contexts to uncover its potential roles in regulating SHP2 signaling networks. This study provides critical molecular insights for the development of efficient SHP2-targeted inhibitors and deepens the understanding of SHP2's pivotal role in disease-related signaling and pathological mechanisms, thereby laying a solid foundation for precision therapeutic strategies in cancer and other SHP2-associated diseases.

4. Materials and Methods

4.1. Simulation System Preparation and Conformational Screening

In this study, we constructed two simulation systems—the wild-type and C459S mutant Mb11-SHP2-PTP complexes—and determined the conformations through a multi-step screening and evaluation process.

First, the initial crystal structure of the Mb13-SHP2-PTP complex (PDB ID: 7TVJ) was obtained from the Protein Data Bank (PDB) [27]. Subsequently, Modeller 10.4 in USCF Chimera was used to add missing non-terminal residues and remove non-protein co-crystallized molecules [34], yielding the complete structures of Mb13, Mb11, and SHP2-PTP. To capture representative conformations, three sets of 1 μs MD simulations were conducted for the unbound structures of Mb11 and SHP2-PTP individually. Protein–protein docking was then performed using ClusPro 2.0 to generate reasonable docking models [30,31], followed by site-directed mutagenesis in UCSF Chimera to construct the corresponding C459S mutants. Finally, three sets of 1 μs MD simulations were conducted for both the wild-type and C459S mutant Mb11-SHP2-PTP systems, and energy evaluation and refinement were carried out using the MM/GBSA method.

4.2. MD Simulations

To investigate the protein dynamics under realistic environmental conditions, MD simulations were performed for both systems using AMBER 20. The initial parameters for energy minimizations and simulations were prepared using the Amber ff14SB force field via the Leap module [35–38]. Both systems were solvated in a truncated octahedral water box filled with transferable intermolecular potential three-point (TIP3P) water molecules, extending 12 Å from the protein surface [39–41]. Counterions were added to neutralize the overall charge of each system, ensuring an electrically neutral simulation environment. All MD simulations utilized the GPU-accelerated PMEMD module for computational efficiency [42]. Energy minimization was conducted in two stages to remove steric clashes and optimize the system. The first stage involved 50,000 steps of minimization, allowing non-protein components to move freely while restraining the protein atoms. In the second stage, all atomic movements were unrestricted, with an additional 100,000 steps performed using a combination of the steepest descent and conjugate gradient algorithms. Following minimization, the systems were gradually heated from 0 K to 310 K over 300 ps using Langevin dynamics in a canonical ensemble (NVT). Equilibration was then performed for 700 ps under the same conditions to stabilize the systems. Subsequently, production runs consisting of three independent 1 μs simulations for each system were carried out in an isothermal-isobaric ensemble (NPT) at 310 K and 1 atm pressure. The simulations employed a time step of 2 fs. Long-range electrostatic interactions were computed using the Particle Mesh Ewald (PME) method with a 10 Å cutoff [43,44]. To maintain system stability, all covalent bonds involving hydrogen atoms were constrained using the SHAKE algorithm [45,46].

4.3. Dynamic Cross-Correlation Matrix Analysis

DCCM analysis was conducted for both systems using the CPPTRAJ module in AMBER20 to evaluate residue–residue correlations. The normalized cross-correlation matrix was computed, with its elements representing cross-correlation coefficients [47], and derived using the following equation:

$$C(i,j) = \frac{c(i,j)}{\sqrt{c(i,i) \times c(j,j)}} \tag{1}$$

In this equation, $C(i,j)$ denotes the cross-correlation coefficient between the fluctuations in the coordinates of Cα atoms i and j, ranging from -1 to 1. $c(i,j)$ represents the covariance of the coordinate fluctuations between Cα atoms i and j. $c(i,i)$ and $c(j,j)$ correspond to the variances of the coordinate fluctuations of Cα atoms i and j, respectively.

4.4. Principal Component Analysis and Free Energy Landscapes

PCA is a widely used statistical method in simulation studies to identify dominant protein dynamics and explain relevant phenomena. By simplifying a large set of interdependent variables (atomic coordinates) into a smaller set of independent variables, PCA facilitates the interpretation of complex motion patterns [48]. This is achieved by solving the eigenvalues and eigenvectors of the covariance matrix, thereby decomposing the observed motions into uncorrelated principal components (PCs). Each PC is represented as a linear combination of correlated variables, expressed by the following equation:

$$PC_i = \alpha_{i1}x_1 + \alpha_{i2}x_2 + \cdots + \alpha_{im}x_m = \sum_{j=1}^{m} \alpha_{ij}x_j \tag{2}$$

Among these eigenvectors, PC1 and PC2 capture the largest proportion of variance, reflecting the dominant motions in the MD simulations [49]. To enable meaningful comparisons across simulations, root mean square (RMS) fitting was performed using a common reference structure, effectively removing translational and rotational motions of the proteins.

The free energy landscape (FEL) serves as a powerful approach for investigating protein conformational ensembles and is derived using the following equation [50–52]:

$$G_i = -k_B T ln \frac{N_i}{N_m} \tag{3}$$

Here, k_B represents the Boltzmann constant, and T is the temperature of the simulation system. N_i and N_m denote the populations of the i-th bin and the most populated bin, respectively. To generate the FELs, PC1 and PC2, RMSD values, and specific distances were utilized as reaction coordinates, with energy levels visualized through color-coded gradients.

4.5. Binding Free Energy Calculations

Binding free energies were calculated using the MM/PBSA method implemented in the MMPBSA.py plugin of AMBER20 [53]. This approach quantifies the Gibbs free energy changes upon ligand binding in binary and ternary systems. The total $\Delta G_{binding}$ was derived as follows:

$$\Delta G_{binding} = \Delta G_{complex} - \Delta G_{ligand} - \Delta G_{receptor} \quad (4)$$

According to thermodynamic principles, the Gibbs free energy change in a solvated system consists of enthalpic (ΔH) and entropic ($-T\Delta S$) contributions. The enthalpic term is further decomposed into molecular mechanical energy (ΔE_{MM}), representing interaction energies between the receptor and ligand, and the solvation free energy (ΔG_{sol}), leading to the following expression:

$$\Delta G_{binding} = \Delta E_{MM} + \Delta G_{sol} - T\Delta S \quad (5)$$

The molecular mechanical energy (ΔE_{MM}) comprises three components: van der Waals (ΔE_{vdW}), electrostatic (ΔE_{ele}), and intramolecular energies (ΔE_{int}) derived from bond, angle, and torsional terms, as follows:

$$\Delta E_{MM} = \Delta E_{vdW} + \Delta E_{ele} + \Delta E_{int} \quad (6)$$

For the solvation term (ΔG_{sol}), a Poisson–Boltzmann continuum solvent model was applied. This term is divided into polar (ΔE_{PB}) and nonpolar ($\Delta E_{nonpolar}$) contributions, as follows:

$$\Delta G_{sol} = \Delta E_{PB} + \Delta E_{nonpolar} \quad (7)$$

The nonpolar component ($\Delta E_{nonpolar}$) was calculated using a simplified linear relation with the solvent-accessible surface area ($SASA$), where γ was set to 0.00542 kcal·mol^{-1}·Å^{-2} and the offset b was assigned a value of 0.92 kcal·mol^{-1}, as follows:

$$\Delta E_{nonpolar} = \gamma * SASA + b \quad (8)$$

Notably, the entropy term ($-T\Delta S$) was excluded from the calculations due to the low conformational fluctuations observed in the systems, as indicated by their low RMSD values. Moreover, estimating the entropy using normal mode analysis or quasi-harmonic approaches would require significantly higher computational resources while introducing potential errors in systems with small conformational variations. The analysis thus focused on the relative free energy contributions of enthalpic and solvation terms, which are sufficient for comparing the binding free energies between similar systems.

4.6. Community Network Analysis

Using the NetworkView plugin in VMD, the correlation coefficient matrix $C(i,j)$ was computed to investigate intercommunity interactions and analyze the community network structure [54]. In this analysis, Cα atoms of both structures were treated as nodes, each corresponding to a residue. Edges were established between nodes if their pairwise distance remained within 4.5 Å for a minimum of 75% of the simulation duration. The edge weight was calculated using the following formula:

$$d_{(i,j)} = -log\left(\left|C_{(i,j)}\right|\right) \quad (9)$$

where i and j denote the respective nodes. The Girvan–Newman divisive algorithm, applied using the gncommunities program, was utilized to segment the network into distinct communities [55]. Intercommunity interactions were evaluated based on betweenness centrality, which quantifies the role of a node as a connector or bridge between separate communities. To enhance the robustness of the analysis, communities comprising fewer than three residues were omitted from the final network visualization.

Supplementary Materials: The following supporting information can be downloaded at: https://www.mdpi.com/article/10.3390/biom15020217/s1, Figure S1: Detailed residue sequence and secondary structure information of Mb11-SHP2-PTP complex; Figure S2: RMSD and RMSF analyses of wild-type and C459S mutant Mb11-SHP2-PTP complexes; Figure S3: PCA eigenvalue distributions for wild-type and C459S mutant Mb11-SHP2-PTP systems; Figure S4: Representative conformations of wild-type and C459S mutant Mb11-SHP2-PTP complexes, along with RMSD heatmaps illustrating the structural variations of the representative conformations; Figure S5: DSSP analysis of residue-specific secondary structure transitions in wild-type and C459S mutant Mb11-SHP2-PTP systems; Figure S6: Heatmap of the Overlap Matrix Between the First 10 PCs of wild-type and C459S mutant complexes; Table S1: Binding free energies (kcal/mol) of Mode 1 and its corresponding C459S mutant, calculated using the MM-GBSA method; Table S2: Binding free energies (kcal/mol) of Mode 2 and its corresponding C459S mutant, calculated using the MM-GBSA method; Table S3: Binding free energies (kcal/mol) of Mode 3 and its corresponding C459S mutant, calculated using the MM-GBSA method; Table S4: Detailed residue composition of Loops M1, M2, P1, and P2.

Author Contributions: Conceptualization, S.L. and Z.C.; methodology, Y.W., X.Q. and R.Z.; validation, Y.W., X.Q. and R.Z.; formal analysis, Y.W., X.Q. and R.Z.; investigation, Y.W. and R.Z.; resources, S.L.; data curation, Y.W.; writing—original draft preparation, Y.W.; writing—review and editing, S.L. and Z.C.; visualization, Y.W., R.Z., L.Z. and Q.Z.; supervision, S.L. and Z.C.; project administration, S.L. and Z.C.; funding acquisition, Z.C. All authors have read and agreed to the published version of the manuscript.

Funding: This research was funded by the National Natural Science Foundation of China (No. 82172846).

Institutional Review Board Statement: Not applicable.

Informed Consent Statement: Not applicable.

Data Availability Statement: Data are contained within the article and Supplementary Materials.

Conflicts of Interest: The authors declare no conflicts of interest.

References

1. Zhang, E.E.; Chapeau, E.; Hagihara, K.; Feng, G.-S. Neuronal Shp2 Tyrosine Phosphatase Controls Energy Balance and Metabolism. *Proc. Natl. Acad. Sci. USA* **2004**, *101*, 16064–16069. [CrossRef] [PubMed]
2. Matozaki, T.; Murata, Y.; Saito, Y.; Okazawa, H.; Ohnishi, H. Protein Tyrosine Phosphatase SHP-2: A Proto-Oncogene Product That Promotes Ras Activation. *Cancer Sci.* **2009**, *100*, 1786–1793. [CrossRef] [PubMed]
3. Chan, R.J.; Feng, G.-S. PTPN11 Is the First Identified Proto-Oncogene That Encodes a Tyrosine Phosphatase. *Blood* **2006**, *109*, 862. [CrossRef] [PubMed]
4. Hof, P.; Pluskey, S.; Dhe-Paganon, S.; Eck, M.J.; Shoelson, S.E. Crystal Structure of the Tyrosine Phosphatase SHP-2. *Cell* **1998**, *92*, 441–450. [CrossRef] [PubMed]
5. Anselmi, M.; Calligari, P.; Hub, J.S.; Tartaglia, M.; Bocchinfuso, G.; Stella, L. Structural Determinants of Phosphopeptide Binding to the N-Terminal Src Homology 2 Domain of the SHP2 Phosphatase. *J. Chem. Inf. Model.* **2020**, *60*, 3157. [CrossRef]

6. Pluskey, S.; Wandless, T.J.; Walsh, C.T.; Shoelson, S.E. Potent Stimulation of SH-PTP2 Phosphatase Activity by Simultaneous Occupancy of Both SH2 Domains. *J. Biol. Chem.* **1995**, *270*, 2897–2900. [CrossRef]
7. Marasco, M.; Kirkpatrick, J.; Nanna, V.; Sikorska, J.; Carlomagno, T. Phosphotyrosine Couples Peptide Binding and SHP2 Activation via a Dynamic Allosteric Network. *Comput. Struct. Biotechnol. J.* **2021**, *19*, 2398. [CrossRef]
8. Tajan, M.; de Rocca Serra, A.; Valet, P.; Edouard, T.; Yart, A. SHP2 Sails from Physiology to Pathology. *Eur. J. Med. Genet.* **2015**, *58*, 509–525. [CrossRef]
9. Tartaglia, M.; Martinelli, S.; Stella, L.; Bocchinfuso, G.; Flex, E.; Cordeddu, V.; Zampino, G.; van der Burgt, I.; Palleschi, A.; Petrucci, T.C.; et al. Diversity and Functional Consequences of Germline and Somatic PTPN11 Mutations in Human Disease. *Am. J. Hum. Genet.* **2006**, *78*, 279–290. [CrossRef]
10. Keilhack, H.; David, F.S.; McGregor, M.; Cantley, L.C.; Neel, B.G. Diverse Biochemical Properties of Shp2 Mutants. Implications for Disease Phenotypes. *J. Biol. Chem.* **2005**, *280*, 30984–30993. [CrossRef]
11. Aceto, N.; Sausgruber, N.; Brinkhaus, H.; Gaidatzis, D.; Martiny-Baron, G.; Mazzarol, G.; Confalonieri, S.; Quarto, M.; Hu, G.; Balwierz, P.J.; et al. Tyrosine Phosphatase SHP2 Promotes Breast Cancer Progression and Maintains Tumor-Initiating Cells via Activation of Key Transcription Factors and a Positive Feedback Signaling Loop. *Nat. Med.* **2012**, *18*, 529–537. [CrossRef] [PubMed]
12. Zhou, X.; Coad, J.; Ducatman, B.; Agazie, Y.M. SHP2 Is Up-Regulated in Breast Cancer Cells and in Infiltrating Ductal Carcinoma of the Breast, Implying Its Involvement in Breast Oncogenesis. *Histopathology* **2008**, *53*, 389–402. [CrossRef] [PubMed]
13. Nagamura, Y.; Miyazaki, M.; Nagano, Y.; Tomiyama, A.; Ohki, R.; Yanagihara, K.; Sakai, R.; Yamaguchi, H. SHP2 as a Potential Therapeutic Target in Diffuse-Type Gastric Carcinoma Addicted to Receptor Tyrosine Kinase Signaling. *Cancers* **2021**, *13*, 4309. [CrossRef] [PubMed]
14. Kim, J.S.; Shin, O.R.; Kim, H.K.; Cho, Y.S.; An, C.H.; Lim, K.W.; Kim, S.S. Overexpression of Protein Phosphatase Non-Receptor Type 11 (PTPN11) in Gastric Carcinomas. *Dig. Dis. Sci.* **2010**, *55*, 1565–1569. [CrossRef]
15. Mainardi, S.; Mulero-Sánchez, A.; Prahallad, A.; Germano, G.; Bosma, A.; Krimpenfort, P.; Lieftink, C.; Steinberg, J.D.; De Wit, N.; Gonçalves-Ribeiro, S.; et al. SHP2 is required for growth of KRAS-mutant non-small-cell lung cancer in vivo letter. *Nat. Med.* **2018**, *24*, 961–967. [CrossRef]
16. Xia, L.; Yang, F.; Wu, X.; Li, S.; Kan, C.; Zheng, H.; Wang, S. SHP2 Inhibition Enhances the Anticancer Effect of Osimertinib in EGFR T790M Mutant Lung Adenocarcinoma by Blocking CXCL8 Loop Mediated Stemness. *Cancer Cell Int.* **2021**, *21*, 337. [CrossRef]
17. Xu, R.; Yu, Y.; Zheng, S.; Zhao, X.; Dong, Q.; He, Z.; Liang, Y.; Lu, Q.; Fang, Y.; Gan, X.; et al. Overexpression of Shp2 Tyrosine Phosphatase Is Implicated in Leukemogenesis in Adult Human Leukemia. *Blood* **2005**, *106*, 3142. [CrossRef]
18. Dong, L.; Han, D.; Meng, X.; Xu, M.; Zheng, C.; Xia, Q. Activating Mutation of SHP2 Establishes a Tumorigenic Phonotype Through Cell-Autonomous and Non-Cell-Autonomous Mechanisms. *Front. Cell Dev. Biol.* **2021**, *9*, 630712. [CrossRef]
19. Fedele, C.; Li, S.; Teng, K.W.; Foster, C.J.R.; Peng, D.; Ran, H.; Mita, P.; Geer, M.J.; Hattori, T.; Koide, A.; et al. SHP2 Inhibition Diminishes KRASG12C Cycling and Promotes Tumor Microenvironment Remodeling. *J. Exp. Med.* **2021**, *218*, e20201414. [CrossRef]
20. Hui, E.; Cheung, J.; Zhu, J.; Su, X.; Taylor, M.J.; Wallweber, H.A.; Sasmal, D.K.; Huang, J.; Kim, J.M.; Mellman, I.; et al. T Cell Costimulatory Receptor CD28 Is a Primary Target for PD-1-Mediated Inhibition. *Science* **2017**, *355*, 1428–1433. [CrossRef]
21. Xu, X.; Hou, B.; Fulzele, A.; Masubuchi, T.; Zhao, Y.; Wu, Z.; Hu, Y.; Jiang, Y.; Ma, Y.; Wang, H.; et al. PD-1 and BTLA Regulate T Cell Signaling Differentially and Only Partially through SHP1 and SHP2. *J. Cell Biol.* **2020**, *219*, e201905085. [CrossRef] [PubMed]
22. Li, J.; Jie, H.-B.; Lei, Y.; Gildener-Leapman, N.; Trivedi, S.; Green, T.; Kane, L.P.; Ferris, R.L. PD-1/SHP-2 Inhibits Tc1/Th1 Phenotypic Responses and the Activation of T Cells in the Tumor Microenvironment. *Cancer Res.* **2015**, *75*, 508–518. [CrossRef] [PubMed]
23. Kleffel, S.; Posch, C.; Barthel, S.R.; Mueller, H.; Schlapbach, C.; Guenova, E.; Elco, C.P.; Lee, N.; Juneja, V.R.; Zhan, Q.; et al. Melanoma Cell-Intrinsic PD-1 Receptor Functions Promote Tumor Growth. *Cell* **2015**, *162*, 1242. [CrossRef]
24. Asmamaw, M.D.; Shi, X.-J.; Zhang, L.-R.; Liu, H.-M. A Comprehensive Review of SHP2 and Its Role in Cancer. *Cell. Oncol.* **2022**, *45*, 729–753. [CrossRef]
25. Tartaglia, M.; Mehler, E.L.; Goldberg, R.; Zampino, G.; Brunner, H.G.; Kremer, H.; van der Burgt, I.; Crosby, A.H.; Ion, A.; Jeffery, S.; et al. Mutations in PTPN11, Encoding the Protein Tyrosine Phosphatase SHP-2, Cause Noonan Syndrome. *Nat. Genet.* **2001**, *29*, 465–468. [CrossRef]
26. Yu, Z.-H.; Xu, J.; Walls, C.D.; Chen, L.; Zhang, S.; Zhang, R.; Wu, L.; Wang, L.; Liu, S.; Zhang, Z.-Y. Structural and Mechanistic Insights into LEOPARD Syndrome-Associated SHP2 Mutations. *J. Biol. Chem.* **2013**, *288*, 10472. [CrossRef]
27. Sha, F.; Kurosawa, K.; Glasser, E.; Ketavarapu, G.; Albazzaz, S.; Koide, A.; Koide, S. Monobody Inhibitor Selective to the Phosphatase Domain of SHP2 and Its Use as a Probe for Quantifying SHP2 Allosteric Regulation. *J. Mol. Biol.* **2023**, *435*, 168010. [CrossRef]

28. Li, M.; Lan, X.; Shi, X.; Zhu, C.; Lu, X.; Pu, J.; Lu, S.; Zhang, J. Delineating the Stepwise Millisecond Allosteric Activation Mechanism of the Class C GPCR Dimer mGlu5. *Nat. Commun.* **2024**, *15*, 7519. [CrossRef]
29. Zhang, M.; Ao, J.; Liu, N.; Chen, T.; Lu, S. Exploring the Constitutive Activation Mechanism of the Class A Orphan GPR20. *Acta Pharmacol. Sin.* **2025**, *46*, 500–511. [CrossRef]
30. Jones, G.; Jindal, A.; Ghani, U.; Kotelnikov, S.; Egbert, M.; Hashemi, N.; Vajda, S.; Padhorny, D.; Kozakov, D. Elucidation of Protein Function Using Computational Docking and Hotspot Analysis by ClusPro and FTMap. *Acta Cryst. D* **2022**, *78*, 690–697. [CrossRef]
31. Desta, I.T.; Porter, K.A.; Xia, B.; Kozakov, D.; Vajda, S. Performance and Its Limits in Rigid Body Protein-Protein Docking. *Structure* **2020**, *28*, 1071–1081.e3. [CrossRef] [PubMed]
32. Genheden, S.; Ryde, U. The MM/PBSA and MM/GBSA Methods to Estimate Ligand-Binding Affinities. *Expert Opin. Drug Discov.* **2015**, *10*, 449–461. [CrossRef] [PubMed]
33. Newman, M.E.J. Modularity and Community Structure in Networks. *Proc. Natl. Acad. Sci. USA* **2006**, *103*, 8577–8582. [CrossRef]
34. Sali, A. Comparative Protein Modeling by Satisfaction of Spatial Restraints. *Mol. Med. Today* **1995**, *1*, 270–277. [CrossRef]
35. Lee, T.-S.; Allen, B.K.; Giese, T.J.; Guo, Z.; Li, P.; Lin, C.; McGee, T.D.; Pearlman, D.A.; Radak, B.K.; Tao, Y.; et al. Alchemical Binding Free Energy Calculations in AMBER20: Advances and Best Practices for Drug Discovery. *J. Chem. Inf. Model.* **2020**, *60*, 5595–5623. [CrossRef]
36. Maier, J.A.; Martinez, C.; Kasavajhala, K.; Wickstrom, L.; Hauser, K.E.; Simmerling, C. ff14SB: Improving the Accuracy of Protein Side Chain and Backbone Parameters from ff99SB. *J. Chem. Theory Comput.* **2015**, *11*, 3696–3713. [CrossRef]
37. Wang, J.; Wolf, R.M.; Caldwell, J.W.; Kollman, P.A.; Case, D.A. Development and Testing of a General Amber Force Field. *J. Comput. Chem.* **2004**, *25*, 1157–1174. [CrossRef]
38. Zhang, H.; Chu, G.; Wang, G.; Yao, M.; Lu, S.; Chen, T. Mechanistic Understanding of the Palmitoylation of Go Protein in the Allosteric Regulation of Adhesion Receptor GPR97. *Pharmaceutics* **2022**, *14*, 1856. [CrossRef]
39. Jorgensen, W.L.; Chandrasekhar, J.; Madura, J.D.; Impey, R.W.; Klein, M.L. Comparison of Simple Potential Functions for Simulating Liquid Water. *J. Chem. Phys.* **1983**, *79*, 926–935. [CrossRef]
40. Nayar, D.; Agarwal, M.; Chakravarty, C. Comparison of Tetrahedral Order, Liquid State Anomalies, and Hydration Behavior of mTIP3P and TIP4P Water Models. *J. Chem. Theory Comput.* **2011**, *7*, 3354–3367. [CrossRef]
41. Zhuang, H.; Ji, D.; Fan, J.; Li, M.; Tao, R.; Du, K.; Lu, S.; Chai, Z.; Fan, X. Mechanistic Insights into the Protection Effect of Argonaute–RNA Complex on the HCV Genome. *Biomolecules* **2022**, *12*, 1631. [CrossRef] [PubMed]
42. Götz, A.W.; Williamson, M.J.; Xu, D.; Poole, D.; Le Grand, S.; Walker, R.C. Routine Microsecond Molecular Dynamics Simulations with AMBER on GPUs. 1. Generalized Born. *J. Chem. Theory Comput.* **2012**, *8*, 1542–1555. [CrossRef] [PubMed]
43. Darden, T.; York, D.; Pedersen, L. Particle Mesh Ewald: An N·log(N) Method for Ewald Sums in Large Systems. *J. Chem. Phys.* **1993**, *98*, 10089–10092. [CrossRef]
44. York, D.M.; Wlodawer, A.; Pedersen, L.G.; Darden, T.A. Atomic-Level Accuracy in Simulations of Large Protein Crystals. *Proc. Natl. Acad. Sci. USA* **1994**, *91*, 8715–8718. [CrossRef]
45. Lee, S.-H.; Palmo, K.; Krimm, S. A Comparative Study of Molecular Dynamics in Cartesian and in Internal Coordinates: Dynamical Instability in the Latter Caused by Nonlinearity of the Equations of Motion. *J. Comput. Chem.* **2007**, *28*, 1107–1118. [CrossRef]
46. Ryckaert, J.-P.; Ciccotti, G.; Berendsen, H.J.C. Numerical Integration of the Cartesian Equations of Motion of a System with Constraints: Molecular Dynamics of *n*-Alkanes. *J. Comput. Phys.* **1977**, *23*, 327–341. [CrossRef]
47. Roe, D.R.; Cheatham, T.E. PTRAJ and CPPTRAJ: Software for Processing and Analysis of Molecular Dynamics Trajectory Data. *J. Chem. Theory Comput.* **2013**, *9*, 3084–3095. [CrossRef]
48. Zhang, H.; Ni, D.; Fan, J.; Li, M.; Zhang, J.; Hua, C.; Nussinov, R.; Lu, S. Markov State Models and Molecular Dynamics Simulations Reveal the Conformational Transition of the Intrinsically Disordered Hypervariable Region of K-Ras4B to the Ordered Conformation. *J. Chem. Inf. Model.* **2022**, *62*, 4222–4231. [CrossRef]
49. Amadei, A.; Linssen, A.B.M.; Berendsen, H.J.C. Essential Dynamics of Proteins. *Proteins Struct. Funct. Bioinform.* **1993**, *17*, 412–425. [CrossRef]
50. Li, X.; Wang, C.; Peng, T.; Chai, Z.; Ni, D.; Liu, Y.; Zhang, J.; Chen, T.; Lu, S. Atomic-Scale Insights into Allosteric Inhibition and Evolutional Rescue Mechanism of Streptococcus Thermophilus Cas9 by the Anti-CRISPR Protein AcrIIA6. *Comput. Struct. Biotechnol. J.* **2021**, *19*, 6108–6124. [CrossRef]
51. Lu, S.; He, X.; Yang, Z.; Chai, Z.; Zhou, S.; Wang, J.; Rehman, A.U.; Ni, D.; Pu, J.; Sun, J.; et al. Activation pathway of a G protein-coupled receptor uncovers conformational intermediates as targets for allosteric drug design. *Nat. Commun.* **2021**, *12*, 4721. [CrossRef] [PubMed]
52. Ni, D.; Wei, J.; He, X.; Rehman, A.U.; Li, X.; Qiu, Y.; Pu, J.; Lu, S.; Zhang, J. Discovery of cryptic allosteric sites using reversed allosteric communication by a combined computational and experimental strategy. *Chem. Sci.* **2021**, *12*, 464–476. [CrossRef] [PubMed]

53. Wang, E.; Sun, H.; Wang, J.; Wang, Z.; Liu, H.; Zhang, J.Z.H.; Hou, T. End-Point Binding Free Energy Calculation with MM/PBSA and MM/GBSA: Strategies and Applications in Drug Design. *Chem. Rev.* **2019**, *119*, 9478–9508. [CrossRef] [PubMed]
54. Eargle, J.; Luthey-Schulten, Z. NetworkView: 3D Display and Analysis of protein·RNA Interaction Networks. *Bioinformatics* **2012**, *28*, 3000–3001. [CrossRef]
55. Girvan, M.; Newman, M.E.J. Community Structure in Social and Biological Networks. *Proc. Natl. Acad. Sci. USA* **2002**, *99*, 7821–7826. [CrossRef]

biomolecules

MDPI

Article

CellGAT: A GAT-Based Method for Constructing a Cell Communication Network Integrating Multiomics Information

Tianjiao Zhang [1], Zhenao Wu [1], Liangyu Li [1], Jixiang Ren [1], Ziheng Zhang [1], Jingyu Zhang [2] and Guohua Wang [1,3,*]

1 College of Computer and Control Engineering, Northeast Forestry University, Harbin 150040, China; tianjiaozhang@nefu.edu.cn (T.Z.); 2022112562@nefu.edu.cn (Z.W.); liangyuli@nefu.edu.cn (L.L.); 1241126773@nefu.edu.cn (J.R.); zzhjs@nefu.edu.cn (Z.Z.)
2 Department of Radiology, The Second Affiliated Hospital of Harbin Medical University, Harbin 150040, China; hmuzjy@hrbmu.edu.cn
3 Faculty of Computing, Harbin Institute of Technology, Harbin 150001, China
* Correspondence: ghwang@nefu.edu.cn; Tel.: +86-13946094199

Abstract: The growth, development, and differentiation of multicellular organisms are primarily driven by intercellular communication, which coordinates the activities of diverse cell types. This cell-to-cell signaling is typically mediated by various types of protein–protein interactions, including ligand–receptor; receptor–receptor, and extracellular matrix–receptor interactions. Currently, computational methods for inferring ligand–receptor communication primarily depend on gene expression data of ligand–receptor pairs and spatial information of cells. Some approaches integrate protein complexes; transcription factors; or pathway information to construct cell communication networks. However, few methods consider the critical role of protein–protein interactions (PPIs) in intercellular communication networks, especially when predicting communication between different cell types in the absence of cell type information. These methods often rely on ligand–receptor pairs that lack PPI evidence, potentially compromising the accuracy of their predictions. To address this issue, we propose CellGAT, a framework that infers intercellular communication by integrating gene expression data of ligand–receptor pairs, PPI information, protein complex data, and experimentally validated pathway information. CellGAT not only builds a priori models but also uses node embedding algorithms and graph attention networks to build cell communication networks based on scRNA-seq (single-cell RNA sequencing) datasets and includes a built-in cell clustering algorithm. Through comparisons with various methods, CellGAT accurately predicts cell–cell communication (CCC) and analyzes its impact on downstream pathways; neighboring cells; and drug interventions.

Keywords: cell–cell interactions; scRNA-seq; deep learning; graph convolutional neural network; graph attention networks

1. Introduction

Intercellular interactions are crucial in many biological processes, including cell growth, division, differentiation, tissue or organ development, and even play significant roles in disease progression. The study of single ligand–receptor pairs and actual cell–cell communication represents a progressive exploration from the molecular to the systemic level. Research on single pairs provides foundational data for understanding cell–cell communication, while the analysis of cell–cell communication further enhances our comprehension of the overall functionality of biological systems. The signaling of CCC is typically mediated by various types of PPIs, including ligand–receptor (LR), receptor–receptor, and extracellular matrix–receptor interactions. Understanding multicellular organisms and their functions requires a deep understanding of cellular activities, often achieved through studying signaling events-induced responses and downstream effects [1]. These signaling events include the binding of secreted ligands to homologous receptors, as well as intercellular [2] or intracellular ligand–receptor interactions. Understanding the phenotypic effects of these cell–cell communication patterns under healthy, perturbed, and diseased conditions is increasingly recognized [3]. Changes in intercellular communication are crucial for understanding tissue function and disease progression [4], and understanding changes in intercellular communication is crucial for deciphering tissue function and disease progression, particularly in tumor advancement, where direct cell–cell communication plays a critical role [5]. Therefore, using computational tools to simulate this cell–cell communication role in tumors can help identify biomarkers for disease severity in patients and improve drug responses in treatment approaches [6].

PPIs on the cell surface facilitate the efficient exchange of information between cells or between cells and their environment, occurring in a highly dynamic, environment, time, and space-dependent manner [7]. However, identifying interactions in native microenvironments often requires expensive experiments or extensive domain knowledge to detect active interactions [8]. To seek relatively facile methods for studying CCC, researchers have developed numerous computational approaches such as CellPhoneDB [9], Crosstalkr [10], NicheNet [11], and CellChat [12]. These methods are all gene ligand–receptor expression information and downstream target analysis, paving new paths for exploring cell–cell communication at the single-cell level. Building upon these methods, researchers have extensively applied approaches to studying CCC in different models, for instance, spatial transcriptomic data, where cellular coordinate data can better visualize CCC predictions and provide the basis for spatial adjacency. However, few methods are considering the importance of PPIs in intercellular communication networks. The absence of PPI information may lead these methods to inaccuracies, as abnormal gene expression of certain genes might not directly translate to loss of activity in the encoded proteins, affecting the accuracy of predictions.

To achieve more accurate predictions of CCC and enhance the biological relevance of its results, there are still several limitations that need to be addressed. For instance, there lacks a unified ground truth for validated LR interactions, and many datasets lack annotations such as protein complex information and pathway information involved in LR interactions. Previous methods have shown that integrating both ligand and receptor expression information with context-specific transcriptomic data affects the accuracy of predicting LR interactions based solely on expression values. Therefore, this study integrates gene expression information of ligands and receptors, PPI information, protein complex information, and experimentally validated pathway information to improve the accuracy of existing methods in predicting CCC. To address the challenge of annotating multiple large dataset inputs and projecting them onto smaller test sets, this study proposes the use of graph attention neural networks as potential solutions for predicting CCC. The introduction of this new CCC prediction framework will aid in the discovery of new ligand–receptor interactions, which may reveal novel cellular functions or roles under different environments.

To address these challenges, this paper proposes CellGAT, a deep learning approach based on graph attention neural networks for predicting cell–cell communication in scRNA-seq data. CellGAT integrates ligand–receptor annotations (such as protein complexes, PPIs, cell type, and pathway information) along with gene expression values. To better infer CCC, we constructed an a priori model. Through a feedforward mechanism, CellGAT constructs a cell communication network and learns the optimal data representation of the model to predict the probability of CCC interactions. CellGAT assigns numerical values to relationships between cell populations, ligands, and receptors, and ranks CCC activities using these embedded communication probabilities. This deep learning framework allows for capturing rich information about cellular signaling networks and transcriptomic data while prioritizing multiple interactions simultaneously. To validate the utility of CellGAT, we apply the model to single-cell transcriptomic data from Drosophila, human heart and lung cancer cell lines, and mouse brain. It is demonstrated that our model can be used to identify downstream activated pathways and biologically relevant changes in CCC in cancer cells following drug interventions and predict cell–cell interactions in spatial microenvironments.

2. Materials and Methods

2.1. Overview of CellGAT

Figure 1 provides an overview of the development and testing of CellGAT. CellGAT integrates existing databases' ligand–receptor information, complex protein information, relevant pathway information, and gene expression information from single-cell transcriptomic data to infer biologically significant known and unknown ligand–receptor interactions between different cell types or cell clusters and visualizes these interactions for analysis. The prediction of cell–cell interactions by CellGAT involves two main steps: (1) constructing a prior model to extract features based on existing ligand–receptor-related information in databases and single-cell sequencing gene expression information (Figure 1A); (2) using transcriptomic information and features output from (1) to predict cell–cell communication in the single-cell dataset (Figure 1B). Finally, based on the computed communication probabilities from (2), CellGAT analyzes the top ten ranked ligand–receptor interactions in terms of communication probability in each single-cell sequencing dataset using Circos and Sankey plots (Figure 1C). CellGAT utilizes gene expression information from single-cell sequencing or spatial transcriptomics data as an input. Furthermore, the output of CellGAT is not a fixed ranked list of interactions predefined by Omnipath between specific cell types. Instead, it is based on the likelihood of ligand–receptor interactions predicted by the model and their corresponding rankings. In this manner, we generate a prediction matrix that reflects potential intercellular interactions. Subsequently, we use the Omnipath database as a validation set to assess the accuracy of the CellGAT predictions.

It is crucial to infer cell–cell communication based on biologically validated ligand–receptor pair-related information. Therefore, in step (1), CellGAT utilizes the OmniPath database, which contains over 30,000 experimentally validated intracellular interactions, more than 3000 intercellular interactions, as well as comprehensive protein information and related pathway information. These pieces of information have been successfully applied in previous methods such as LIANA [13] and CellChat to predict known and unknown cell–cell interactions. CellGAT extracts all significantly expressed ligands and receptors from the input single cell sequencing gene expression matrix and constructs a directed graph using ligand–receptor pairs existing in the OmniPath database. Additionally, CellGAT provides protein complex and KEGG pathway annotation information in the form of node embeddings. For any pair of ligand and receptor nodes in the directed graph, CellGAT assigns corresponding values to represent the relevance and complexity similarity of ligands/receptors with pathways. Based on these steps, the information in the directed graph is updated, and then feature learning is performed using the Node2Vec [14] framework. This framework updates the representation information of each node based on experimentally validated linkage information.

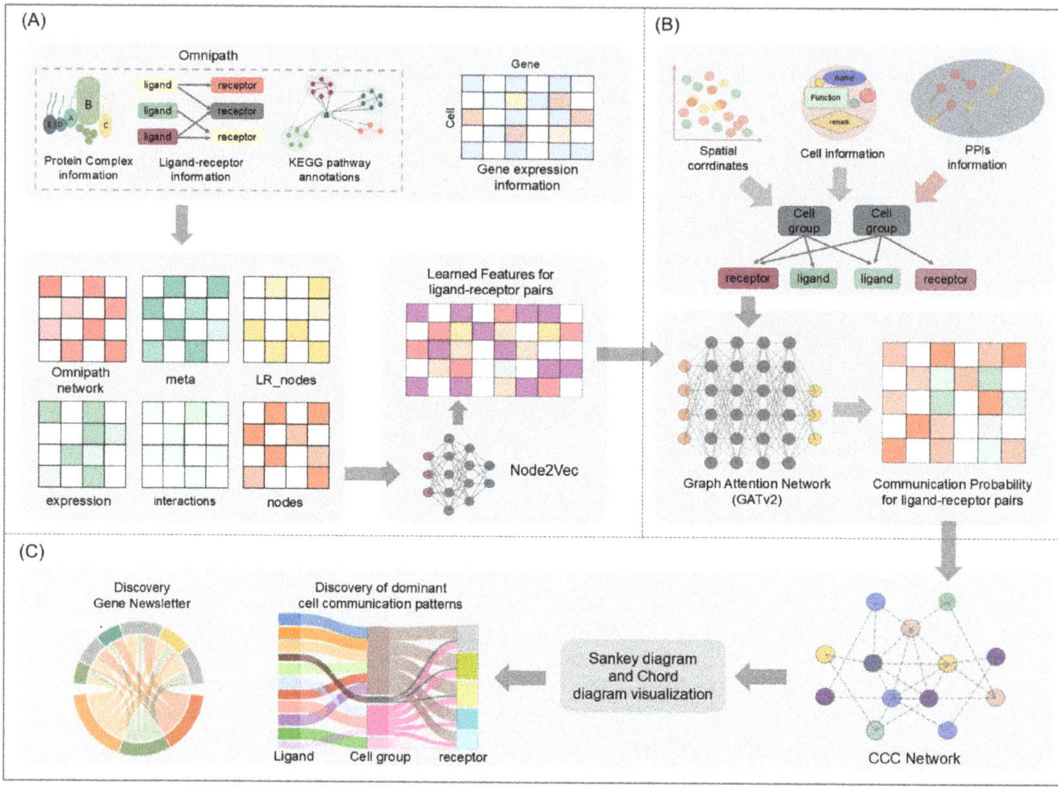

Figure 1. This figure outlines the entire workflow, including data input, model architecture, and analysis steps for predicting intercellular communication. (**A**) Feature construction of CellGAT model input. (**B**) CellGAT predicts interactions. (**C**) CellGAT analyzes prediction results and performs visual analysis.

Gene co-expression is crucial for inferring cell–cell communication. Therefore, in step (2), when computing the communication probability for all possible ligand–receptor pairs, CellGAT employs the Leiden [15] clustering method to identify cell types within a given single-cell sequencing dataset, thereby obtaining cell clusters. For each cluster, CellGAT identifies differentially expressed ligands and receptors. Based on these cell clusters, ligands, and receptors, a new directed graph is constructed. If a ligand/receptor is significantly upregulated in a cell cluster, an edge is drawn from the given cell cluster to the ligand/receptor. Additionally, the previously obtained positional embeddings and new contextual information are further annotated. The newly identified directed graph is then input into the graph attention network (GATv2) [16] for training. GATv2 continuously updates the information of all nodes in a direction favorable to the ligand–receptor interactions based on the learned facts. Finally, the values obtained by GATv2 through inner product calculation serve as the potential communication probabilities for the corresponding ligand–receptor pairs in the single-cell sequencing dataset.

Due to the lack of a universally accepted gold standard for evaluating the accuracy of predicted cell–cell communication, we used the biologically validated OmniPath dataset as a benchmark for assessment. OmniPath, a plugin designed to help researchers analyze and integrate complex molecular and cellular interaction networks, has been extensively studied by Ceccarelli et al., following its creation by Türei et al. These studies have demonstrated the effectiveness of OmniPath in validating communication network analysis. This allowed us to leverage experimentally confirmed LR interactions as a reference point. In addition, we publicly integrated ligand–receptor interaction pairs based on previous scientific studies to validate the performance of CellGAT.

2.2. Datasets

CellGAT underwent benchmark testing on four different datasets covering three distinct biological environments: developmental, perturbation, and fibrosis settings. The Drosophila developmental dataset consists of 5835 cells, clustered using the Leiden algorithm (GEO accession number: GSE95025). The PC9 lung cancer cell line perturbation dataset comprises 6778 cells on day 0, 4770 cells on day 7 (replicate 1), and 3515 cells on day 7 (replicate 2), also clustered using the Leiden algorithm (GEO accession number: GSE150949). The mouse brain dataset consists of 2305 cells, clustered using the Leiden algorithm https://data.caltech.edu/records/s0vdx-0k302 (accessed on 25 February 2025). The 10X Visium dataset includes spatial slides of the left ventricle of human myocardial infarction patients, comprising 3100 cells with 9 cell types https://zenodo.org/record/6578047 (accessed on 25 February 2025). To ensure the uniqueness of interaction pairs, we remove duplicates for each ligand–receptor pair within the dataset.

2.3. Ligand–Receptor Basic Truth Database

Correctly assigning the roles of signaling molecules and understanding their interactions are crucial for predicting biologically meaningful cell-to-cell communication. Therefore, CellGAT utilizes the OmniPath (1.0.6) [17] database curated by Martina Kutmon et al., which provides all scientifically validated protein–protein interactions (31729 pairs), intercellular interactions (3599 pairs), protein complexes (8022 complexes), and KEGG pathway annotations (7534 entries) [17–19] as the experimentally verified relationships to measure the accuracy of the method. These data are presented in tabular form in the OmniPath package in Python (3.11), with ligands and receptors identified using HGCN symbols. We divided the dataset into a training set and a validation set, with the validation set comprising 20% of the original dataset. This ensures that the validation set is a representative subset of the entire dataset, allowing us to assess the performance of our model on unseen data.

2.4. Cell Clustering

CellGAT provides a built-in functionality for classifying cells into different cell clusters. Specifically, CellGAT first computes the adjacency relationships between cells, a process that utilizes dimensionality reduction techniques (such as principal component analysis or nearest neighbor embedding) to map high-dimensional gene expression data into a lower-dimensional space. The dimensionality reduction technique used in CellGAT is Node2vec [20], an embedding method specifically designed for graph data. Node2vec not only preserves the neighborhood information of nodes but also retains the structural features of the network, providing an effective tool for calculating the adjacency relationships between cells. Next, the low-dimensional data are subjected to the Leiden community detection algorithm to cluster cells into clusters. The Leiden clustering algorithm automatically determines the number of categories based on the features of the data and assigns cells to the respective categories.

2.5. Mapping PPIs to LR Interactions

Given gene expression data $G = \{g_1, g_2, \ldots, g_u\}$, where g_i is the identifier of the i-th gene and u is the total number of genes, and protein–protein interaction data $P = \{p_1, p_2, \ldots, p_v\}$, where p_j is the identifier of the j-th protein and v is the total number of proteins, the mapping relationship between ligand or receptor genes and proteins can be represented as a $u \times v$ binary matrix D, where D_{ij} indicates whether gene g_i is associated with protein p_j. Principle of prioritizing gene co-expression between cells:

$$D_{ij} = \begin{cases} 1 & \text{if } g_i \text{ and } p_j \text{ are related} \\ 0 & \text{otherwise} \end{cases}, \tag{1}$$

The goal of mapping genes to proteins is to determine the elements D_{ij} in the matrix D. This can be achieved by comparing the gene names or identifiers from the gene expression data with the protein names or identifiers from the protein–protein interaction data. Specifically, for each gene g_i in the gene expression data, it can be checked whether it is associated with any protein p_j in the PPI data. If it is, then D_{ij} is set to 1; otherwise, D_{ij} is set to 0.

2.6. Data Preprocessing

CellGAT first transposes the gene expression matrix of the input scRNA-seq data and creates an AnnData object. After filtering and normalization of the gene expression information, it calculates the pairwise distances between cells to construct an adjacency matrix. In the process of cell communication, the interactions between cells often rely on the coordinated expression of genes. To adhere to the principle of prioritizing gene co-expression between cells, CellGAT employs the Leiden algorithm for cell clustering on the cleaned data:

$$\vartheta = \frac{1}{2J} \sum_c \left(e_c - \gamma \frac{K_c^2}{2J} \right), \tag{2}$$

Among them, J represents the total number of borders, e_c represents the number of edges in community node c, and K_c represents the degree of node c. In experiments involving single-cell sequencing data, the cell type annotations are often not provided directly. The Leiden algorithm is widely used in biological network analysis.

For each clustered cell cluster ϑ, the average expression level of genes in each cluster is computed. Finally, information regarding cell types, PPIs, ligand–receptor interactions, and protein complexes is extracted from the OmniPath database. Based on this information, data frames for nodes and interactions are constructed. Subsequently, the data are further organized and processed to generate a dataset containing cells and interactions, which serves as the input for Node2vec node embedding.

2.7. Interaction Inference

CellGAT employs two key steps to predict intercellular communication: firstly, it constructs a prior model using the OmniPath database, which encompasses information such as cell types, PPIs, LR interactions, and protein complexes; secondly, it utilizes a graph attention network to infer intercellular communication. The integration of these two steps enables CellGAT to demonstrate outstanding performance in predicting intercellular interactions and offers a comprehensive understanding of the complex communication network among cells. A single ligand–receptor pair refers to the interaction between a specific ligand (a signaling molecule) and its corresponding receptor in the context of cell-to-cell communication. In contrast, true cell-to-cell communication occurs in a complex multicellular environment, where cells interact through multiple signaling pathways and mechanisms. Our study focuses on cell-to-cell interactions within complex multicellular environments, aiming to comprehensively analyze the communication networks between cells, consistent with the definitions of intercellular interactions in the Omnipath database. The interactions considered in this research are directed; for example, we specifically analyze scenarios where one cell type (e.g., cell type A) binds to another cell type (e.g., cell type B) through a receptor–ligand combination (P/L).

2.8. Construction of Prior Model

After identifying the ligands and receptors present in the single-cell RNA dataset, cross-referencing is performed using the OmniPath database to determine the valid ligand–receptor interactions present in the dataset. But it is not limited to the connections between each cell group in the cluster. Using the PyTorch [21] Geometric library, a directed graph is constructed with ligands and receptors as nodes, and validation edges are built based on the relationships between ligands and receptors. This directed graph is accompanied by a matrix with dimensions (number of nodes) × (number of nodes). For each cell in the matrix, it represents the connectivity between any two nodes, with its value being

$$W(m,n) = z_{m,n} + a_{m,n}, \quad (3)$$

where W represents a cell, and m and n represent a ligand and its corresponding receptor, respectively. $W(m,n)$ denotes the value of that cell in the feature matrix, $Z_{m,n}$ is the number of protein complexes where both nodes m and n are subunits, and $a_{m,n}$ is the number of KEGG pathway members where both nodes m and n are included.

Next, CellGAT employs the Node2Vec model to perform random walks, training the labeled feature matrix for 100 iterations to generate two-dimensional embeddings. Upon training completion, it extracts the two-dimensional spatial coordinates for each node. Subsequently, it utilizes the inner product of these node positions to construct a position matrix of size equal to the number of ligands multiplied by the number of receptors, thereby forming a spatial distribution map of cell–cell interactions. The closer the distance, the more likely communication exists. Spatial information helps validate that our model favors spatial proximity.

2.9. Inference of Intercellular Communication by Graph Attention Network

Before using CellGAT for the prediction of single-cell RNA sequencing data, data undergo logarithmic transformation and normalization. Next, genes with non-negative average expression levels across all cells within a given cell type are identified to determine significantly expressed genes. For each cell group, a directed graph is constructed using Pytorch Geometric, where each cell group points to its respective ligands and receptors. This series of steps lays the groundwork for subsequent predictions of intercellular communication.

The directed graph of the cell group is similar to the graph used when constructing the prior model. Before inference, a feature matrix of size (number of nodes) × (number of nodes) is used to label the directed graph. For nodes m and n, the value in each cell represents the connection status between any two nodes, as follows:

If node m is a ligand/receptor, or node n is a ligand/receptor:

$$X(m,n) = Y_m Y_n + A_{m,n}, \quad (4)$$

where Y_m represents the average expression level of ligand/receptor m, Y_n represents the average expression level of ligand/receptor n with respect to A_m, and n corresponds to the value of the matrix generated by the previous step.

If node m is a cell group and node n is a ligand/receptor:

$$X(m,n) = Y_{m,n}, \quad (5)$$

In the formula, $Y_{m,n}$ is the average expression of ligand/receptor n in cell group m.

If nodes m and n both represent cell types, embeddings are provided only when spatial information indicating cell-specific information is included in the dataset. Cell spatial coordinates are an example of this. In such cases, the embedding values are as follows:

$$X(m,n) = Z_{m,n}, \tag{6}$$

where $Z_{m,n}$ is a selection function similar to describing the positional information between cell types m and n. For example, for spatial coordinates, $Z_{m,n}$ is computed as the minimum Euclidean distance between cells of type m and cells of type n.

$$l(h_m, h_n) = b^T LeakyReLU(O \cdot [h_m || h_n]), \tag{7}$$

In the equation $b \in R^{2d'}$, $W \in R^{d' \times d}$ are learnable parameters. For each node m, after computing scores with all neighbors, normalization is performed using softmax:

$$\alpha_{mn} = softmax_n(l(h_m, h_n)) = \frac{\exp(l(h_m, h_n))}{\sum_{n' \in \eta_m} \exp(l(h_m, h_{n'}))}, \tag{8}$$

Finally, GATv2 uses weights for aggregation:

$$h'_m = \sigma\left(\sum_{n \in \eta_m} \alpha_{mn} \cdot O h_n\right), \tag{9}$$

For semi-supervised classification tasks such as GATv2, the commonly used loss function is Cross-Entropy Loss, which is formulated as follows:

$$L_{CE} = -\sum_{i=1}^{N} f_i \log(q_i), \tag{10}$$

Here, f_i represents the true label probability distribution of the i-th sample, and q_i is the predicted probability that the model assigns to the i-th sample being a positive class. This loss function is suitable for situations where nodes have categorical labels, aiming to minimize the difference between the true labels and the predicted labels.

During the training process, the negative log-likelihood loss function is utilized. Upon completion of training, a single forward pass is conducted to obtain two-dimensional embeddings for all nodes in the graph. To identify the top-ranked validated ligand–receptor links, an inner product operation is performed on the embeddings of ligands and receptors, resulting in a matrix of shape (ligands) × (receptors). Subsequently, this matrix is flattened, and the ligand–receptor links are ranked in descending order based on the inner product. To determine the source and destination cell types, CellGAT computes the inner product of cell type embeddings with ligand and receptor embeddings, respectively. This yields a matrix of shape (number of ligands × number of receptors), assigning communication probabilities to each potential edge. The interactions are then sorted based on the inner product values. To avoid randomness of results and verify the stability of the model, we conducted experiments on the data under different default parameters and compared the outcomes.

To delve deeper into the predicted interactions by CellGAT in single-cell transcriptomic data, circular plots and Sankey diagrams were employed to illustrate the top ten results. These graphical tools offer an intuitive display of cellular communication patterns, aiding in unraveling the complexity and significance of intercellular interactions. We visualized the top 10 results, and no impact on accuracy was observed when considering only the top-1, top-3, or top-5 results. The Circos plot focuses on illustrating the overall topology of the cell communication network, highlighting global relationships between different cell types, while the Sankey diagram emphasizes the direction and weight of signal transmission, making it suitable for analyzing dynamic changes in specific pathways. The Circos plot helps identify strong connections among key cell populations, whereas the Sankey diagram traces the detailed flow of signals. The two visualizations complement each other, providing both a global perspective and an in-depth analysis of specific interactions.

All abbreviations used in the text are listed in Abbreviations section.

3. Results

3.1. Comparison with Other Methods

To better predict intercellular communication, we integrated publicly available LR interaction databases to validate the accuracy of CellGAT. These databases include human and mouse LR interactions based on reliable and validated scientific studies. We prioritized interactions supported in the literature to enhance the accuracy and reliability of predicting intercellular communication. Additionally, we accounted for the presence of multi-subunit complexes, which are critical functional units in complex LR interactions within cellular processes. By considering these factors, we ensured that the predicted interactions reflect the complexity of actual biological processes. Finally, all data were consolidated, and redundant LR pairs were removed to ensure the uniqueness and accuracy of the dataset. As a result, CellGAT retained 5786 validated human LR interactions and 4806 validated mouse LR interactions (Supplementary Data S1 and S2).

To generate negative LR samples, we randomly shuffled ligand and receptor gene lists to create new pairings, ensuring that these randomly generated LR pairs did not overlap with known true interactions. Redundant pairs were eliminated, resulting in a negative sample set for comparative experiments to further validate the model's performance [22]. Using this approach, we generated negative LR sample sets for the PC9 lung adenocarcinoma and Mouse brain datasets and compared CellGAT with other intercellular communication methods.

The PC9 lung adenocarcinoma dataset is a comprehensive resource for studying high-throughput spatial gene expression in lung adenocarcinoma. Analyzing intercellular communication within this dataset provides insights into cancer cells' spatial distribution in the lungs or other metastatic target organs, migration pathways, and interactions with other cell types in the microenvironment. This is crucial for uncovering the complex dynamics of the tumor microenvironment, cancer cell metastasis mechanisms, and potential therapeutic targets. We compared the performance of CellGAT with three other methods on this dataset, using their default thresholds and evaluating performance through ACC. As shown in Table 1, CellGAT achieved the highest ACC value.

Table 1. Comparison of CellGAT and other three methods ACC.

	CellGAT	CellPhoneDB	CellChat	Connectome
PC9_lung_adenocarcinoma	**0.91**	0.85	0.80	0.87
Mouse_brain	**0.85**	0.78	0.81	0.79

Next, we evaluated CellGAT on the Mouse brain dataset, a classic resource for studying intercellular communication and signal transduction in the nervous system. Similar to the PC9 dataset, we assessed performance using ACC, and as shown in Table 1, CellGAT again achieved the highest ACC value. These results demonstrate that CellGAT efficiently identifies relevant LR pairs across diverse datasets, consistently exhibiting high accuracy. Furthermore, as shown in Figure 2, CellGAT demonstrates superior performance in predicting intercellular communication, as evidenced by its high AUC values and statistically significant improvements over other models. Its robust framework for integrating validated interaction data is likely the key to its success across diverse biological contexts.

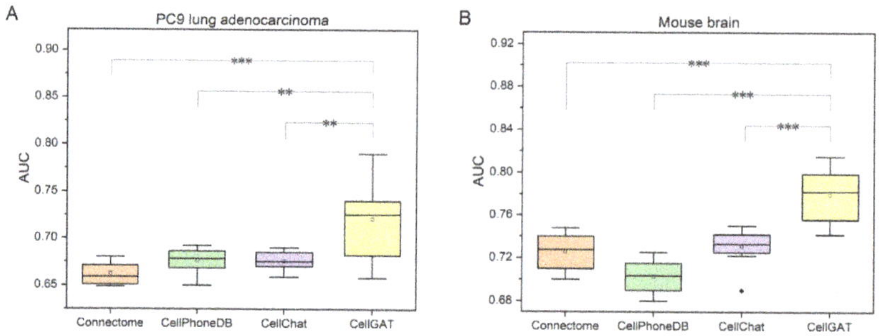

Figure 2. Boxplots comparing the performance of Connectome, CellPhoneDB, CellChat, and CellGAT in predicting intercellular communication. (**A**) Performance on PC9 lung adenocarcinoma dataset. (**B**) Performance on mouse brain dataset. CellGAT achieves significantly higher AUC values compared to other models. (** ($p \leq 0.01$), *** ($p \leq 0.001$), the discrete points in the figure are one of the results of the test).

3.2. Predicting Pathway Activity in Virtual In Situ Hybridization

To validate CellGAT's ability to capture spatial communication patterns in situ hybridization, we applied it to a single-cell sequencing dataset of stage 6 Drosophila embryos, focusing on the pathway activity of CCC (Figure 3A) [23]. Previous studies have identified the expression patterns of the Hippo signaling pathway in stage 6 Drosophila embryos by querying spatially expressed ligands and receptors and reconstructing their spatial expression patterns. These studies further revealed the correlation between Hippo pathway activity and the process of mitosis [24].

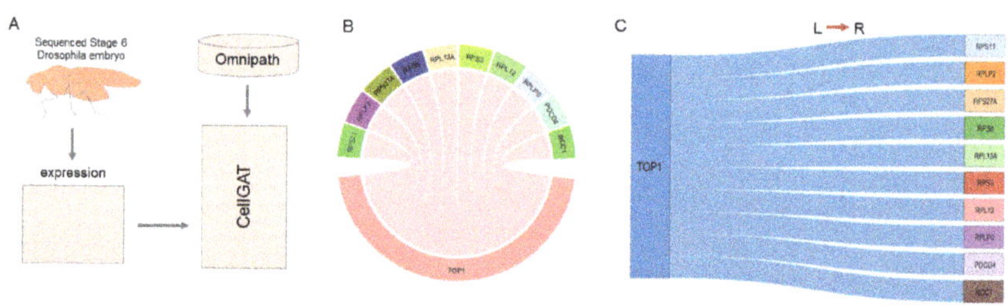

Figure 3. CellGAT applied to sequenced stage 6 Drosophila embryos. (**A**) Sequenced stage 6 Drosophila embryo application diagram. (**B**) Circle diagram visualization of the top ten interaction pairs predicted by CellGAT for sequenced stage 6 Drosophila embryos. (**C**) Sankey diagram visualization of the top ten interaction pairs predicted by CellGAT for sequenced stage 6 Drosophila embryos.

CellGAT analyzed the top-ranked CCC interactions in stage 6 Drosophila embryos and visualized them (Figure 3B,C), with the main contribution coming from TOP1. In previous research by Professor Qianwen Sun and colleagues, TOP1 activity was found to influence R-loop levels, and the topological organization of the genome is highly correlated with activities such as DNA replication, transcription, and repair. CellGAT not only identified CCC interactions with members of the Hippo pathway as ligands/receptors but also discovered interactions with many other pathways, including animal mitosis, MAPK signaling, the cell cycle [25], phospholipase D signaling, and the Notch signaling pathway. MAPK, phospholipase D, and Notch signaling have been confirmed to participate in Drosophila embryonic development. While CellGAT's results cannot determine the downstream effects of ligand–receptor interactions, the identified interaction between the cell cycle and the Hippo signaling pathway may be relevant to previous research findings.

Finally, CellGAT compared the coverage of the Hippo pathway in the top 1000 interactions with other CCC methods. Among the three CCC methods—CellPhoneDB [9], Connectome [26], and CellChat [12]—none captured any interactions involving the Hippo pathway. The reason for this outcome might be that other CCC methods only utilize intercellular interactions, whereas CellGAT was trained on all intra- and intercellular interactions. Although previous studies have identified interactions associated with the ErbB pathway and found its involvement in the Hippo pathway in Drosophila [27], experimental evidence suggests that CellGAT uniquely predicts previously published cell–cell communication results and provides more intricate interactions.

3.3. Predicting Changes in Intercellular Communication Caused by Non-Genetic and Pharmacological Perturbations

We applied CellGAT to study CCC in a dataset covering PC9 lung adenocarcinoma cells before and after drug treatment (Figure 4A), aiming to explore CellGAT's potential in revealing the effects of drug treatment on pathways and CCC [28]. CellGAT inferred cell–cell communication in populations of cells treated with osimertinib (a tyrosine kinase inhibitor) [29] at days 0, 7, and 14, and visualized them, respectively, for day 0 (Figure 4B,C), day 7 (Figure 4E,F), and day 14 (Figure 4H,I). To validate CellGAT's ability to further identify drug targets/mechanisms of action [30], we categorized these data into two classes: (1) the same biological sample under different conditions (day 0, 7/14); (2) the same biological replicate sample (day 7, 14). As shown in Figure 4, CellGAT yielded results on three independent datasets under two different conditions, including one single-cell RNA dataset for day 0 and two biological replicate single-cell RNA datasets for day 7.

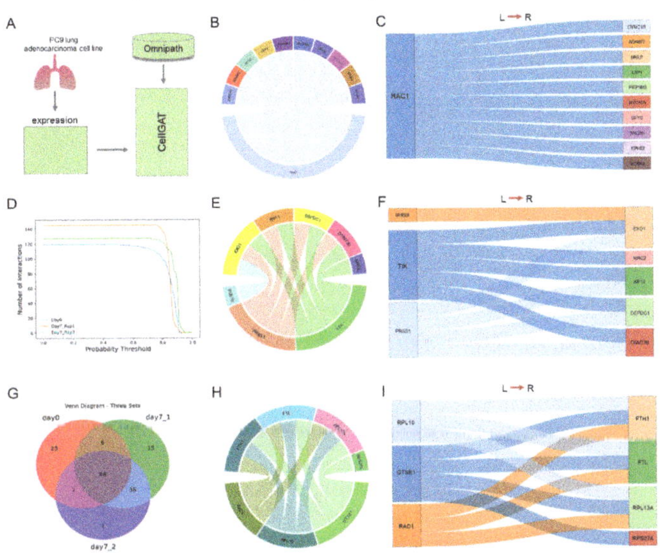

Figure 4. CellGAT applied to PC9 lung adenocarcinoma cell line dataset. (**A**) PC9 lung adenocarcinoma cell line dataset application diagram. (**B**,**C**) Visualization of the circle diagram and Sankey diagram of the top ten interaction pairs predicted by CellGAT for day 0 dataset. (**D**) Changes in the interaction logarithm of day 0, day 7, and day 14 datasets with different probability thresholds. (**E**,**F**) Visualization of the circle diagram and Sankey diagram of the top ten interaction pairs predicted by CellGAT for day 7 dataset. (**G**) Venn diagram visualizes the overlap of predicted interaction pairs by CellGAT in three datasets of day 0, day 7, and day 14 dataset. (**H**,**I**) Visualization of the circle diagram and Sankey diagram of the top ten interaction pairs predicted by CellGAT for day 14 dataset.

Based on the LR interactions validated in the OmniPath database and known pathway information annotated through KEGG, CellGAT predicted the number of ligand–receptor pairs in the two classes of datasets, as shown in Figure 4D. In the dataset for day 7, the second-ranked ligand was found to be TNF, a multifunctional pro-inflammatory cytokine belonging to the tumor necrosis factor (TNF) superfamily. In previous scientific research, elevated levels of TNF expression have been observed in lung cancer and may play a tumorigenic role [31,32]. We analyzed the top 100 validated CCCs in each of the three datasets, and the overlap of the predicted results for the three datasets is shown in in Figure 4G. In both classes of datasets, CellGAT discovered not only different interactions in the same biological sample before and after treatment but also differences in the predicted results for the same biological replicate sample after treatment. This discrepancy demonstrates CellGAT's ability to predict new interactions in the same dataset, which may suggest new findings regarding drug sensitivity or chemotherapy resistance in PC9 [33].

To evaluate the ability of the method to explore novel interactions, we compared the proportion of non-overlapping results across the three datasets. In comparison with other methods, CellGAT captured a relatively higher proportion of unique interactions among the top 100 biologically validated interactions, as shown in Figure 5B. This indicates that CellGAT is better able to identify unique interactions before and after disease treatment. Therefore, CellGAT can identify unique interactions before and after tumor treatment, potentially providing new insights for drug therapy.

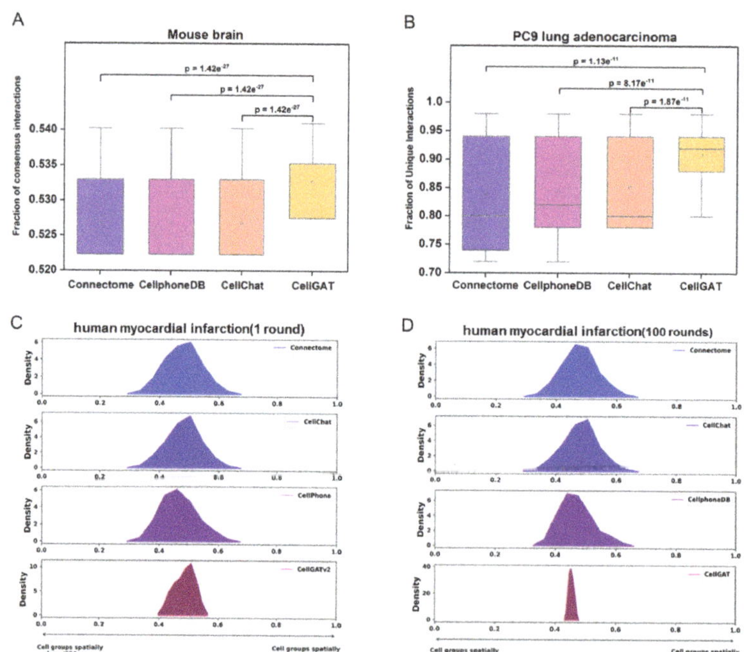

Figure 5. CellGAT compared with other methods. (**A**) Comparison of the performance of CellGAT with Connectome, CellphoneDB, and CellChat in predicting verified cell–cell interaction pairs in the Mouse brain dataset. (The means are 0.527, 0.527, 0.527, and 0.533, respectively.) (**B**) Comparison of the performance of CellGAT with Connectome, CellphoneDB, and CellChat in predicting validated interaction pairs in the PC9 lung adenocarcinoma dataset. (The means are 0.80, 0.82, 0.81, and 0.91, respectively.) (**C**,**D**) Comparison of the spatial proximity of CellGAT in predicting validated interaction pairs with Connectome, CellphoneDB, and CellChat in the human myocardial infarction dataset.

3.4. Predicting Neuronal Cell Interactions

Communication between neurons plays a critical role in cognitive processes and behavioral performance. By delving into the communication networks among neurons, we can uncover patterns of connectivity between different brain regions and understand how these connections influence cognitive function and behavior. To validate the ability of CellGAT to capture communication patterns between neurons, we applied it to a mouse brain dataset (Figure 6A). We curated a validation set, called consensus_mouse, based on interaction pairs and pathway information from the OmniPath database in this dataset.

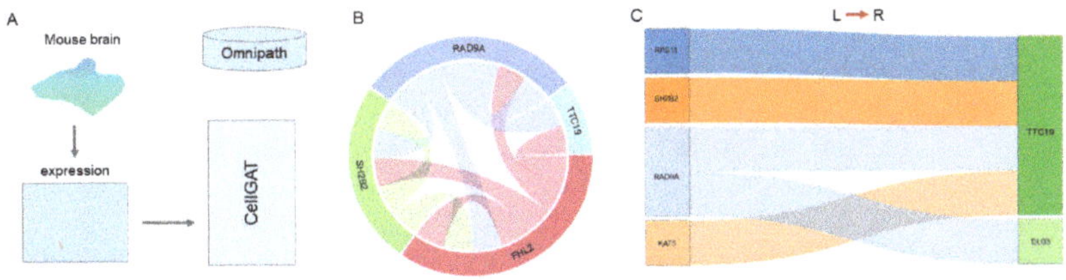

Figure 6. CellGAT applied to Mouse brain dataset. (**A**) Mouse brain dataset application diagram. (**B**) Circle diagram visualization of the top ten interaction pairs predicted by CellGAT for Mouse brain dataset. (**C**) Sankey diagram visualization of the top ten interaction pairs predicted by CellGAT for Mouse brain dataset.

Therefore, CellGAT further analyzed the top-ranked interaction pairs and visualized them (Figure 6B,C). We discovered an interaction between RPS11 and TTC19, where previous research has demonstrated RPS11 as an antibody, while mutations in TTC19 lead to neural damage [34]. Additionally, in Burkitt's lymphoma cell lines, SH2B2 is phosphorylated on tyrosine upon B-cell receptor stimulation. SH2 domains typically bind ligands relative to phosphotyrosine. Its association with stimulation and Grb2 is independent, and it appears to play a role in signaling pathways from the receptor to Shc/Grb2 upon receptor activation. These interactions are likely associated with the pathogenesis of certain diseases. Previous studies have indicated that the interaction between SH2B2 and TTC19 involves the coupling of immune receptors with the Ras signaling pathway. Upon activation of immune receptors during immune response, a cascade of signaling events, including activation of the Ras signaling pathway, is initiated. SH2B2, as a crucial signaling adaptor protein, likely mediates the signal transduction of immune receptors by binding to phosphorylated tyrosine residues on active receptors and interacts with TTC19 during this process. TTC19, as a protein, may regulate the formation or function of this complex, further impacting the progression of signal transduction. This interaction likely plays a pivotal role in modulating cellular immune responses, especially in regulating the Ras signaling pathway. Therefore, the findings from CellGAT provide a novel perspective on inter-neuronal communication, further enhancing our understanding of the regulatory mechanisms underlying neuronal function.

Based on a meticulously planned validation set, we compared the performance of CellGAT with other methods on the Mouse brain dataset (Figure 5A). The results demonstrate that CellGAT outperforms current methods used to predict interactions between neurons and exhibits greater stability. This suggests that CellGAT effectively identifies interactions between neurons, potentially providing opportunities for discovering new drug targets. "Verified interaction pairs" refer to the biologically validated dataset from OmniPath. OmniPath is a resource designed to assist researchers in analyzing and integrating complex molecular and cellular interaction networks. Following its creation by Türei et al., it has been extensively studied by Ceccarelli et al., demonstrating its effectiveness in validating communication network analyses. Additionally, in our methodology, "interaction pairs" specifically denote the intercellular interaction pairs predicted by CellGAT, which align with the known biological interaction information from the OmniPath database. By comparing our model's predictions against existing, experimentally validated interaction pairs in OmniPath, we validate the accuracy and reliability of our predictions.

3.5. Predict Spatially Adjacent CCC

To assess CellGAT's ability to predict interactions consistently with other modalities and to analyze new directions in CCC, CellGAT was applied to a dataset of human myocardial infarction, which includes spatial transcriptomic data from different regions affected by myocardial infarction (such as the ischemic zone, remote zone, border zone, and fibrotic zone), as well as unaffected control regions (Figure 7A) [35]. The human myocardial infarction 10X Visium spatial transcriptomics dataset we used includes samples from the fibrotic zone (FZ). The primary objective of this benchmark test is to demonstrate CellGAT's ability to learn graph structures consistent with spatial microenvironments, namely cellular localization and proximity. This provides a new perspective for comparing its prediction results with background knowledge of intercellular communication. For the sequencing data from various regions of the heart, cellular expression data are accompanied by spatial coordinates. Besides validating expected ligand–receptor pairs, CellGAT identifies potential spatial proximity between ligand cell populations and receptor cell populations, influencing the likelihood of these interactions.

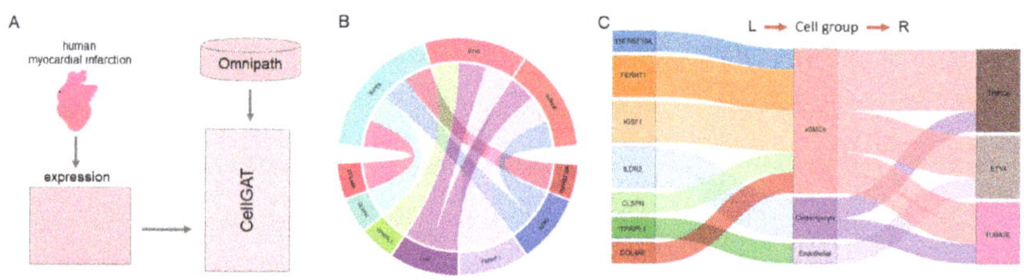

Figure 7. CellGAT applied to human myocardial infarction dataset. (**A**) Human myocardial infarction dataset application diagram. (**B**) Circle diagram visualization of the top ten interaction pairs predicted by CellGAT for human myocardial infarction dataset. (**C**) Sankey diagram visualization of the top ten interaction pairs predicted by CellGAT for human myocardial infarction dataset.

We applied a single spatial transcriptomic data sample from the fibrotic zone (FZ) to infer intercellular communication using CellGAT (Figure 7B,C). The dataset contains spatial coordinates for all cells, used to construct maps organized by cell types. We utilized this map to partition subsets of cell types involved in interactions and visualize their spatial proximity (Figure 5C,D). In the prediction results, we identified a ranked-sixth interaction involving vascular smooth muscle cells (vSMCs) and myeloid cells [36]. This interaction has been studied in the carotid artery region, involving interactions between CXCL12 and CXCR4. Previous studies have extensively analyzed this interaction, particularly to understand its role in myocardial infarction, demonstrating its high biological relevance to the dataset [37]. Additionally, we identified lower-ranked cell–cell interactions involving peripheral cells and neurons. Within these cell populations, we observed a ligand–receptor interaction between RELN and ITGB1. This validated interaction primarily occurred in glioblastoma cells. RELN is a critical signaling protein known for its role in brain neurodevelopment and neuronal migration. ITGB1 is an integrin protein that serves as a key link between the extracellular matrix and cells. It is speculated that RELN may mediate cell–cell signaling, cell migration, or other related biological processes in glioblastoma cells by binding to its receptor ITGB1.

When comparing the spatial proximity of the top 10 predicted interactions by CellGAT with other methods in the human myocardial infarction dataset, CellGAT exhibits lower mean and variance values. These findings suggest that, based on accurate information, CellGAT not only prioritizes biologically meaningful and validated interactions but also prioritizes interactions between cell types that are spatially adjacent.

4. Discussion

In this study, we introduce a novel bioinformatics method called CellGAT, designed to infer intercellular communication from single-cell RNA sequencing data and validate it using transcriptomic data. This method integrates single-cell gene expression data with ligand–receptor pair databases and previous biological pathway annotations to achieve a more comprehensive and accurate inference of intercellular communication. The scope of the model extends beyond connections between cell groups within individual clusters to encompass broader interactions across the entire cellular network. This interconnectedness is crucial for understanding the complex dynamics of cell communication, as it enables signal propagation and coordination of cellular activities throughout the tissue or organism. By considering these broader interactions, the model aims to provide a comprehensive perspective on cell communication patterns, uncovering new insights and patterns that may not be evident within individual clusters.

CellGAT utilizes a directed graph representation based on graph neural networks to depict the relationships between cell types and identified ligand–receptor pairs. Additionally, it prioritizes information on cell–cell interactions and interprets aspects such as cell localization, pathway annotations, and protein complexes. During the training phase, CellGAT utilizes PPIs obtained from the OmniPath database to explain CCC occurring between cells/cell types and internally. By training on both intra- and inter-cellular PPIs obtained from the OmniPath database, CellGAT can predict CCC between cell types and internally. To validate CellGAT's ability to predict intercellular communication, we utilized four datasets and compared them with three other methods. In the fly dataset, CellGAT's inferred intercellular interactions were compared with validated results, demonstrating its predictive capability for intercellular communication. For the other three datasets, we focused on intercellular communication to reflect the definition used in previous studies and emphasized more diverse patterns occurring between different cell types/populations. We were able to demonstrate that CellGAT can reproduce intercellular communication results seen through spatial reconstruction, identify unique interactions in cell populations after drug treatment, and prioritize interactions occurring between spatially adjacent cell types.

Our model effectively integrates prior knowledge from databases such as OmniPath, enabling accurate inference of cell–cell communication based on known ligand–receptor pairs. By combining multi-omics data with advanced computational methods, such as node embedding and graph attention networks, we provide a robust platform capable of predicting interactions across various biological contexts. However, the model's reliance on existing databases like OmniPath introduces certain limitations, particularly when predicting novel cell–cell interactions. To address this issue, researchers are exploring more data-driven approaches, leveraging large-scale single-cell and spatial transcriptomics datasets to reduce dependency on existing databases, discover novel ligand–receptor interactions, and enhance the model's comprehensiveness.

Exploring intercellular communication offers new insights into biomedical research, deepening our understanding of interactions within and between cells. By delving into the interactions between different cell types, developmental stages, disease states, or external perturbations, we can uncover crucial information closely related to cell state recognition and functional regulation. Although many methods exist to predict potential communication activities between cells using validated ligand–receptor databases and transcriptomic data, these approaches often struggle to fully capture the complexity of cellular social networks and face limitations in integrating different modalities. Preliminary research findings suggest that for more accurate prediction of intercellular communication, we need to consider the complexity of intercellular interaction networks. This involves a comprehensive understanding of intercellular signal transduction and how to integrate this information with prior knowledge of ligand–receptor interactions. Therefore, studying CCC requires more comprehensive and in-depth methods to better understand the complex interaction networks between cells.

Our model effectively integrates prior knowledge from existing databases such as OmniPath, enabling highly accurate inference of cell–cell communication based on known ligand–receptor pairs. However, differences in cell communication across tissue samples may exist, and certain unique interaction pairs in cancer tissues could limit the model's accuracy when applied to other datasets. Additionally, the lack of standardized datasets to measure the accuracy of these predictions restricts the direct comparison between methods, as well as the reproducibility and comparability of their performance. The development of multi-omics datasets in the future may help address this limitation. Furthermore, there is currently no concrete definition of cell communication. Communication types include intracellular and extracellular, long-range, and short-range signaling. Integrating a unified definition of cell communication could provide researchers with clearer directions, allowing them to overcome existing limitations and further advance the field.

Furthermore, in the inference of intercellular communication, the exacerbation due to the lack of foundational facts has been previously recognized, emphasizing the critical importance of speculative analysis based on foundational facts. can lead to the discovery of more effective CCC interactions in different contexts, thereby improving the prediction and evaluation of ligand–receptor interactions. Based on these facts, we developed a graph-based deep learning method, CellGAT, capable of accurately capturing ligand–receptor communication in single-cell transcriptomic data. Additionally, when studying spatially regulated intercellular communication, integrating different omics data and multimodal datasets (such as 10× Multiome and Digital Spatial Profiling) can provide more comprehensive, multidimensional information, enhancing our understanding of cell interactions and regulatory mechanisms. In such cases, reliable computational models are needed to accurately integrate multimodal data and perform inference.

5. Conclusions

In summary, CellGAT combines gene expression data with protein–protein interactions (PPIs) and protein complex information from the OmniPath database. This approach helps to mitigate the loss of predictive accuracy due to the inactivation of ligands or receptors caused by certain gene inactivation. Consequently, CellGAT demonstrates high robustness and accuracy when handling data with abnormal gene expression. CellGAT integrates node embedding algorithms with graph attention networks, excelling at capturing both local and global graph structural features within the data. The node embedding algorithms effectively extract graph structure information, while the graph attention networks adeptly process these features, making CellGAT highly effective in predicting cell interactions.

You can access the open-source Python implementation of CellGAT on GitHub at https://github.com/wuzhenao/CellGAT (accessed on 25 February 2025). This version provides tutorials and examples for the following analyses: ligand activity analysis based on a gene set of interest, single-cell ligand activity analysis, and visualization of ligand-to-target signaling pathways. Additionally, we offer tutorials for model validation, including training the prior model with databases and evaluating target gene and ligand activity.

Users can utilize these tutorials to compare their methods with CellGAT and build models tailored to their specific needs (for example, using their own data sources to explore more cell–cell communications).

Supplementary Materials: The following supporting information can be downloaded at https://www.mdpi.com/article//10.3390/biom15030342/s1, Supplementary S1: Supplementary_Data_1.csv, Supplementary S2: Supplementary_Data_2.csv.

Author Contributions: T.Z.: conceptualization, methodology, software; Z.W.: writing—original draft preparation; L.L.: formal analysis; J.R.: visualization; Z.Z.: software, validation; J.Z.: data curation; G.W.: writing—reviewing and editing. All authors have read and agreed to the published version of the manuscript.

Funding: This work was supported by the National Natural Science Foundation of China [62473094], the Natural Science Foundation of Heilongjiang Province, China [LH2024F003] and the National Science Foundation for Distinguished Young Scholars of China [62225109].

Institutional Review Board Statement: Not applicable.

Informed Consent Statement: Not applicable.

Data Availability Statement: Sequencing data for the sixth stage of Drosophila embryo can be found in the GEO database under accession code GsSE95025. The data for PC9 cell line treated with osimertinib before and after treatment are available under the GEO registration code GSE150949. The Mouse brain dataset can be obtained from "https://data.caltech.edu/records/s0vdx-0k302 accessed on 25 February 2025". Additionally, spatial transcriptomics data can be accessed from "https://zenodo.org/record/6578047 accessed on 25 February 2025".

Acknowledgments: The authors thank the anonymous reviewers for their constructive suggestions.

Conflicts of Interest: The authors declare that they have no conflicts of interest.

Abbreviations

CCC	cell-cell communication
LR	ligand-receptor
PPIs	protein-protein interactions

References

1. Zhou, X.; Franklin, R.A.; Adler, M.; Jacox, J.B.; Bailis, W.; Shyer, J.A.; Flavell, R.A.; Mayo, A.; Alon, U.; Medzhitov, R. Circuit design features of a stable two-cell system. *Cell* **2018**, *172*, 744–757.e717. [CrossRef] [PubMed]
2. Handly, L.N.; Pilko, A.; Wollman, R. Paracrine communication maximizes cellular response fidelity in wound signaling. *eLife* **2015**, *4*, e09652. [CrossRef] [PubMed]
3. Armingol, E.; Officer, A.; Harismendy, O.; Lewis, N.E. Deciphering cell–cell interactions and communication from gene expression. *Nat. Rev. Genet.* **2021**, *22*, 71–88. [CrossRef] [PubMed]
4. Leimkühler, N.B.; Gleitz, H.F.; Ronghui, L.; Snoeren, I.A.; Fuchs, S.N.; Nagai, J.S.; Banjanin, B.; Lam, K.H.; Vogl, T.; Kuppe, C. Heterogeneous bone-marrow stromal progenitors drive myelofibrosis via a druggable alarmin axis. *Cell Stem Cell* **2021**, *28*, 637–652.e638. [CrossRef] [PubMed]
5. Dominiak, A.; Chełstowska, B.; Olejarz, W.; Nowicka, G. Communication in the cancer microenvironment as a target for therapeutic interventions. *Cancers* **2020**, *12*, 1232. [CrossRef]
6. Liu, Z.; Hu, S.; Yun, Z.; Hu, W.; Zhang, S.; Luo, D. Using dynamic cell communication improves treatment strategies of breast cancer. *Cancer Cell Int.* **2021**, *21*, 275. [CrossRef] [PubMed]
7. Wojtowicz, W.M.; Vielmetter, J.; Fernandes, R.A.; Siepe, D.H.; Eastman, C.L.; Chisholm, G.B.; Cox, S.; Klock, H.; Anderson, P.W.; Rue, S.M. A human IgSF cell-surface interactome reveals a complex network of protein-protein interactions. *Cell* **2020**, *182*, 1027–1043.e1017. [CrossRef]
8. Sharma, V.; Ranjan, T.; Kumar, P.; Pal, A.K.; Jha, V.K.; Sahni, S.; Prasad, B.D. Protein–protein interaction detection: Methods and analysis. In *Plant Biotechnology, Volume 1*; Apple Academic Press: Palm Bay, FL, USA, 2017; pp. 391–411.

9. Efremova, M.; Vento-Tormo, M.; Teichmann, S.A.; Vento-Tormo, R. CellPhoneDB: Inferring cell–cell communication from combined expression of multi-subunit ligand–receptor complexes. *Nat. Protoc.* **2020**, *15*, 1484–1506. [CrossRef] [PubMed]
10. Weaver, D.T.; Scott, J.G. Crosstalkr: An open-source R package to facilitate drug target identification. *bioRxiv* **2023**. [CrossRef]
11. Browaeys, R.; Saelens, W.; Saeys, Y. NicheNet: Modeling intercellular communication by linking ligands to target genes. *Nat. Methods* **2020**, *17*, 159–162. [CrossRef]
12. Jin, S.; Guerrero-Juarez, C.F.; Zhang, L.; Chang, I.; Ramos, R.; Kuan, C.-H.; Myung, P.; Plikus, M.V.; Nie, Q. Inference and analysis of cell-cell communication using CellChat. *Nat. Commun.* **2021**, *12*, 1088. [CrossRef] [PubMed]
13. Dimitrov, D.; Türei, D.; Garrido-Rodriguez, M.; Burmedi, P.L.; Nagai, J.S.; Boys, C.; Ramirez Flores, R.O.; Kim, H.; Szalai, B.; Costa, I.G. Comparison of methods and resources for cell-cell communication inference from single-cell RNA-Seq data. *Nat. Commun.* **2022**, *13*, 3224. [CrossRef]
14. Grover, A.; Leskovec, J. node2vec: Scalable feature learning for networks. In Proceedings of the 22nd ACM SIGKDD International Conference on Knowledge Discovery and Data Mining, San Francisco, CA, USA, 13–17 August 2016; pp. 855–864.
15. Traag, V.A.; Waltman, L.; Van Eck, N.J. From Louvain to Leiden: Guaranteeing well-connected communities. *Sci. Rep.* **2019**, *9*, 5233. [CrossRef]
16. Hu, Y.; Shi, W.; Yeh, C.-H. Spatiotemporal convolution sleep network based on graph attention mechanism with automatic feature extraction. *Comput. Methods Programs Biomed.* **2024**, *244*, 107930. [CrossRef]
17. Türei, D.; Korcsmáros, T.; Saez-Rodriguez, J. OmniPath: Guidelines and gateway for literature-curated signaling pathway resources. *Nat. Methods* **2016**, *13*, 966–967. [CrossRef]
18. Kanehisa, M.; Sato, Y.; Kawashima, M.; Furumichi, M.; Tanabe, M. KEGG as a reference resource for gene and protein annotation. *Nucleic Acids Res.* **2016**, *44*, D457–D462. [CrossRef]
19. Tsitsiridis, G.; Steinkamp, R.; Giurgiu, M.; Brauner, B.; Fobo, G.; Frishman, G.; Montrone, C.; Ruepp, A. CORUM: The comprehensive resource of mammalian protein complexes–2022. *Nucleic Acids Res.* **2023**, *51*, D539–D545. [CrossRef] [PubMed]
20. Zhang, Y.; Tang, M. A Theoretical Analysis of DeepWalk and Node2vec for Exact Recovery of Community Structures in Stochastic Blockmodels. *IEEE Trans. Pattern Anal. Mach. Intell.* **2023**, *46*, 1065–1078. [CrossRef] [PubMed]
21. Imambi, S.; Prakash, K.B.; Kanagachidambaresan, G. PyTorch. In *Programming with TensorFlow: Solution for Edge Computing Applications*; Springer Nature: Cham, Switzerland, 2021; pp. 87–104.
22. Zhang, T.; Zhang, X.; Wu, Z.; Ren, J.; Zhao, Z.; Zhang, H.; Wang, G.; Wang, T. VGAE-CCI: Variational graph autoencoder-based construction of 3D spatial cell–cell communication network. *Brief. Bioinform.* **2025**, *26*, bbae619. [CrossRef] [PubMed]
23. Karaiskos, N.; Wahle, P.; Alles, J.; Boltengagen, A.; Ayoub, S.; Kipar, C.; Kocks, C.; Rajewsky, N.; Zinzen, R.P. The Drosophila embryo at single-cell transcriptome resolution. *Science* **2017**, *358*, 194–199. [CrossRef] [PubMed]
24. Staley, B.K.; Irvine, K.D. Hippo signaling in Drosophila: Recent advances and insights. *Dev. Dyn.* **2012**, *241*, 3–15. [CrossRef] [PubMed]
25. Vergassola, M.; Deneke, V.E.; Di Talia, S. Mitotic waves in the early embryogenesis of Drosophila: Bistability traded for speed. *Proc. Natl. Acad. Sci. USA* **2018**, *115*, E2165–E2174. [CrossRef] [PubMed]
26. Raredon, M.S.B.; Yang, J.; Garritano, J.; Wang, M.; Kushnir, D.; Schupp, J.C.; Adams, T.S.; Greaney, A.M.; Leiby, K.L.; Kaminski, N. Computation and visualization of cell–cell signaling topologies in single-cell systems data using connectome. *Sci. Rep.* **2022**, *12*, 4187. [CrossRef] [PubMed]
27. Snigdha, K.; Gangwani, K.S.; Lapalikar, G.V.; Singh, A.; Kango-Singh, M. Hippo signaling in cancer: Lessons from Drosophila models. *Front. Cell Dev. Biol.* **2019**, *7*, 85. [CrossRef]
28. Oren, Y.; Tsabar, M.; Cuoco, M.S.; Amir-Zilberstein, L.; Cabanos, H.F.; Hütter, J.-C.; Hu, B.; Thakore, P.I.; Tabaka, M.; Fulco, C.P. Cycling cancer persister cells arise from lineages with distinct programs. *Nature* **2021**, *596*, 576–582. [CrossRef] [PubMed]
29. Bollinger, M.K.; Agnew, A.S.; Mascara, G.P. Osimertinib: A third-generation tyrosine kinase inhibitor for treatment of epidermal growth factor receptor-mutated non-small cell lung cancer with the acquired Thr790Met mutation. *J. Oncol. Pharm. Pract.* **2018**, *24*, 379–388. [CrossRef]
30. AlMusawi, S.; Ahmed, M.; Nateri, A.S. Understanding cell-cell communication and signaling in the colorectal cancer microenvironment. *Clin. Transl. Med.* **2021**, *11*, e308. [CrossRef]
31. Gong, K.; Guo, G.; Beckley, N.; Zhang, Y.; Yang, X.; Sharma, M.; Habib, A.A. Tumor necrosis factor in lung cancer: Complex roles in biology and resistance to treatment. *Neoplasia* **2021**, *23*, 189–196. [CrossRef]
32. Xu, G.; Liu, C.; Liang, T.; Zhang, Z.; Jiang, J.; Chen, J.; Xue, J.; Zeng, H.; Lu, Z.; Zhan, X. Gene expression profile and bioinformatics analysis revealed key molecular characteristics of chordoma—Before and after TNF-a treatment. *Medicine* **2020**, *99*, e18790. [CrossRef]
33. Ramirez, M.; Rajaram, S.; Steininger, R.J.; Osipchuk, D.; Roth, M.A.; Morinishi, L.S.; Evans, L.; Ji, W.; Hsu, C.-H.; Thurley, K. Diverse drug-resistance mechanisms can emerge from drug-tolerant cancer persister cells. *Nat. Commun.* **2016**, *7*, 10690. [CrossRef] [PubMed]

34. Radaelli, E.; Assenmacher, C.-A.; Verrelle, J.; Banerjee, E.; Manero, F.; Khiati, S.; Girona, A.; Lopez-Lluch, G.; Navas, P.; Spinazzi, M. Mitochondrial defects caused by PARL deficiency lead to arrested spermatogenesis and ferroptosis. *eLife* **2023**, *12*, e84710. [CrossRef]
35. Kuppe, C.; Ramirez Flores, R.O.; Li, Z.; Hayat, S.; Levinson, R.T.; Liao, X.; Hannani, M.T.; Tanevski, J.; Wünnemann, F.; Nagai, J.S. Spatial multi-omic map of human myocardial infarction. *Nature* **2022**, *608*, 766–777. [CrossRef]
36. Xue, Y.; Luo, M.; Hu, X.; Li, X.; Shen, J.; Zhu, W.; Huang, L.; Hu, Y.; Guo, Y.; Liu, L. Macrophages regulate vascular smooth muscle cell function during atherosclerosis progression through IL-1β/STAT3 signaling. *Commun. Biol.* **2022**, *5*, 1316. [CrossRef] [PubMed]
37. Wang, Y.; Dembowsky, K.; Chevalier, E.; Stüve, P.; Korf-Klingebiel, M.; Lochner, M.; Napp, L.C.; Frank, H.; Brinkmann, E.; Kanwischer, A. CXC motif chemokine receptor 4 blockade promotes tissue repair after myocardial infarction by enhancing regulatory T cell mobilization and immune-regulatory function. *Circulation* **2019**, *139*, 1798–1812. [CrossRef] [PubMed]

biomolecules

MDPI

Article

IL-6-Inducing Peptide Prediction Based on 3D Structure and Graph Neural Network

Ruifen Cao [1], Qiangsheng Li [1], Pijing Wei [2], Yun Ding [3], Yannan Bin [2,*] and Chunhou Zheng [3,*]

[1] Information Materials and Intelligent Sensing Laboratory of Anhui Province, School of Computer Science and Technology, Anhui University, Hefei 230601, China; rfcao@ahu.edu.cn (R.C.); e22301211@stu.ahu.edu.cn (Q.L.)
[2] Institutes of Physical Science and Information Technology, Anhui University, Hefei 230601, China; weipj@ahu.edu.cn
[3] School of Artificial Intelligence, Anhui University, Hefei 230601, China; yunding@ahu.edu.cn
* Correspondence: ynbin@ahu.edu.cn (Y.B.); zhengch99@ahu.edu.cn (C.Z.)

Abstract: Interleukin-6 (IL-6) is a potent glycoprotein that plays a crucial role in regulating innate and adaptive immunity, as well as metabolism. The expression and release of IL-6 are closely correlated with the severity of various diseases. IL-6-inducing peptides are critical for the development of immunotherapy and diagnostic biomarkers for some diseases. Most existing methods for predicting IL-6-induced peptides use traditional machine learning methods, whose feature selection is based on prior knowledge. In addition, none of these methods take into account the three-dimensional (3D) structure of peptides, which is essential for their functional properties. In this study, we propose a novel IL-6-inducing peptide prediction method called DGIL-6, which integrates 3D structural information with graph neural networks. DGIL-6 represents a peptide sequence as a graph, where each amino acid is treated as a node, and the adjacency matrix, representing the relationships between nodes, is derived from the predicted residue contact graph of the peptide sequence. In addition to commonly used amino acid representations, such as one-hot encoding and position encoding, the pre-trained model ESM-1b is employed to extract amino acid features as node features. In order to simultaneously consider node weights and information updates, a dual-channel method combining Graph Attention Network (GAT) and Graph Convolutional Network (GCN) is adopted. Finally, the extracted features from both channels are merged for the classification of IL-6-inducing peptides. A series of experiments including cross-validation, independent testing, ablation studies, and visualizations demonstrate the effectiveness of the DGIL-6 method.

Keywords: interleukin-6; IL-6-inducing peptides; graph attention network; graph convolutional network; ESM-1b

1. Introduction

Interleukin-6 (IL-6), a genetically encoded 184-amino acid quadruple helical protein, is a key pro-inflammatory cytokine [1,2]. It is also commonly referred to as interferon- β2 (IFN- β2), B-cell-stimulating factor-2, and hybridoma/plasmacytoma growth factor. As a pleiotropic cytokine, IL-6 is produced by various cell types, including T cells, fibroblasts, monocytes, endothelial cells, macrophages, and dendritic cells [3–5]. This cytokine plays a critical role in adaptive and innate immune responses, hematopoiesis, acute phase response, inflammatory diseases, and organ development [6]. Furthermore, cytokine burst or cytokine release syndrome is closely related to the progression of COVID-19. In patients

 https://doi.org/10.3390/biom15010099

with COVID-19, significantly elevated levels of IL-6 and other pro-inflammatory cytokines, such as IL-7, IL-8, and IL-12, are strongly related to disease progression [7]. It should be noted that the *SARS-CoV-2* virus can selectively induce IL-6 overproduction, leading to lymphocyte depletion [8]. IL-6-inducing peptides play a crucial role in the regulation of IL-6 expression and activity, and their interactions significantly impact immune balance and disease progression [9]. Therefore, accurate identification of IL-6-inducing peptides is essential to deepen our understanding of IL-6-mediated mechanisms and provide valuable information for the development of diagnostic tools and immunotherapy strategies.

Several machine learning approaches have been developed in recent years to predict IL-6-inducing peptides. Dhall et al. introduced IL-6Pred, which uses 15 peptide feature descriptors to map sequences to feature vectors, and then input them into a random forest model for prediction [10]. To further improve predictive performance, Charoenkwan et al. proposed StackIL6, a stacked ensemble model that combines three primary peptide features with five well-known traditional machine learning models [11]. By fine-tuning the hyperparameters of individual classifiers, StackIL6 significantly enhances both recognition accuracy and generalization capabilities. However, machine learning-based methods heavily rely on extensive human driven feature engineering and domain expertise. Deep learning approaches have also advanced the prediction of IL-6-inducing peptides. MVIL6, a deep learning-based predictor, leverages peptide sequence information by combining the Transformer architecture with the pre-trained MG-BERT model, effectively extracting feature representations and improving recognition performance [12]. Despite these promising advancements, existing studies largely overlook the structural features of IL-6-inducing peptides, which are crucial for their functional roles [13–16].

To address the limitations of existing methods, we propose a novel tool for IL-6-inducing peptide prediction, named DGIL-6, which leverages graph neural networks and 3D structural information. DGIL-6 models a peptide sequence as a graph, where each node represents an amino acid, and edges are defined by the adjacency matrix based on the predicted 3D peptide structure. In addition to conventional amino acid feature extraction methods, such as one-hot encoding and position encoding, DGIL-6 employs a pre-trained model, ESM-1b [17], to extract amino acid features as node attributes. After obtaining the graph representation of a peptide, a dual-channel graph network architecture is utilized to extract features. Specifically, the graph attention network (GAT) [18] is applied to update node features, while the graph convolutional network (GCN) [19] focuses on updating node weights. The features obtained from both channels are subsequently fused to classify IL-6-inducing peptides effectively.

2. Materials and Methods

2.1. Data Collection

To effectively compare our method with previous studies, we used the same baseline dataset as Dhall et al. (2020) [10]. This dataset was selected due to its rigorous validation and widespread use in research on IL-6-inducing peptide recognition. DHall et al. extracted 583 experimentally validated IL-6-inducing peptides from the Immune Epitope Database (IEDB) [20]. Identical peptides and those longer than 25 amino acids were excluded, as most IL-6-inducing peptides are shorter than 25 amino acids. After filtering, 365 IL-6-inducing peptides (referred to as positive data) were retained; these had been tested in human or mouse hosts. For the negative dataset, Dhall et al. extracted experimentally validated peptides from IEDB that induced other cytokines, such as IL-1 α, IL-1 β, TNF α, IL-8, IL-12, IL-17, and IL-18, but not IL-6. These peptides, known as non-IL-6-inducing peptides, were used as negative samples, since they do not induce IL-6. The negative dataset consists of sequences have been tested either in human or mouse hosts. As with the positive peptides, identical peptides and those longer than 25 amino acids were excluded. Ultimately, 2991 non-IL-6-inducing peptides were retained. This dataset consists of 365 IL-6-inducing peptides and 2991 non-IL-6-inducing peptides, serving as the benchmark for studies on IL-6-inducing peptide prediction tools. The imbalance between positive and negative samples may affect the model's performance with regard to the minority class. Moreover, peptides with IL-6 inducing activity are generally much smaller than those without IL-6 induction in real-world datasets. To address this issue, we employed balancing techniques during training, such as weighting the positive and negative samples, to improve the model's ability to recognize the minority class (IL-6-inducing peptides). Several studies have previously used an 80:20 split ratio for training and testing datasets [21–24]. Following this standard protocol, 80% of the data were used for training, and 20% were reserved for testing. The training set contained 292 IL-6-inducing peptides and 2393 non-IL-6-inducing peptides, while the test set comprised 73 IL-6-inducing peptides and 598 non-IL-6-inducing peptides. In order to better understand the distribution of samples in the training and testing sets, we conducted amino acid composition analysis (Figure 1A) and amino acid length analysis (Figure 1B) on them. The results showed comparable amino acid compositions between the two sets, with no significant biases in residue frequencies. Similarly, the sequence length distributions were consistent, confirming the fairness of the dataset split. The specific details of the dataset are provided in Table 1.

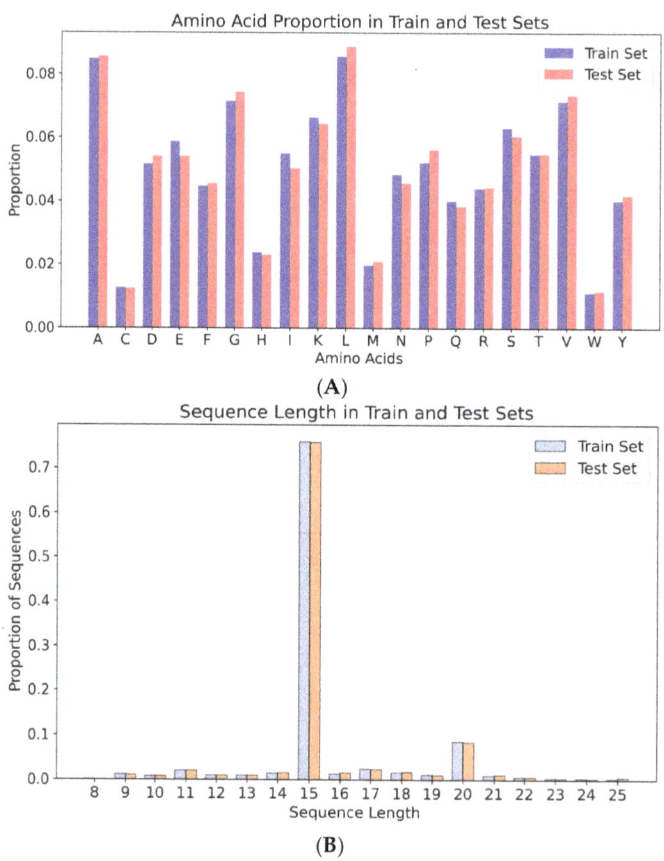

Figure 1. (**A**) Amino Acid proportion in training and testing sets. (**B**) Sequence-length proportion distribution in training and testing sets.

Table 1. An overview of the dataset.

Dataset	Positive Data	Negative Data	Ratio
Training set	292	2393	0.122
Test set	73	598	

2.2. Overview of DGIL-6

The framework of the DGIL-6 method is depicted in Figure 2 and consists of three main components: graph representation, a dual-channel feature extraction module, and a classification module. Graph representation: Amino acids in the peptide sequence are encoded as node features using one-hot encoding, position encoding, and features extracted by the pre-trained ESM-1b model. Structural features are derived from the peptide's structural information, and an adjacency matrix is constructed based on the residue contact map. Together, these elements form a graph representation of the peptide sequence, where nodes represent amino acids, and edges reflect their structural relationships. Dual-channel feature extraction module: The graph representation is input into a dual-channel feature extraction module. One channel utilizes a multi-head Graph Attention Network (GAT) to update node features, while the other channel employs a Graph Convolutional Network (GCN) to update node weights. This dual-channel design captures complementary information, enabling more robust feature extraction. Classification module: The features extracted from the dual-channel module are fused and passed to the classification module, which predicts whether a peptide induces IL-6. Details of each component are provided in the subsequent sections.

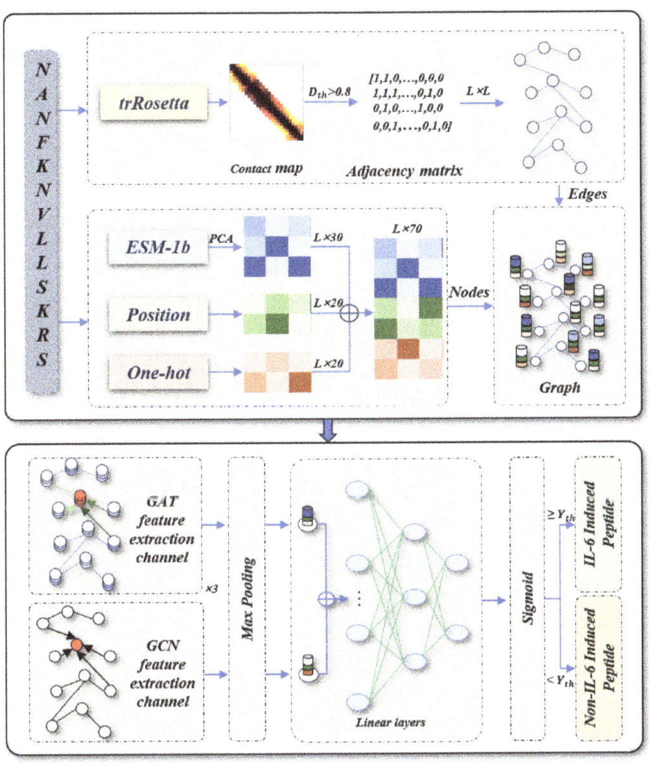

Figure 2. The network architecture of DGIL-6.

2.3. Graph Representation

To investigate the role of peptide structural information for identifying IL-6-inducing peptides, we represent each peptide as a graph $G = (V, E)$, where V represents the sets of nodes (amino acids) and E denotes the set of edges (amino acid contacts), respectively. Representing peptides as graphs enables the model to capture the structural relationships between amino acids in peptide sequences. Each amino acid is treated as a node in the graph, and the edges between nodes represent the interactions or spatial relationships between amino acids. This graph-based representation incorporates three-dimensional structural information, providing a more comprehensive understanding of peptide features. The features of each graph node are generated by concatenating one-hot encoding, position encoding, and ESM-1b extracting features. For a graph containing N nodes, the node feature is $x_1, x_2, ..., x_N, \mathbf{x}_i \in \mathbb{R}^D, i \in (1, N)$ and it D is the dimensions of the features. The spatial relationships between nodes are encoded in an adjacency matrix A, where $A_{ij} = 1$ if there is an edge between nodes i and j; otherwise, $A_{ij} = 0$. The construction of node features and adjacency matrices used in this study will be described in detail in the following sections.

2.3.1. Node Feature Embedding

Amino Acid Composition Representation Based on One-Hot Encoding

One-hot encoding is employed to represent the 20 standard amino acids as numerical vectors. In this encoding scheme, a value of one is assigned to the position corresponding to a given amino acid, while all other positions in the vector are set to zero. For instance, the encoding of the amino acid "A" is represented as $[1, 0, 0, 0, \ldots, 0]_{20}$, which means each amino acid is represented by a 20-dimensional vector. This encoding method maps discrete feature values to points in Euclidean space, where each unique value corresponds to a specific position. By placing discrete features in Euclidean space, one-hot encoding facilitates more meaningful distance calculations.

Residue Position Encoding

Position encoding in many applications is considered an effective descriptor [25] calculated as follows:

$$P(\text{pos}, 2i) = \sin\left(\frac{\text{pos}}{b^{2i/d}}\right) \tag{1}$$

$$P(\text{pos}, 2i + 1) = \cos\left(\frac{\text{pos}}{b^{2i/d}}\right) \tag{2}$$

Let pos (where $0 \leq pos < L$) denote the position of the amino acid in a length of L, and let $2i$ and $2i + 1$ (where $0 \leq i < 10$) represent the individual positions in the corresponding position vector. The values at even positions are calculated using the sine function, while those at odd positions are calculated using the cosine function. Alternating between sine and cosine functions enhances the model's robustness and performance. Here, b and d are constants, and are set to 1000 and 20, respectively. Each amino acid is represented by a 20-dimensional vector.

ESM-1b Feature Encoding

To comprehensively capture the features of each amino acid, this study utilizes the pre-trained ESM-1b model. ESM-1b is a high-capacity transformer model, optimized through hyperparameter tuning, that takes protein sequences as input. The embeddings generated by ESM-1b not only encode direct information about protein sequences but also implicitly capture deeper biological insights, including structural, functional, and evolutionary knowledge. The pre-trained ESM-1b model can uncover information that is absent in one-hot encoding or residue position encoding. The features from these encoding methods are combined to represent each amino acid [17]. The feature matrix derived from

ESM-1b has a size of $L \times 1280$ (where L is the length of a peptide sequence), while both one-hot encoding and position encoding feature matrices have a size of $L \times 20$. Due to the substantial dimensionality difference between these feature types, directly concatenating the ESM-1b features with the other two could result in the ESM-1b features dominating, leading to poor performance. Thus, it is necessary to reduce the dimensionality of the ESM-1b features to match the scale of the other feature sets. Principal Component Analysis (PCA) [26] was used for dimensionality reduction, with the optimal dimension selected through experiments. Further details are provided in Section 3.3.

2.3.2. Adjacency Matrix Construction

The adjacency matrix represents the relationships between nodes in the graph, which, in this study, corresponds to the relationship between two amino acids. To more accurately capture this relationship, we construct the adjacency matrix using predicted contact maps from trRosetta [27]. First, we query the sequence against the uniclust30_2018_08 database using the HHblits tool [28], generating a multiple sequence alignment (MSA). The MSA is then input into the trRosetta tool to predict the distributions of distances and orientations between residue pairs. The distance information represents the spatial relationship between two atoms in the contact map. Distances ranging from 2 Å to 20 Å are divided into 36 segments, with each representing a 0.5 Å increment, with an additional segment for residues that do not come into contact. TrRosetta predicts the contact probability for residues within each segment. In previous work, the segment with the highest probability among all equally spaced segments was used to represent residue distance; when the distance between two atoms in the contact map is less than or equal to a predefined threshold, it indicates that the residues are in contact [27]. However, this approach is not optimal. Therefore, the method proposed in this study sums the probabilities of all segments corresponding to distances between 2 and 8 Å to represent the contact probability between residues, and uses this to calculate the adjacency matrix A_{ij} of the peptide. The calculation can be expressed as follows:

$$C_{ij} = \sum_{n=1}^{13} dist_{ij} \tag{3}$$

As shown in Figure 3, $dist_{ij}$ represents the probability distribution C_{ij}, where $n = 1$ and $n = 13$ correspond to distances of 2 Å and 8 Å, respectively. Residue pairs with contact probabilities C_{ij} between residues greater than $C_{ij} = 0.8$ are added to the graph as edge information:

$$A_{ij} = \begin{cases} 1, & \text{if } C_{ij} \geq D_{\text{th}} \text{ or } i = j, \\ 0, & \text{otherwise.} \end{cases} \tag{4}$$

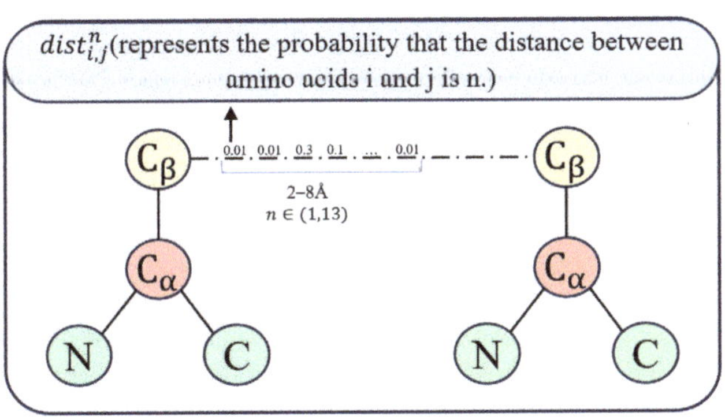

Figure 3. Calculation method for amino acid contact probability.

Here, C_{ij} represents the contact probability between amino acids i and j, and D_{th} denotes the contact threshold used in this study. As discussed in Section 3.4.3, setting this threshold to 0.8 yields the best performance, likely because it helps filter out improbable spatial interactions, thereby improving the precision of protein structure predictions.

2.4. Dual-Channel Feature Extraction

2.4.1. The GAT Feature Extraction Channel

In this study, we use a multi-head GAT to model the weights of adjacent nodes and learn the hidden representations of nodes within the graph. The GAT utilizes a masked self-attention mechanism to assign weights to neighboring nodes, reflecting the importance of their connections. Specifically, GAT is employed to capture local correlations between residues and their neighbors. Due to errors in polypeptide structure prediction, the actual contact probabilities may deviate from the predicted ones. GAT can adaptively assign lower weights to these noisy data, thereby minimizing their negative impact on structural prediction outcomes. The process of using multiple iterations of the multi-head attention layer to calculate node features for the subsequent layer is as follows:

$$H_i' = ||_{k=1}^{K}\sigma\left(\sum_{v_j \in N(v_i)} \alpha_{ij}^{(k)} W^{(k)} H_j\right) \tag{5}$$

where H_i' represents the feature representation of node i, K denotes the number of heads subjected to multi-head attention, σ the symbol stands for the activation function, $N(v_i)$ represents the neighboring nodes of node i, and $W^{(k)}$ denotes the projection transformation matrix of the node features of the head i. $\alpha_{ij}^{(k)}$ is computed as follows:

$$\alpha_{ij}^{(k)} = \frac{exp\left(g\left(a^{(k)T}\left[W^{(k)}H_i||W^{(k)}H_j\right]\right)\right)}{\sum_{u \in N_{(i)}} exp\left(g\left(a^{(k)T}\left[W^{(k)}H_i||W^{(k)}H_u\right]\right)\right)} \tag{6}$$

where $a(\cdot)$ denotes a vector of learnable parameters, $g(\cdot)$ denotes a ReLU [29] activation function, and $||$ denotes a cascade.

In this study, the graph information is input into a three-layer GAT with eight attention heads to obtain node weights. Layer normalization is applied to reduce covariate shift between layers, enhancing both training efficiency and the generalization of the model. ReLU is used as the activation function for the GAT layers. Following this, a global max pooling layer is introduced to select the global maximum after the GAT layers, thereby capturing the most representative information from the entire graph. In addition, a dropout layer [30] is employed to randomly discard neurons, further improving the generalization of the model.

2.4.2. The GCN Feature Extraction Channel

The GCN fully aggregates information from adjacent nodes along with its own node information, enabling the node features to be updated. Therefore, the graph constructed from peptide sequences is input into the GCN feature-learning module developed in this study, allowing the model to focus on updating node features and learning effective representations. The GCN takes two primary inputs: the node feature matrix X and the adjacency matrix A. For a graph with N nodes, each node has D-dimensional features, which can be organized into an $N \times D$ matrix X. Additionally, the relational connections between nodes are represented by an $N \times N$ adjacency matrix A.

The weight update process in the GCN can be described by the following formula:

$$H' = \sigma\left(\widetilde{D}^{-1/2}\widetilde{A}\widetilde{D}^{-1/2}HW\right) \tag{7}$$

Let $H \in R^{N \times D}$, where N represents the number of nodes in the graph, and D is the dimensionality of the vector used to represent each node. The input layer consists of H and the adjacency matrix $\widetilde{A}$, which includes self-connections. $\widetilde{D}$ denotes the degree matrix of $\widetilde{A}$, and W represents the trainable weight parameters. The symbol σ refers to the ReLU activation function. By applying this formula, the output graph representation H' is obtained. After the graph convolution, a global Max-pooling layer is used to select the global maximum, which captures the most informative feature of the entire graph.

2.5. Classification Module

The classification module uses the fused features obtained from the dual-channel feature extraction module, employing four fully connected layers and a sigmoid function to achieve the final classification of IL-6-inducing peptides. The specific implementation process is as follows:

$$x_0 = concatenate\left(maxpooling\left(H_i'\right), maxpooling\left(H'\right)\right) \tag{8}$$

$$x_i = W_i x_{i-1} + b_i \tag{9}$$

$$y_i = \frac{1}{1 + e^{-x_i}} \tag{10}$$

Let W_i denote the weight matrix of the ith linear layer, while b_i represents the bias of the ith linear layer. Dropout and ReLU activation functions are applied after each linear layer to process the features. The model's output, y_i, is obtained through the sigmoid activation function. Finally, a threshold $Y_{th} = 0.35$ is set to determine whether the input sequence is an IL-6-inducing peptide.

The BCEWithLogitsLoss function, which combines a sigmoid function and a binary cross-entropy loss, is used during training. The formula of BCEWithLogitsLoss can be expressed as follows:

$$\text{Loss} = \frac{1}{N}\sum_{i=1}^{N}(y_i \cdot \log(\sigma(p_i)) + (1 - y_i) \cdot \log(1 - \sigma(p_i))) \tag{11}$$

where N denotes the total number of samples, $\sigma(\cdot)$ denotes a sigmoid function, and p_i represents the probability that the ith sample is positive.

2.6. Evaluation Metrics

To enable a more objective and comprehensive comparison between the proposed model and previous studies, we select the same evaluation metrics used in prior research. These include Balanced Accuracy (BACC), Sensitivity (SN), Specificity (SP), Matthews Correlation Coefficient (MCC), and Area Under the ROC Curve (AUC). The AUC is particularly valuable, as it is independent of specific threshold settings, providing a more robust assessment of the model's generalization performance. Balanced Accuracy (BACC) is also an important metric, as it accounts for both positive and negative samples, thereby mitigating potential bias caused by class imbalance. The formulas for calculating these metrics are as follows:

$$BACC = \frac{SN + SP}{2} \tag{12}$$

$$SN = \frac{TP}{TP + FN} \tag{13}$$

$$SP = \frac{TN}{TP + FN} \quad (14)$$

$$MCC = \frac{TP \times TN - FP \times FN}{\sqrt{(TP \times FN)(TP \times FP)(TN \times FP)(TN \times FN)}} \quad (15)$$

$$AUC : Area \quad under \quad the \quad ROC \quad Curve \quad (16)$$

In this context, TP and FP denote true positives and false positives, respectively. True positives refer to the number of samples that are correctly predicted as positive, while false positives represent the samples incorrectly predicted as positive. Similarly, TN and FN represent true negatives and false negatives. True negatives correspond to the samples correctly predicted as negative, whereas false negatives refer to the samples incorrectly predicted as negative.

2.7. Model Implementation

The model is constructed using PyTorch and its Geometric framework [31]. ADAM is employed as the optimizer, with a training batch size of 256. To address the class imbalance between positive and negative samples, the weights for positive and negative samples are set at a ratio of 10:1. The initial learning rate is set to 0.001, and it is reduced by 5% every 5 epochs. The model is trained for a total of 40 epochs. For hyperparameter optimization, a grid search strategy is employed, with the best combinations selected based on the AUC from the validation set.

3. Results

3.1. Pre-Trained Model Selection Experiment

In the process of selecting the pre-trained models, we conducted a five-fold cross-validation and compared the performance of BERT [32], ProtT5 [33], and ESM-1b models. Since these models extract feature vectors of different dimensions, we performed dimensionality reduction on the features before conducting the comparative experiments. As shown in Figure 4A, the ESM-1b model outperformed the others in terms of evaluation metrics such as AUC, BACC, and SN. This can be attributed to the fact that, although BERT and ProtT5 excel in capturing long-range dependencies and contextual information, the relatively short sequences in our study may limit their performance. In contrast, ESM-1b is capable of capturing the position and relative importance of amino acids in three-dimensional structures by considering their deep features. This ability not only enhances our understanding of how amino acids interact with each other but also provides insights into their functional roles within peptide segments. The deep features enable the model to identify the importance of key amino acid positions for biological activity, thereby improving prediction accuracy. After carefully considering both the experimental results and the model characteristics, we ultimately selected ESM-1b as the pre-trained model for our study.

3.2. Threshold Analysis for Classification

Due to the substantial imbalance between positive and negative samples in the dataset, with a ratio of 0.122, selecting an appropriate classification threshold was crucial. To determine the optimal threshold, BACC was used as the evaluation metric, and five-fold cross-validation was performed on the training set. The results, as shown in Figure 5, indicate that setting the classification threshold to 0.35 yields the highest BACC, with the model performing best when it comes to distinguishing between negative and positive samples. Specifically, if the predicted probability of a positive sample, obtained from the softmax function, exceeds 0.35, the sample is classified as positive; otherwise, it is classified as negative. Thus, the classification threshold was set to 0.35 in this study.

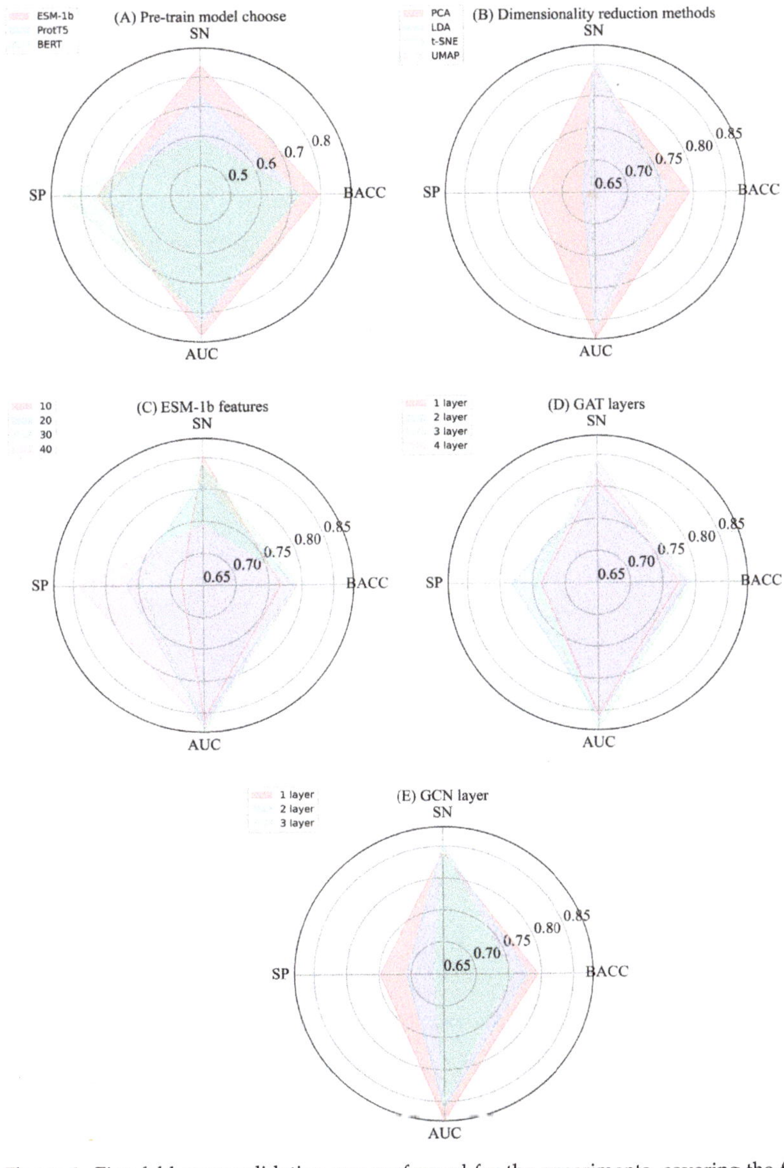

Figure 4. Five-fold cross-validation was performed for the experiments, covering the following aspects: (**A**) selection of the pre-trained model, (**B**) selection of dimensionality reduction methods, (**C**) selection of dimensions for ESM-1b features, (**D**) selection of GAT layers, and (**E**) selection of GCN layers.

3.3. Dimension Selection of ESM-1b Features

To effectively integrate the features obtained from ESM-1b with the other two feature types and ensure optimal prediction performance, it is essential to reduce the dimensionality of the 1280-dimensional ESM-1b features to a smaller number of dimensions [34]. In this study, several dimensionality reduction methods were compared, including linear discriminant analysis (LDA) [35], t-distributed stochastic neighbor embedding (t-SNE) [36], and uniform manifold approximation and projection (UMAP) [37]. These methods were evaluated by reducing the dimensionality of the ESM-1b features to the same dimension, as shown in Figure 4B. After comparison, PCA was chosen as the dimensionality reduction method. PCA effectively projects high-dimensional data into a lower-dimensional space while preserving as much variance as possible, thereby maintaining the primary structure of the data and reducing the number of features [26]. To ensure consistency and comparability across models, the same dimensionality reduction method was applied to all models. This approach allows for fair evaluation of performance differences between models. To determine the optimal dimensionality, experiments were conducted on the training set, testing four dimensions in intervals of 10, ranging from 10 to 40, with AUC as the evaluation metric. The dimension corresponding to the best result was selected for further analysis. Figure 4C presents the average results of five-fold cross-validation for each experiment. The results show that the best and most stable performance is achieved when the ESM-1b features are reduced to 30 dimensions. At this dimension, the difference between the ESM-1b features and those extracted from position encoding and one-hot encoding becomes minimal, and the reduction in dimensionality has little impact on the model's performance. This helps retain the essential feature information, leading to the best results.

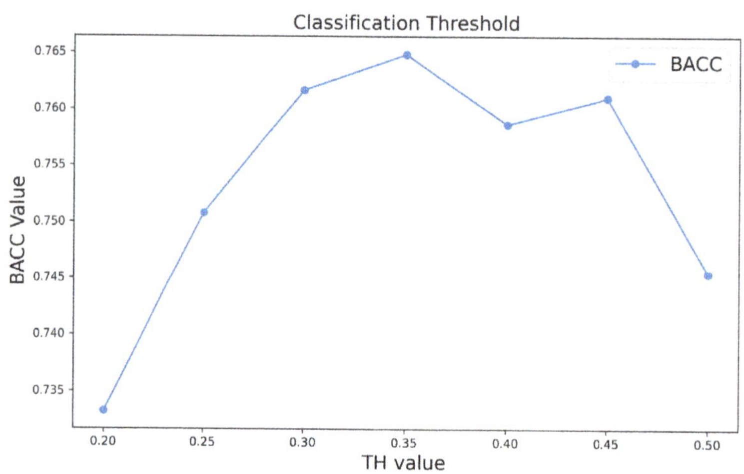

Figure 5. Comparison of the model's performance across various classification thresholds using five-fold cross-validation.

3.4. Analysis of GAT and GCN Layers

3.4.1. Experiments on GAT Layers Selection

Using a five-fold cross-validation approach on the training set, we systematically evaluated various numbers of GAT layers to determine the optimal configuration. We tested different combinations of GAT layers, ranging from one to four, to examine how the number of layers affects model performance. As shown in Figure 4D, the three-layer GAT demonstrated significant improvements, with increases of 0.2% to 4.9% in SN and 1.7% to 2.3% in AUC, indicating that the three-layer GAT outperformed others in predicting IL-6-inducing peptides. The superior performance of the three-layer GAT may be attributed to its ability to capture and represent the complex structure and relationships between nodes in the graph more effectively by increasing the network depth and leveraging the flexibility of the attention mechanism. Although a four-layer GAT could theoretically offer greater model capacity, the limited volume of data in this study suggests that a four-layer

GAT might overcomplicate the model, increase training difficulty, and potentially lead to overfitting. Considering both the model's performance and training feasibility, we ultimately selected the three-layer GAT as a key module in the dual-channel architecture. This choice aims to balance predictive power with training efficiency, ensuring that the model captures the essential data characteristics while avoiding unnecessary complexity.

3.4.2. Experiments on GCN Layer Selection

By evaluating the results of five-fold cross-validation on the training set, we determined the optimal number of GCN layers. One-layer, two-layer, and three-layer GCNs were tested. As shown in Figure 4E, the one-layer GCN outperformed the others. Specifically, compared to the two-layer GCN, the one-layer GCN improved the BACC on the training set by 1.6% to 3.9%, the MCC by 1.6%, and the AUC by 2% to 2.6%. Given the relatively short sequence length and simple graph structure in our dataset, a one-layer GCN was sufficient to capture valuable information from the graph. Increasing the number of layers did not yield further performance gains and may have unnecessarily increased the model's complexity. Therefore, considering both performance and model simplicity, we chose to use a one-layer GCN as the second channel in the dual-channel feature extraction module. This decision ensured that the model maintained efficient performance while preserving structural simplicity.

3.4.3. Additional Parameter Selection Experiments

In this section, we discuss the impact of model parameters on the performance of DGIL-6, specifically focusing on the amino acid contact threshold, the number of GAT attention heads, and the embedding dimension of hidden layers. These parameters were selected based on AUC and SN from the results of five-fold cross-validation.

The selection of the amino acid contact threshold significantly influences the composition stage. Firstly, we tested thresholds of 0.6, 0.7, 0.8, and 0.9. As shown in Figure 6A, a lower threshold generates excessive edge information, negatively impacting the model's performance, while a higher threshold may exclude key edge information. Therefore, we selected a contact threshold of 0.8. Next, the multi-head attention mechanism in the GAT layer was evaluated. We tested different numbers of attention heads (4, 8, 12, and 16). As shown in Figure 6B, DGIL-6 performed better as the number of attention heads increased, but its performance declined when the number exceeded eight. Consequently, we chose eight attention heads for the GAT layer. Finally, we examined the impact of the embedding dimension of hidden layers on model performance. Values of 32, 64, 128, and 256 were tested. As shown in Figure 6C, DGIL-6 achieved optimal performance with an embedding dimension of 64.

3.5. *Ablation Experiments*

3.5.1. Ablation Experiments of Node Features

To evaluate the impact of different node characteristics on model performance, we conducted a series of ablation experiments on the testing set, with the results presented in Table 2. The experimental results demonstrate that combining ESM-1b-extracted features with position encoding and one-hot encoding significantly improved the model's predictive performance, particularly in BACC and AUC metrics. This feature fusion strategy enhanced the model's representation capability: one-hot encoding supplied categorical information about nodes, position encoding facilitated the capture of amino acid positions within peptide sequences, and ESM-1b features enabled the model to capture biological signals related to protein functionality. Notably, the ESM-1b-extracted features significantly enhanced the model's performance, with AUC increasing by 5%, MCC by 10.5%, BACC by 6.7%, and SN by 8.2%. The relative positions of amino acids in peptide sequences directly influenced their functions and interactions. However, graph neural networks often struggle to capture sequential positions when processing sequence data. The introduction of position encoding compensates for this limitation. By providing precise positional information for each node, position encoding enabled the model to better interpret node positions within peptide sequences [38]. The experimental results show that the AUC increased by 2.5%, MCC by 1.1%, the SN increased by 9.6%, and the BACC increased by 2.1%. Additionally, while one-hot encoding is less effective than ESM-1b features, the amino acid composition information it provides remains vital to the model's performance. It resulted in the AUC increasing by 3.1%, MCC increasing by 4.5%, SP increasing by 3.4%, SN increasing by 1.6%, and BACC increasing by 2.4%. One-hot encoding allows the model to effectively capture the diversity of amino acids and their influence on protein functionality, thereby driving these improvements. These findings underscore the distinct contributions and synergistic effects of various features in enhancing the model's predictive capability.

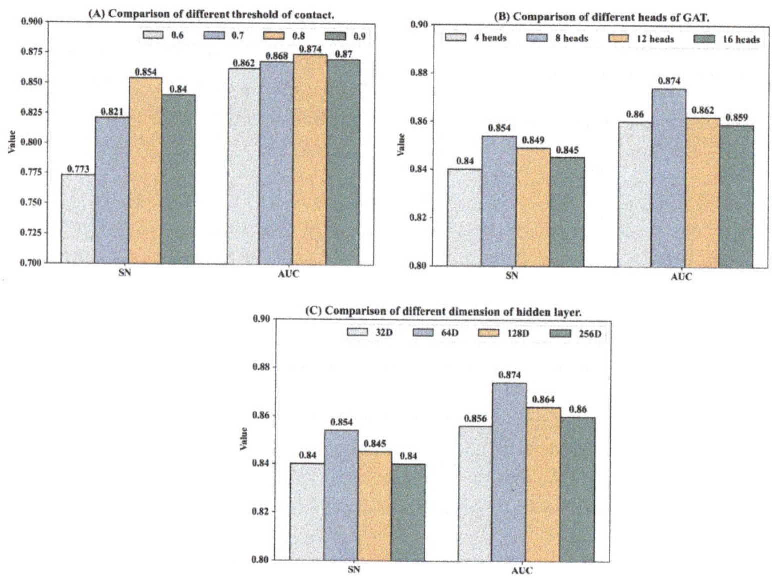

Figure 6. Five-fold cross-validation was conducted in the experiments. (**A**) Selection of the amino acid contact threshold. (**B**) Selection of the number of GAT attention heads. (**C**) Selection of the hidden layer embedding dimension.

Table 2. Ablation experiment results of node features.

Node Features	BACC	SN	SP	MCC	AUC
No-One-hot	0.814	0.863	0.765	0.426	0.871
No-Position	0.817	0.780	**0.854**	0.460	0.877
No-ESM-1b	0.771	0.794	0.749	0.366	0.852
DGIL-6	**0.838**	**0.876**	0.799	**0.471**	**0.902**

Note: The best value for each indicator is shown in bold.

3.5.2. Ablation Experiments of Dual-Channel Feature Extraction

Ablation experiments were conducted on the testing set to evaluate the effectiveness of the dual-channel feature extraction module and determine which component had the greatest impact on the performance of the model. As presented in Table 3, "No-GAT" refers to the model with the GAT module removed, while "No-GCN" represents the model with the GCN module excluded.

Table 3. Ablation experiment results of dual-channel feature extraction.

Dual-Channel	BACC	SN	SP	MCC	AUC
No-GAT	0.805	0.863	0.645	0.321	0.841
No-GCN	0.821	0.794	**0.849**	0.453	0.876
DGIL-6	**0.838**	**0.876**	0.799	**0.471**	**0.902**

Note: The best value for each indicator is shown in bold.

When the GAT module was removed from the model, a 6.1% decrease in AUC, a substantial 15% decrease in MCC, a 3.3% decrease in BACC, and a 15.4% decrease in SP were observed. Interestingly, SN increased by 1.3%, indicating that while GAT excels at predicting negative samples, it is relatively less effective at identifying positive samples. In contrast, removing the GCN module led to a 2.6% decrease in AUC, a 2.8% decrease in MCC, a 1.7% decrease in BACC, an 8.2% decrease in SN, and a 5% increase in SP. These results suggest that GCNs are more effective when it comes to predicting positive samples. The observed differences may stem from the unique characteristics of the two channels: GCN leverages local neighborhood information to learn node features, whereas GAT uses an attention mechanism to assign varying importance weights to neighbors, capturing more nuanced relationships between nodes. By integrating these two mechanisms, the model can comprehensively capture both global and local features within the graph, enhancing its overall predictive performance. These findings highlight the complementary roles of GAT and GCN modules and their critical contributions to improving the prediction accuracy of IL-6-inducing peptides.

3.6. Comparison Experiments with Existing Methods

To compare the proposed model against existing prediction tools, we conducted comparative experiments on the testing set. The evaluated IL-6-inducing peptide prediction models include IL-6Pred, StackIL6, and MVIL6. Performance metrics of IL-6Pred, StackIL6, and MVIL6 were derived from prior studies, with references specifically citing the MVIL6 study. The comparative results of these methods are summarized in Table 4.

Table 4. Comparative experimental results with existing methods.

Tool	BACC	SN	SP	MCC	AUC
IL-6Pred	0.743	0.753	0.732	-	0.830
StackIL6	0.795	0.849	0.741	0.393	0.841
MVIL6	0.834	0.863	**0.804**	0.465	0.883
DGIL-6	**0.838**	**0.876**	0.799	**0.471**	**0.902**

Note: The best value for each indicator is shown in bold; "-" represents a null value.

As shown in Table 4, DGIL-6 showed improvements across several key performance metrics: AUC increased by 1.9% to 7.2%, BACC increased by 0.4% to 9.5%, MCC increased by 0.6%to 7.8%, SP increased by 0.5% to 7.2%, and SN increased significantly, by 1.3% to 12.3%. These advancements can be largely attributed to the utilization of the pre-trained ESM-1b model for extracting amino acid features, which enriches the model with more comprehensive biological information, thereby enhancing prediction accuracy and reliability. The DGIL-6 achieves complementary features by integrating 3D structure-based graph representations with amino acid-based depth features, resulting in superior performance in IL-6-inducing peptide prediction.

3.7. Hierarchical Clustering Analysis of Dual-Channel Features

To visually demonstrate the features extracted by different channels of the DGIL-6 model, we conducted a hierarchical clustering analysis after concatenating the features from the GAT and GCN channels. Hierarchical clustering was selected because it effectively groups similar features into a hierarchical structure, facilitating a clear observation of feature similarity. This method is particularly well suited for exploring the intrinsic structure of data and identifying feature patterns. The visualization helps to quickly pinpoint similar features and provides insight into their distribution and significance across the entire dataset. The results of the hierarchical clustering analysis [39] are presented in Figure 7. As shown in the characteristic distribution of IL-6-inducing peptides and non-IL-6-inducing peptides in Figure 7, regions with larger eigenvalues tend to form clusters. This finding suggests that features derived from the model are similar for sequences with similar functions but significantly different for sequences with distinct functions. It is noted that larger eigenvalues indicate that these features play a more significant role in distinguishing different categories.

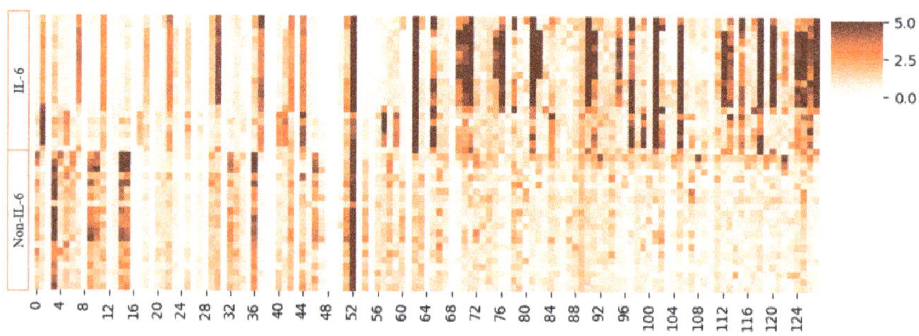

Figure 7. The feature visualization results of two channels presented via hierarchical clustering on positive and negative samples. The vertical axis represents different individual samples. The horizontal axis indicates the feature indices, with features 0–63 derived from the GAT channel and features 64–127 derived from the GCN channel. The heatmaps utilize color gradients to visualize normalized feature values, where darker colors indicate higher values and lighter colors indicate lower values.

From the feature distributions extracted by GAT and GCN in Figure 7, it can be observed that the two channels capture distinct features of the data. GAT focuses on neighboring nodes that exert the greatest impact on classification, enhancing the model's sensitivity to critical features. In contrast, GCN effectively smooths node features, bringing similar nodes closer in feature space, making it well suited for capturing overall graph structural information. This complementary understanding enables the model to better identify structural patterns in peptides. These distinct contributions improve the classification of IL-6-inducing peptides, further validating the effectiveness of the dual-channel feature extraction module. The clustering results in Figure 7 demonstrate that the extracted features effectively separate IL-6-inducing peptides from non-IL-6-inducing peptides, highlighting their strong discriminative capacity. However, some features exhibit low values in both IL-6 inducing and non-IL-6-inducing peptides, suggesting these features may be shared between the two groups, making it difficult for the model to distinguish them. Further optimization of these features could enhance the model's performance and reduce false positives.

4. Conclusions

We developed a prediction tool, DGIL-6, for identifying IL-6-inducing peptides using graph neural networks and predicted structural information. DGIL-6 represents each peptide sequence as a graph, where amino acids serve as nodes, and the adjacency matrix, representing relationships between nodes, is computed based on the predicted residue contact graph of the peptide sequence. The graph's node features are composed of one-hot encoding, positional encoding, and features extracted using ESM-1b. To extract multi-view features, a dual-channel feature extraction module was implemented, combining a three-layer multi-head GAT with a one-layer GCN. The effectiveness of the proposed method was validated through various experiments, including comparison studies, ablation analysis, and feature visualizations. Although we achieved promising results, there is still room for improvement. Future work could address the challenge of insufficient positive samples using transfer learning or enhance graph construction performance by leveraging structural information of protein sequences obtained from advanced large-scale models.

Author Contributions: Conceptualization, R.C. and C.Z.; methodology, R.C., P.W., Y.D., Y.B. and Q.L.; software, R.C. and Q.L.; writing—original draft preparation, R.C. and Q.L.; writing—review and editing, R.C., P.W., Y.D. and Y.B.; supervision, R.C. and C.Z.; funding acquisition, R.C. All authors have read and agreed to the published version of the manuscript.

Funding: This work was supported by the National Natural Science Foundation of China (62373001 and 62272004), the National Key Research and Development Program of China (2020YFA0908700), the State Key Laboratory of Pathogenesis, Prevention and Treatment of High Incidence Diseases in Central Asia Fund (SKL-HIDCA-2024-AH4), and the Education Department of Anhui Province (2023AH050089).

Institutional Review Board Statement: Not applicable.

Informed Consent Statement: Not applicable.

Data Availability Statement: The source code and dataset used in this study are available at https://github.com/IL-6Li/DGIL-6 (accessed on 7 January 2025).

Conflicts of Interest: The authors declare no conflicts of interest.

References

1. Hirano, T.; Yasukawa, K.; Harada, H.; Taga, T.; Watanabe, Y.; Matsuda, T.; Kashiwamura, S.; Nakajima, K.; Koyama, K.; Iwamatsu, A.; et al. Complementary DNA for a novel human interleukin (BSF-2) that induces B lymphocytes to produce immunoglobulin. *Nature* **1986**, *324*, 73–76. [CrossRef] [PubMed]
2. Lu, R.; Zhao, X.; Li, J.; Niu, P.; Yang, B.; Wu, H.; Wang, W.; Song, H.; Huang, B.; Zhu, N.; et al. Genomic characterisation and epidemiology of 2019 novel coronavirus: Implications for virus origins and receptor binding. *Lancet* **2020**, *395*, 565–574. [CrossRef] [PubMed]
3. Kishimoto, T. IL-6: From its discovery to clinical applications. *Int. Immunol.* **2010**, *22*, 347–352. [CrossRef] [PubMed]
4. Li, Y.S.; Ren, H.C.; Cao, J.H. Roles of Interleukin-6-mediated immunometabolic reprogramming in COVID-19 and other viral infection-associated diseases. *Int. Immunopharmacol.* **2022**, *110*, 109005. [CrossRef] [PubMed]
5. Tanaka, T.; Narazaki, M.; Kishimoto, T. Interleukin (IL-6) Immunotherapy. *Cold Spring Harb. Perspect. Biol.* **2018**, *10*, a028456. [CrossRef]
6. Hong, D.S.; Angelo, L.S.; Kurzrock, R. Interleukin-6 and its receptor in cancer. *Cancer* **2007**, *110*, 1911–1928. [CrossRef]
7. Ujiie, H.; Tomida, M.; Akiyama, H.; Nakajima, Y.; Okada, D.; Yoshino, N.; Takiguchi, Y.; Tanzawa, H. Serum Hepatocyte Growth Factor and Interleukin-6 Are Effective Prognostic Markers for Non-small Cell Lung Cancer. *Anticancer. Res.* **2012**, *32*, 3251–3258.
8. Tang, Y.; Liu, J.; Zhang, D.; Xu, Z.; Ji, J.; Wen, C. Cytokine storm in COVID-19: The current evidence and treatment strategies. *Front. Immunol.* **2020**, *11*, 1708. [CrossRef]
9. Sakuta, H.; Inaba, K.; Muramatsu, S. Calcitonin gene-related peptide enhances cytokine-induced IL-6 production by fibroblasts. *Cell. Immunol.* **1995**, *165*, 20–25. [CrossRef]
10. Dhall, A.; Patiyal, S.; Sharma, N.; Usmani, S.S.; Raghava, G.P. Computer-aided prediction and design of IL-6 inducing peptides: IL-6 plays a crucial role in COVID-19. *Briefings Bioinform.* **2020**, *22*, 936–945. [CrossRef]

11. Charoenkwan, P.; Chiangjong, W.; Nantasenamat, C.; Hasan, M.M.; Manavalan, B.; Shoombuatong, W. StackIL6: A stacking ensemble model for improving the prediction of IL-6 inducing peptides. *Briefings Bioinform.* **2021**, *22*, bbab172. [CrossRef] [PubMed]
12. Wang, R.; Feng, Y.; Sun, M.; Jiang, Y.; Li, Z.; Cui, L.; Wei, L. MVIL6: Accurate identification of IL-6-induced peptides using multi-view feature learning. *Int. J. Biol. Macromol.* **2023**, *246*, 125412. [CrossRef]
13. Yang, J.; Anishchenko, I.; Park, H.; Peng, Z.; Ovchinnikov, S.; Baker, D. Improved protein structure prediction using predicted interresidue orientations. *Proc. Natl. Acad. Sci. USA* **2020**, *117*, 1496–1503. [CrossRef] [PubMed]
14. Lai, B.; Xu, J. Accurate protein function prediction via graph attention networks with predicted structure information. *Briefings Bioinform.* **2022**, *23*, bbab502. [CrossRef] [PubMed]
15. Gligorijević, V.; Renfrew, P.D.; Kosciolek, T.; Leman, J.K.; Berenberg, D.; Vatanen, T.; Chandler, C.; Taylor, B.C.; Fisk, I.M.; Vlamakis, H.; et al. Structure-based protein function prediction using graph convolutional networks. *Nat. Commun.* **2021**, *12*, 3168. [CrossRef]
16. Yan, K.; Lv, H.; Guo, Y.; Peng, W.; Liu, B. sAMPpred-GAT: Prediction of antimicrobial peptide by graph attention network and predicted peptide structure. *Bioinformatics* **2023**, *39*, btac715. [CrossRef]
17. Rives, A.; Meier, J.; Sercu, T.; Goyal, S.; Lin, Z.; Liu, J.; Guo, D.; Ott, M.; Zitnick, C.L.; Ma, J.; et al. Biological structure and function emerge from scaling unsupervised learning to 250 million protein sequences. *Proc. Natl. Acad. Sci. USA* **2021**, *118*, e2016239118. [CrossRef]
18. Veličković, P.; Cucurull, G.; Casanova, A.; Romero, A.; Lio, P.; Bengio, Y. Graph attention networks. *arXiv* **2017**, arXiv:1710.10903.
19. Kipf, T.N.; Welling, M. Semi-supervised classification with graph convolutional networks. *arXiv* **2016**, arXiv:1609.02907.
20. Vita, R.; Overton, J.A.; Greenbaum, J.A.; Ponomarenko, J.; Clark, J.D.; Cantrell, J.R.; Wheeler, D.K.; Gabbard, J.L.; Hix, D.; Sette, A.; et al. The immune epitope database (IEDB) 3.0. *Nucleic Acids Res.* **2015**, *43*, D405–D412. [CrossRef]
21. Agrawal, P.; Bhalla, S.; Chaudhary, K.; Kumar, R.; Sharma, M.; Raghava, G.P. In silico approach for prediction of antifungal peptides. *Front. Microbiol.* **2018**, *9*, 323. [CrossRef] [PubMed]
22. Nagpal, G.; Chaudhary, K.; Agrawal, P.; Raghava, G.P. Computer-aided prediction of antigen presenting cell modulators for designing peptide-based vaccine adjuvants. *J. Transl. Med.* **2018**, *16*, 181. [CrossRef] [PubMed]
23. Qureshi, A.; Thakur, N.; Kumar, M. VIRsiRNApred: A web server for predicting inhibition efficacy of siRNAs targeting human viruses. *J. Transl. Med.* **2013**, *11*, 305. [CrossRef]
24. Patiyal, S.; Agrawal, P.; Kumar, V.; Dhall, A.; Kumar, R.; Mishra, G.; Raghava, G.P. NAGbinder: An approach for identifying N-acetylglucosamine interacting residues of a protein from its primary sequence. *Protein Sci.* **2020**, *29*, 201–210. [CrossRef]
25. Vaswani, A.; Shazeer, N.; Parmar, N.; Uszkoreit, J.; Jones, L.; Gomez, A.N.; Kaiser, Ł.; Polosukhin, I. Attention is all you need. In Proceedings of the 31st Conference on Neural Information Processing Systems (NIPS 2017), Long Beach, CA, USA, 4–9 December 2017; Volume 30.
26. Abdi, H.; Williams, L.J. Principal component analysis. *Wiley Interdiscip. Rev. Comput. Stat.* **2010**, *2*, 433–459. [CrossRef]
27. Du, Z.; Su, H.; Wang, W.; Ye, L.; Wei, H.; Peng, Z.; Anishchenko, I.; Baker, D.; Yang, J. The trRosetta server for fast and accurate protein structure prediction. *Nat. Protoc.* **2021**, *16*, 5634–5651. [CrossRef]
28. Remmert, M.; Biegert, A.; Hauser, A.; Söding, J. HHblits: Lightning-fast iterative protein sequence searching by HMM-HMM alignment. *Nat. Methods* **2012**, *9*, 173–175. [CrossRef]
29. Nair, V.; Hinton, G.E. Rectified linear units improve restricted boltzmann machines. In Proceedings of the 27th International Conference on Machine Learning (ICML-10), Haifa, Israel, 21–24 June 2010; pp. 807–814.
30. Srivastava, N.; Hinton, G.; Krizhevsky, A.; Sutskever, I.; Salakhutdinov, R. Dropout: A simple way to prevent neural networks from overfitting. *J. Mach. Learn. Res.* **2014**, *15*, 1929–1958.
31. Fey, M.; Lenssen, J.E. Fast graph representation learning with PyTorch Geometric. *arXiv* **2019**, arXiv:1903.02428.
32. Devlin, J.; Chang, M.W.; Lee, K.; Toutanova, K. Bert: Pre-training of deep bidirectional transformers for language understanding. *arXiv* **2018**, arXiv:1810.04805.
33. Weissenow, K.; Heinzinger, M.; Rost, B. Protein language-model embeddings for fast, accurate, and alignment-free protein structure prediction. *Structure* **2022**, *30*, 1169–1177. [CrossRef] [PubMed]
34. Nguyen, L.D.; Lin, D.; Lin, Z.; Cao, J. Deep CNNs for microscopic image classification by exploiting transfer learning and feature concatenation. In Proceedings of the 2018 IEEE International Symposium on Circuits and Systems (ISCAS), Florence, Italy, 27–30 May 2018; pp. 1–5.
35. Xanthopoulos, P.; Pardalos, P.M.; Trafalis, T.B.; Xanthopoulos, P.; Pardalos, P.M.; Trafalis, T.B. Linear discriminant analysis. In *Robust Data Mining*; Springer: New York, NY, USA, 2013; pp. 27–33.
36. Van der Maaten, L.; Hinton, G. Visualizing data using t-SNE. *J. Mach. Learn. Res.* **2008**, *9*, 2579–2605.
37. McInnes, L.; Healy, J.; Melville, J. Umap: Uniform manifold approximation and projection for dimension reduction. *arXiv* **2018**, arXiv:1802.03426.

38. Liu, Y.; Yuan, H.; Cai, L.; Ji, S. Deep learning of high-order interactions for protein interface prediction. In Proceedings of the 26th ACM SIGKDD International Conference on Knowledge Discovery & Data Mining, Virtual, 6–10 July 2020; pp. 679–687.
39. Johnson, S.C. Hierarchical clustering schemes. *Psychometrika* **1967**, *32*, 241–254. [CrossRef]

biomolecules

MDPI

Article

PInteract: Detecting Aromatic-Involving Motifs in Proteins and Protein-Nucleic Acid Complexes

Dong Li [1,2], **Fabrizio Pucci** [1,2] **and Marianne Rooman** [1,2,*]

1 Computational Biology and Bioinformatics, Université Libre de Bruxelles, 1050 Brussels, Belgium; dong.li@ulb.be (D.L.); fabrizio.pucci@ulb.be (F.P.)
2 Interuniversity Institute of Bioinformatics in Brussels, Université Libre de Bruxelles, 1050 Brussels, Belgium
* Correspondence: marianne.rooman@ulb.be

Abstract: With the recent development of accurate protein structure prediction tools, virtually all protein sequences now have an experimental or a modeled structure. It has therefore become essential to develop fast algorithms capable of detecting non-covalent interactions not only within proteins but also in protein-protein, protein-DNA, protein-RNA, and protein-ligand complexes. Interactions involving aromatic compounds, particularly their π molecular orbitals, hold unique significance among molecular interactions due to the electron delocalization, which is known to play a key role in processes such as protein aggregation. In this paper, we present PInteract, an algorithm that detects π-involving interactions in input structures based on geometric criteria, including π-π, cation-π, amino-π, His-π, and sulfur-π interactions. In addition, it is capable of detecting chains and clusters of π interactions as well as particular recurrent motifs at protein-DNA and protein-RNA interfaces, called stair motifs, consisting of a particular combination of π-π stacking, cation/amino/His-π and H-bond interactions.

Keywords: stacking geometry; T-shaped geometry; protein stability; protein affinity; solubility; aggregation.

1. Introduction

Among the various types of noncovalent interactions that stabilize the 3-dimensional (3D) structures of proteins and protein-involving complexes, interactions where one of the partners carries an aromatic ring have a status apart [1,2]. Indeed, their London dispersion contribution to the free energy is particularly important, due to the strength of π-electron delocalization. In particular, interactions between two aromatic moieties, called π-π interactions [3], have large dispersion energies especially if the aromatic rings are parallel. When the rings are in T-shaped conformations, the electrostatic contributions are increased at the expense of dispersion contributions. Both T-shaped and stacked conformations are energetically favorable; their relative preference depends on the particular type of extracyclic atoms and on the environment. Here, we considered π-π interactions in which one of the partners is an aromatic amino acid (Phe, Tyr, Trp) and the other is either another aromatic amino acid or a nucleic acid base (Gua, Ade, Cyt, Thy, Ura).

In addition to π-π interactions, we considered cation-π, amino-π, and His-π interactions, where the π-partner is the ring of an aromatic amino acid or nucleic acid base. In cation-π interactions [4–6], the other partner is a positively charged amino acid Arg or Lys. In Lys-π interactions, the electrostatic contribution is predominant, whereas in Arg-π interactions the dispersion energy is also important due to the charge delocalization on the planar guanidinium group of the Arg side chain. In amino-π interactions [5,7], sometimes also called amide-π, the second partner is the neutral side chain of Asn or Gln, which is the planar and polarizable formamide group; the electrostatic contribution is here small and the dispersion energy is the main stabilizing force. The His-π interaction [8] is, in essence, contained in other subtypes according to the protonation state of the His: when neutral, His-π can be considered as a π-π interaction, and when protonated, as a cation-π interaction. This makes His-π a special and versatile pH-dependent interaction that plays a unique role in function and molecular recognition [9,10]. We chose to treat it as a separate form of π interaction.

The last π-involving type of interactions we considered is sulfur-π interaction [11,12], also called thiol-π, which links an aromatic moiety of an amino acid or nucleic acid base with a sulfur atom of Met or Cys side chains. Such interactions have been shown to be stabilizing, though less than other π interactions [13], and to be sometimes of functional importance [14].

Different π interactions are often found to combine into larger motifs, such as chains or clusters of π-π interactions [15] and of cation/amino/His-π interactions [16]. For example, a positively charged ion can be sandwiched between two aromatic rings, or an aromatic ring can be sandwiched between two positively charged groups. Moreover, recurrent motifs called stair motifs, which mix π-π interactions and H-bonds with cation-π, amino-π, or His-π interactions, are also commonly found at protein-DNA interfaces [16,17].

There has been great success in studying the importance of aromatic-involving interactions in biological processes [18–21]. For example, cation-π and π-π interactions have been shown to dictate the phase transition in liquid-liquid phase separation [22]. Furthermore, by mutating some residues in histones into Trp, thus increasing the number of cation-π interactions, researchers managed to improve the affinity between histones and their reader proteins, which further enhances downstream genetic regulations [23]. Moreover, a mutation from Phe to Thr in BanLec, a banana protein which is regarded as a potential antiviral agent and works by binding to glycan molecules via π-π stacking, was demonstrated to reduce the protein's antiviral competence as the mutation disrupted the capacity of glycan binding of BanLec. Sulfur-π interactions, despite a relatively lower appearance frequency, have been pointed out for contribution to protein stability [24]. Finally, His-π interactions, being interchangeable among different electrostatic states, has been shown to impact protein activity according to the environmental salt concentration.

Several databases of aromatic interactions in proteins have been publicly released, including CAD [25] and A^{2}ID [26], each encompassing different types of aromatic interactions. Despite these advancements, there is still a lack of publicly released computational tools that allow the identification of different types of aromatic interactions. To fill this gap, we developed PInteract, a program written in C, which is able to quickly and accurately localize aromatic-involving interactions in input structures using criteria that combine both distances and angles, which is more chemically reasonable than distance-only tools [27]. With a wide range of supported types of input structures, including protein monomers, multimers, and protein-nucleic acid complexes, we expect PInteract to be an efficient tool in the calculation and analysis of aromatic interactions in biological molecules.

2. Materials and Methods

2.1. Types of Interactions

In the interactions we considered, the first partner is an aromatic moiety of an amino acid (Phe, Tyr, Trp, His) or a nucleic acid (Gua, Ade, Cyt, Thy, Ura). The second partner is an amino acid that is either aromatic (Phe, Tyr, Trp), contains a sulfur atom (Met, Cys), carries a net positive charge (Lys, Arg), possesses a side chain amide group with both a partially positively charged NH_2 group and a partially negatively charged C=O group

(Asn, Gln), or is His, which is both aromatic and sometimes positively charged. According to the type of partner, the aromatic-involving interactions are referred to as π-π, sulfur-π, cation-π, amino-π, and His-π, respectively.

2.2. Design Principles

PInteract detects π interactions using geometric criteria which are in agreement with previously derived energetic principles established through quantum chemical studies; see e.g., [1,2,5,8,28,29]. The interaction energy, defined as $\Delta E = E_{A-B} - (E_A + E_B)$, serves as a unified descriptor for diverse π interactions, where E_{A-B} is the total energy of the interacting partners A and B, and E_A and E_B are the energies of the isolated components. A negative ΔE indicates a stabilizing interaction. Key factors that determine the geometry and strength of π interactions include:

- The distance between the two interacting partners. Sometimes, this distance is defined as the distance between the centroids of the two interacting functional groups. In other studies, like here, it is defined as the closest distance between any two atoms of the two functional groups. We chose this distance definition because the considered aromatic moieties carry various substituents which influence how the ring interacts with other partners and breaks the symmetry around the ring's center. Similarly, the electron delocalization in other functional groups is better captured by selecting the closest atom rather than the functional group's centroid.
- The α angle; it is typically defined as the angle between the vector linking the centroids of both partners and the vector normal to the aromatic ring of the aromatic partner or of one of the aromatic partners in the case of π-π interactions. The complement of α, i.e., $90^\circ - \alpha$, is called elevation angle and is sometimes also used.
- The β angle; it only applies to interactions in which the two functional groups are planar, i.e., contain an aromatic moiety (Phe, Tyr, Trp, His), a guanidinium group (Arg) or an formamide group (Asn, Gln). It is defined as the angle between the vectors normal to the planes of the two interacting functional groups and measures the degree of parallellism between the planes.

The energy values of the interacting partners vary significantly according to their interaction geometries. Optimal ΔE values are observed for inter-partner distances of 2.5–4.0 Å for cation-π and amino-π interactions [5,18,25,30,31], about 6.0 Å for sulfur-π [11,21], and up to about 5.0 Å for π-π interactions [2]. In addition to the distance, the angular parameter α constraints the interaction geometry and modulates the interaction energies. It allows the specification that the functional group of one partner is positioned above (or below) the aromatic moiety of the other, thereby enabling interaction with its π orbital. The β angle further influences the interaction energy by defining whether the interacting molecular planes adopt a stacked or T-shaped geometry.

For PInteract to be able to detect stair motifs at protein-DNA and protein-RNA interfaces, which combine a π-π interaction, an H-bond and a cation-π, amino-π, or His-π interaction between an amino acid and two nucleobases, H-bonds must be identified prior to running PInteract. For this purpose, we used the HBPLUS algorithm [32], which the user must install and execute before using PInteract. Note that all other interaction motifs considered do not require H-bond identification.

2.3. Technical Implementation

2.3.1. Functional Groups

The π interactions between functional groups are defined using geometric criteria. The atoms considered to be part of the functional groups are the following:

- Aromatic moiety: all atoms that make up the aromatic ring or rings, as well as their centers referred to as mC5 or mC6 depending on whether the ring is 5- or 6-membered;
- Positive charge: NH1, NH2, CZ and NE for Arg; and NZ, CE, H1 for Lys; ND1, CE1, NE2 for His if its caonsidered as a cation;
- Partial positive charge: NE2, CD and OE1 for Gln; ND2, CG and OD1 for Asn;
- Sulfur: SG for Cys; SD for Met.

By convention, partner 1 is taken as the aromatic moiety, or one of the aromatic moieties in the case of π-π interactions.

2.3.2. Distance Criterion

The first criterion is a distance threshold, d_{max}, which specifies the maximum allowed separation between atoms involved in the interaction. Specifically, we compute the smallest distance d between any atom of the functional group of partner 1 and any atom of the functional group of partner 2, specified in Section 2.3.1. This is illustrated in Figure 1a. For the interaction to be considered valid, this distance must satisfy $d \leq d_{\text{max}}$ where:

- $d_{\text{max}} = 5.0$ Å for π-π interactions;
- $d_{\text{max}} = 6.0$ Å for sulfur-π interactions;
- $d_{\text{max}} = 4.5$ Å for cation-π, amino-π, and His-π interactions.

These distance thresholds are provided by default, but they can easily be modified by the user. This is particularly important when using low-resolution or modeled protein 3D structures, in which the positioning of the side chain atoms is not sufficiently accurate.

If an aromatic partner consists of a fused double-ring system, one ring with 5 atoms and the other with 6, as is the case for Trp, Ade and Gua, we determine whether the closest distance involves an atom that is closer to the centroid of the 6-atom ring or of the 5-atom ring. In the former case, we focus on the 6-atom ring for checking whether the other geometrical criteria are satisfied and in the latter case, on the 5-atom ring.

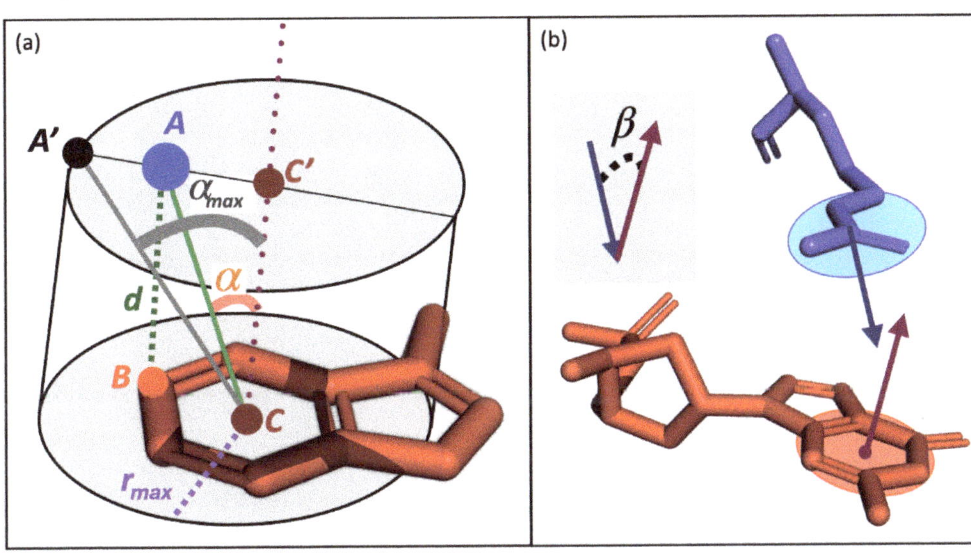

Figure 1. Geometric definition of π-involving interactions. (**a**) The aromatic moiety of partner 1 (here Trp) is in red; the centroid of its 6-atom ring is labeled C. The closest atoms between the two partners are atom B of the 6-atom aromatic ring and atom A of the functional group of partner 2, represented

by a blue ball; their distance is equal to d. A must be inside a cylinder, whose lower base is a circle of radius r_{max} centered on C and its upper base passes through A and is centered on C'. The angle between vector $\overrightarrow{CA}$ and vector $\overrightarrow{CC'}$ which is normal to the cylinder's bases, is called α. For A being inside the cylinder, α must be $\leq \alpha_{max}$, where α_{max} is the angle between the normal vector $\overrightarrow{CC'}$ and vector $\overrightarrow{CA'}$, where A' is on the upper base's edge and on the diameter passing through A and C'. (**b**) Partner 1 (Trp) is in red and partner 2 (Arg), in blue. The planes formed by the 6-atom aromatic ring of Trp and by Arg's guanidinium group are shown in red and blue, respectively, as well as the vectors normal to these planes. The angle between these vectors, called β, taken between 0° and 90°, is shown on a gray background and measures the deviation from the parallelism between the two planes.

2.3.3. Angle Criterion

To ensure that the interacting partner is positioned within the effective electronic range of the π orbital, and thus that the interaction is π-mediated, we apply an angular constraint based on a cylindrical model, illustrated in Figure 1a. The cylinder's base is centered on the geometric center C of the aromatic ring of partner 1 (or of the ring closest to partner 2 in the case of a double aromatic moiety), and its radius is equal to r_{max} taken to be :

- $r_{max} = 2\,r$ for cation-π, His-π, amino-π and sulfur-π interactions;
- $r_{max} = 3\,r$ for π-π interactions.

where r is the radius of the aromatic ring. The height of the cylinder is such that the closest atom of the functional group of partner 2 (noted A) is included in the cylinder's second ("upper") base. If this is impossible, thus if A remains outside the cylinder regardless of its height, the interaction is not considered as a π interaction.

The r_{max} threshold can be translated into an angle criterion. Indeed, for A to be in the cylinder, the angle α formed by vector $\overrightarrow{CA}$, linking the aromatic moiety's center to A, and the vector normal to the aromatic ring (taken in the direction of A so that $\alpha \leq 90°$) must have a value that is lower than a maximum value, called α_{max}. This value is reached when A is on the cylinder's edge, as shown in Figure 1a. Note that the α_{max} threshold value depends on the distance d between the the 2 partners, unlike the r_{max} threshold.

In the case of a π-π interaction, the cylinder is constructed in turn on the aromatic ring of partners 1 and 2, and the configuration yielding the smallest α-angle value is retained. This procedure ensures that the two π groups are treated symmetrically.

The r_{max} thresholds are provided by default in the PInteract program but, like for the distance thresholds, they can easily be changed by the user. Since the asymmetrically distributed electron cloud can extend beyond the aromatic ring's atomic boundaries, using a larger radius ensures inclusion of geometries where the functional group of partner 2 is laterally offset from the ring's centroid of partner 1 but still engages in energetically significant π interactions.

2.3.4. Plane Parallelism Angle

In the case the functional groups of the two partners are planar, as in Arg-π, amino-π, His-π, and π-π interactions (see Section 2.2), we compute the smallest angle β between the vectors that are normal to the planes of the interacting functional groups, as illustrated in Figure 1b for an Arg-Trp interaction. The β value is equal to 0° when the two planes are parallel, or equivalently when the two partners are in stacked conformation, and equal to 90° when the planes are perpendicular or in T-shaped conformation.

2.4. Structure Datasets' Construction

To apply PInteract at proteome scale, we set up several non-redundant and good-resolution structure datasets, which contain:

- $\mathcal{D}_{\text{monomer}}$: 5590 protein monomers;
- $\mathcal{D}_{\text{homodimer}}$: 3940 protein homodimers;
- $\mathcal{D}_{\text{heterodimer}}$: 1037 protein heterodimers;
- $\mathcal{D}_{\text{TCRpMHC}}$: 51 complexes between a T-cell receptor (TCR) and a peptide-bound major histocompatibility complex (pMHC) ;
- $\mathcal{D}_{\text{AbAg}}$: 495 antibody-antigen complexes ;
- $\mathcal{D}_{\text{protNA}}$: 573 protein-DNA or protein-RNA complexes.

The $\mathcal{D}_{\text{monomer}}$ and $\mathcal{D}_{\text{heterodimer}}$ datasets were collected and curated from X-ray structures of protein monomers and heterodimers, respectively, in the Protein DataBank (PDB) [33] using the PISCES server [34]; the $\mathcal{D}_{\text{homodimer}}$ dataset from the HomodimerDB database [35]; the $\mathcal{D}_{\text{TCRpMHC}}$ set from the TCR3d database [36,37]; the $\mathcal{D}_{\text{AbAg}}$ set from the SAbDab database [38–40]; and the $\mathcal{D}_{\text{protNA}}$ set from the DNAproDB and RNAproDB databases [41,42].

To ensure high-quality datasets, we applied the following filters: no missing main-chain atomic coordinates; resolution $\leq$ 2.5 Å; *R*-factor $\leq$ 0.25; and sequence length between 40 and 10,000 residues. Redundancy within each dataset was reduced by applying pairwise sequence identity thresholds. For $\mathcal{D}_{\text{monomer}}$, $\mathcal{D}_{\text{homodimer}}$, $\mathcal{D}_{\text{heterodimer}}$, and $\mathcal{D}_{\text{protNA}}$ (considering only protein chains), the threshold was set to $\leq$30%. For $\mathcal{D}_{\text{AbAg}}$ and $\mathcal{D}_{\text{TCRpMHC}}$, we used sequence identity thresholds of $\leq$70% for antibodies and TCRs (considering complementarity-determining regions only), and $\leq$90% for antigens and MHCs. Non-redundancy was defined using an "or" rule: two antibody-antigen complexes were considered non-redundant if the antibody sequences shared $\leq$70% identity or the antigen sequences shared $\leq$90% identity; the same rule was applied to TCR-pMHC complexes.

To compute the frequency of occurrence of π interactions in these datasets, we divided the number of occurrences of a given type by the total number of interactions. More specifically, for $\mathcal{D}_{\text{monomer}}$, the total number of interactions was estimated as the number of residue pairs in contact, defined as having at least one heavy atom from each residue's side chain within a maximum distance of 5Å. For $\mathcal{D}_{\text{heterodimer}}$ and $\mathcal{D}_{\text{homodimer}}$, the frequencies were obtained in the same way, except that only contacts across the protein-protein interface were considered, and for $\mathcal{D}_{\text{protNA}}$, only contacts across the protein-DNA/RNA interface. In $\mathcal{D}_{\text{AbAg}}$, where the complexes typically consist of three chains, the considered interactions link the antigen and antibody chains. In $\mathcal{D}_{\text{TCRpMHC}}$, where the complexes consist of four or five chains, the interactions taken into account are between the TCR chains and the pMHC molecule.

3. Results

3.1. The PInteract Algorithm

We developed the PInteract algorithm to identify π-involving interactions in proteins, protein-protein complexes, and interfaces between proteins and DNA, RNA, or nucleic acid base-containing ligands. These interactions include π-π, sulfur-π, cation-π, amino-π, and His-π interactions, which are determined based on the geometric criteria detailed in Section 2.3.

The input to PInteract is the name of a directory that contains structure files in PDB format [33]. The output consists of three files: PInteract.csv, a table of individual π interactions; PInteract1.csv, a table of π-chains and stair motifs; and PInteract.txt, a human-readable text file containing all the content from PInteract.csv and PInteract1.csv, with comments and explanations prefixed by # to aid understanding. PInteract is implemented in C and is extremely fast, enabling the detection of π interactions at the proteome scale. The code can be freely downloaded from our GitHub repository: https://github.com/3BioCompBio/PInteract (accessed on 19 August 2025). All technical details are provided in this repository.

3.2. Examples of π-Involving Interactions

Here, we provide examples of PInteract applications to identify specific geometric arrangements of π-involving interactions. The first example of PInteract's application is illustrated in Figure 2a. A cation-π and a π-π interaction were found at the interface between hen egg lysozyme and the heavy chain of an antibody (PDB ID: 2QDJ). These two interactions are generally particularly important in antibody-antigen recognition [43]. In this specific case, for example, their disruption has been experimentally shown to reduce the antibody-antigen binding affinity by at least two orders of magnitude [44].

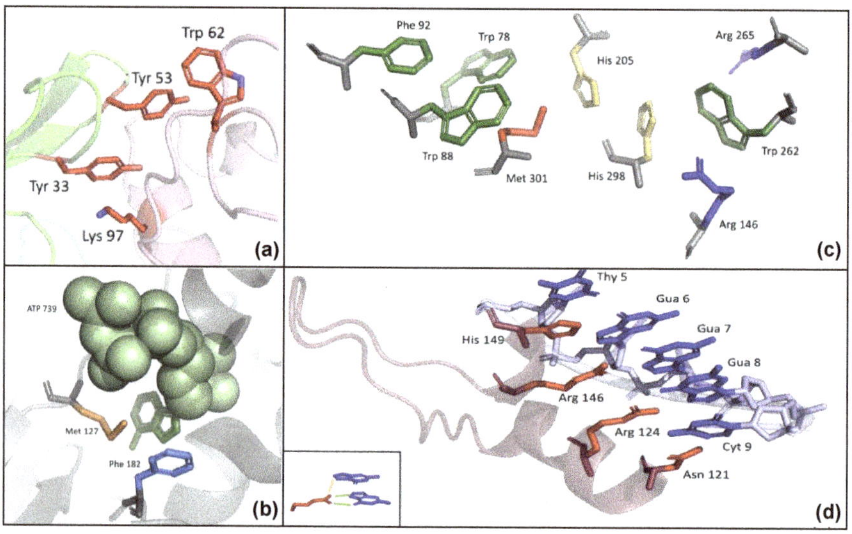

Figure 2. Illustration of π-involving interactions. (**a**) Example of π interactions at a protein-protein interface. Cation-π interaction between Lys Y97 and Tyr H33 and π-π interaction between Tyr H53 and Trp Y62 at the interface between a hen egg lysozyme and its cognate antibody (PDB ID: 2QDJ). Chain Y is the antigen depicted in pink; chain H is the antibody's heavy chain and is shown in green; the light chain is shown in cyan. (**b**) Example of protein-ligand π interactions. ATP molecule C739 (in green spheres with its Ade nucleobase in green sticks), Phe B182 (in blue sticks) and Met B127 (in orange sticks) in an NAD kinase from *Archaeoglobus fulgidus* (PDB ID: 1Z0S). (**c**) Example of π-chains. Residues Trp 78, Trp 88, Phe 92, Arg 146, His 205, Trp 262, Arg 265, His 298, Met 301 in chain A of a proteobacterium's carboxylesterase (PDB ID: 1QLW). Aromatic side chains are in green, sulfur-containing side chains in red, histidines in yellow, and positively charged side chains in blue. Backbone atoms are in gray. (**d**) Example of stair motifs at the protein-DNA interface. Nucleic acid bases Thy B5, Gua B6, Gua B7, Gua B8, Cyt B9 and residues His A149, Arg A146, Arg A124 and Asn A 121 in a zinc finger-DNA complex in *Mus musculus* (PDB ID: 1A1G). Nucleic acid bases are depicted as blue sticks and amino acid side chains, as red sticks. In the lower left corner, the H-bonds are depicted as green lines and the cation-π interaction as an orange line.

PInteract is also able to detect π interactions with ligands that contain a nucleic acid base, such as ATP or GTP. An example of such a protein-ligand π interaction is shown in Figure 2b. A Met and a Phe residue of a NAD kinase (PDB ID: 1Z0S) make a sulfur-π and a π-π interaction, respectively, with the Ade base included in the ATP molecule.

Successive π-involving interactions identified by our algorithm are referred to as π-chains in the output files. These chains represent continuous networks of interacting π-systems, which may play a structural or functional role in stabilizing molecular assemblies. They can be successive interactions of the same type, e.g., clusters of π-π interactions or a double cation-π interaction where the cation is sandwiched between aromatic rings or where an aromatic moiety is sandwiched between two cations. They can also mix different types of π interactions, e.g., a π-π interaction followed by a cation-π interaction. An example is given in Figure 2c, which shows a π-chain formed of three π-π , two cation-π and three sulfur-π interactions in a bacterial esterase from an *Alcaligenes* species (PDB ID: 1QLW).

Another type of mixed interactions identified by PInteract is the stair motif [16,17], recurrently observed at protein-DNA and protein-RNA interfaces. These motifs consist of two consecutive stacked nucleic acid bases that form a π-π interaction, and an amino acid that forms both a cation-π, amino-π or His-π interaction with one of the nucleic acid bases and an H-bond with the next base along the base stack. As mentioned in Section 2.2, for PInteract to be able to detect such motifs, H-bonds must be identified beforehand using the HBPLUS program [32]. An example of four successive stair motifs is shown in Figure 2d, where five stacked nucleobases make H-bonds and His-π, cation-π and amino-π interactions with a His, two Arg and an Asn of a zinc finger protein (PDB ID: 1A1G).

3.3. Large-Scale Analysis of π-Involving Interactions

In order to show the applicability of our tool for large-scale analysis of π-involving interactions, we investigated the relative frequency of each type of π interaction across the proteins in the six $\mathcal{D}$ datasets described in Section 2.4: monomeric proteins, homodimers, heterodimers, antibody-antigen complexes, TCR-pMHC complexes, and protein-DNA and RNA complexes. These frequencies were computed as the number of interactions identified by PInteract divided by the total number of interactions in the protein (see Section 2.4 for details). Note that in doing so, we did not normalize for amino acid frequencies. Our objective was to compare the types of interactions that contribute most to protein and protein complex stabilization in different contexts, independently of sequence composition.

As shown in Table 1 and Figure 3, protein-DNA/RNA interfaces contain, on the average, by far the highest numbers of cation-π and π-π interactions, although with very large standard deviations, and relatively few other types of π-involving interactions. This is largely due to the frequent occurrence of positively charged amino acids in the vicinity of the negatively charged DNA or RNA sugar-phosphate backbone and their recurrent tendency to form stair motifs [16,17]; aromatic amino acids also establish stabilizing interactions with nucleobases or sugars of the sugar-phosphate backbone of DNA or RNA molecules [45]. The large standard deviation arises from the fact that in some complexes, the interactions involve the sugar–phosphate backbone rather than the nucleobases.

Antibody-antigen interfaces rank second for cation-π and π-π interactions and also contain numerous amino-π interactions; the role of these interactions in antibody–antigen binding is well established [43,46]. TCR–pMHC binding interfaces display numerous π-π interactions—fewer than antibody-antigen interfaces, but more than observed in monomers, heterodimers, and homodimers. The amount of cation-π interactions is the same in TCR–pMHC and heterodimer interfaces and higher than in monomers. The proportion of amino-π and His-π is identical in TCR-pMHC and antibody-antigen interfaces and larger than in heterodimer interfaces and in monomers. In contrast, sulfur-π interactions are rare in general, and even rarer in TCR-pMHC interfaces. Notably, the homodimer interfaces contain the lowest number of π-involving interactions in general. These different tendencies are statistically significant, as shown in Supplementary Table S1. However, it should be emphasized that these represent average trends. The standard deviations are very large (Table 1), indicating substantial variability between proteins.

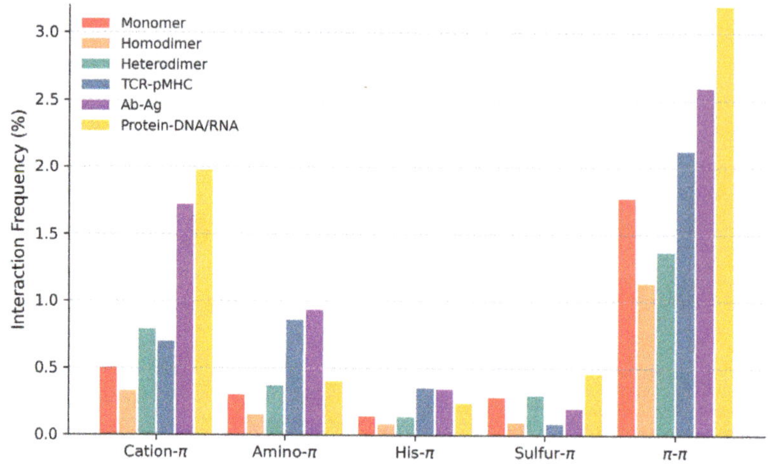

Figure 3. Comparison between the average relative frequency of each type of π interactions in the datasets $\mathcal{D}_{\text{monomer}}$, $\mathcal{D}_{\text{homodimer}}$, $\mathcal{D}_{\text{heterodimer}}$, $\mathcal{D}_{\text{TCRpMHC}}$, $\mathcal{D}_{\text{AbAg}}$ and $\mathcal{D}_{\text{protNA}}$. The relative frequencies were computed as the number of interactions identified by PInteract divided by the total number of interactions in the protein, as defined in Section 2.4. For homo- and heterodimers, only the interactions occurring at the dimer interface were counted; for antibody-antigen complexes, only interactions that link one of the two antibody chains with the antigen; for TCR-pMHC, only interactions between one of the two TCR chains and the pMHC molecule; and for protein-DNA/RNA, only interactions linking the protein with DNA or RNA.

Table 1. Average relative frequencies of π interactions in the set of monomers $\mathcal{D}_{\text{monomer}}$ and in the sets of protein-protein complexes $\mathcal{D}_{\text{homodimer}}$, $\mathcal{D}_{\text{heterodimer}}$, $\mathcal{D}_{\text{TCRpMHC}}$, $\mathcal{D}_{\text{AbAg}}$, and $\mathcal{D}_{\text{protNA}}$, computed as the number of interactions identified by PInteract divided by the total number of interactions in the protein (see Section 2.4). The standard deviation is given in parentheses. The associated p-values are given in Supplementary Table S1.

Dataset	Cation-π	Amino-π	His-π	Sulfur-π	π-π
$\mathcal{D}_{\text{monomer}}$	0.50 (0.46)	0.30 (0.33)	0.14 (0.22)	0.28 (0.37)	1.77 (1.13)
$\mathcal{D}_{\text{homodimer}}$	0.33 (1.14)	0.17 (0.81)	0.09 (0.51)	0.10 (0.56)	1.13 (2.00)
$\mathcal{D}_{\text{heterodimer}}$	0.79 (1.80)	0.37 (1.36)	0.14 (0.73)	0.30 (0.89)	1.37 (2.38)
$\mathcal{D}_{\text{TCRpMHC}}$	0.70 (1.22)	0.86 (1.73)	0.35 (0.84)	0.09 (0.44)	2.12 (2.97)
$\mathcal{D}_{\text{AbAg}}$	1.72 (3.21)	0.93 (1.93)	0.34 (1.15)	0.20 (0.78)	2.59 (3.89)
$\mathcal{D}_{\text{protNA}}$	2.02 (12.67)	0.41 (1.08)	0.23 (0.89)	0.47 (4.29)	3.73 (10.41)

To investigate the characteristic distances between the functional groups involved in π interactions and evaluate the relevance of our predefined distance thresholds d_{max}, we analyzed the distribution of distances d (defined in Section 2.3.2) for cation-π, amino-π, His-π, sulfur-π, and π-π interactions detected in dataset $\mathcal{D}_{\text{monomer}}$. The results are shown in Figure 4. Clearly, as a general trend, the interaction distances consistently exceed 3 Å. The d-distributions exhibit characteristic peaks at different values according to the interaction types: 3.6 Å for cation-π, amino-π, and His-π interactions, and 3.8 Å for sulfur-π and π-π interactions. Interestingly, the peak is less marked for sulfur-π interactions, showing that the geometry of this type of interaction is less constrained. Our analysis also confirms that the chosen thresholds for each interaction type are reasonable, as these are located near the tail of the distribution, corresponding to low-density regions. Note again that the different distance thresholds can be modified by the users.

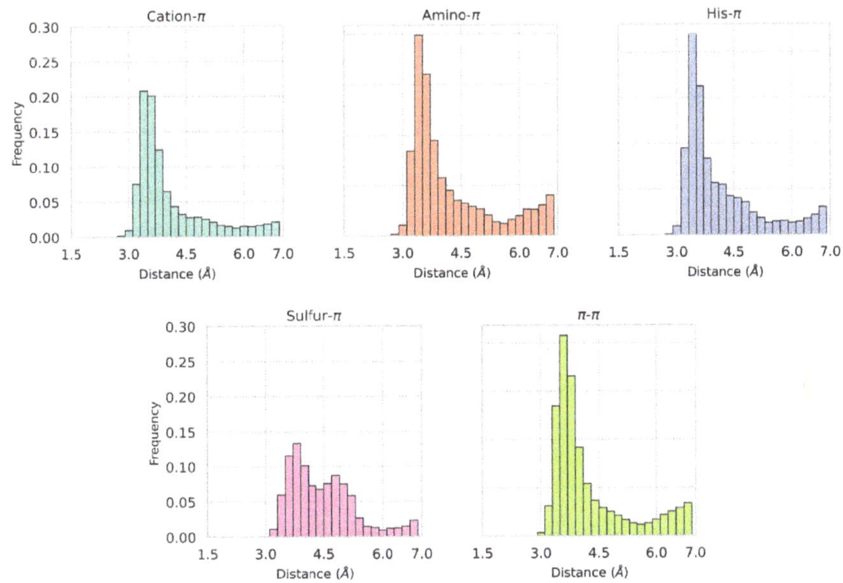

Figure 4. Distribution of the distances d between functional groups of π-involving interactions, defined in Section 2.3.2, for all considered types of π interactions detected in the $\mathcal{D}_{\text{monomer}}$ dataset.

We also analyzed the distribution of α angle values, defined in Section 2.3.3, which represent the position of the functional group of partner 2 above (or below) the aromatic ring of partner 1. A value of $\alpha = 0$ indicates that this functional group is just above (or below) the aromatic ring's center. We plotted the frequency of $(1 - \cos\alpha)$ values, which represent the lateral displacement relative to the ring's center, for the different interaction types observed in $\mathcal{D}_{\text{monomer}}$. As shown in Figure 5, all $(1 - \cos\alpha)$ values have an almost equal frequency up to α angles of about 30° $(1 - \cos 30^\circ \approx 0.13)$ for all interaction types except π-π. This indicates that the formation of π interactions requires the interacting partner to be clearly above (or below) the aromatic plane, where the π-orbital is situated; however, no clear preference for a particular displacement from the ring's center is observed, which reflects a certain plasticity of the interaction geometry. For α angles larger than about 45° $(1 - \cos 45^\circ \approx 0.23)$, the frequency of all non-π-π interactions totally vanishes. The fact that the decrease is gradual is related to the fact that the angle threshold value α_{max} depends on the distance d between interacting partners and is thus not clear-cut.

In contrast, π-π interactions exhibit a clear preference for small displacement values, with α values between 0° and about 18° $(0 \leq 1 - \cos 18^\circ \leq 0.05)$. This means that the closest atom of the aromatic ring of partner 2 is preferably situated above the centroid of the aromatic ring of partner 1. It has however to be noted that these displacements are not always computed between the centroids of the two interacting aromatic rings, but often between an atom of one aromatic ring and the center of the other ring. This can explain the overrepresentation of small α angles. Also, larger displacements are observed for this type of π interactions, up to $(1 - \cos 53^\circ \approx 0.40)$; this is due to the larger α_{max} value accepted for this type of interaction (Section 2.3.3).

The same tendencies are observed in the $(1 - \cos\alpha)$ distributions of antibody-antigen complexes, TCR-pMHC, homodimers, and heterodimers. In contrast, at protein-DNA and protein-RNA interfaces, larger angles are also observed, with displacement values up to $(1 - \cos 60^\circ = 0.5)$ (see Supplementary Figure S1b).

Finally, we computed the distribution of β angles, or rather of the $(1-\cos\beta)$ displacement values, where β is the angle between the planes of the functional groups of the two interacting partners, when these groups are planar (Section 2.3.4). As shown in Figure 6, this distribution varies markedly depending on the type of π interaction. For Arg-involving cation-π interactions, smaller $(1-\cos\beta)$ values—corresponding to parallel, stacked conformations—are strongly preferred. A similar but slightly less marked trend is observed for amino-π and His-π interactions. To an even much lesser extent, this trend is also observed for π-π interactions. However, these interactions tend to accommodate a wide range of conformations, from stacked to T-shaped, with almost no preference, suggesting a high plasticity in interaction geometry. Note that these trends are similar to those observed in the $(1-\cos\beta)$ distributions of all considered sets of protein-protein interfaces. At protein–DNA and protein–RNA interfaces, however, π–π interactions display a stronger preference for stacked conformations, which can be explained by their ability to partially intercalate into the nucleobase stack (see Supplementary Figure S2b).

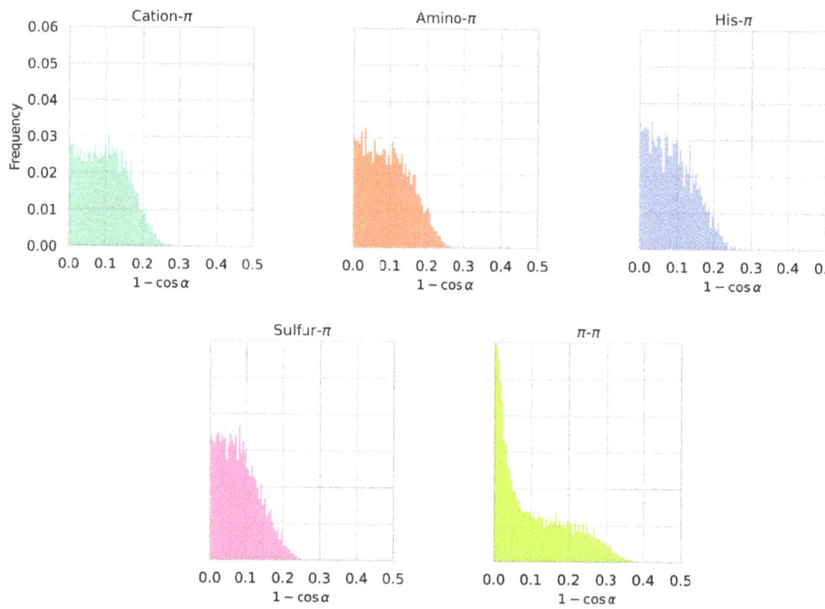

Figure 5. Distribution of the values of the lateral displacement with respect to the aromatic rings' center, $(1-\cos\alpha)$, in which the α angle measures the location of the functional group of interacting partner 2 above (or below) the plane of the aromatic ring of partner 1 (see Section 2.3.3), for different types of π interactions detected in the $\mathcal{D}_{\text{monomer}}$ dataset. The smaller the value of $(1-\cos\alpha)$, the more directly the functional group of partner 2 is positioned above (or below) the center of the aromatic ring. See Supplementary Figure S1 for the corresponding α angle distributions.

To further evaluate whether these observed preferences of α and β angles are preserved in computationally predicted structures, we applied PInteract to selected subsets of the AlphaFold Protein Structure Database (AlphaFoldDB) [47], as described in Supplementary Section S3. As shown in Supplementary Figures S3 and S4, the resulting distributions of α and β angles have a high degree of consistency with those observed in experimentally solved crystal structures, confirming that models predicted by AlphaFold [48] largely retain the geometric characteristics of natural π interactions. A minor exception is the attenuated peak of the β angle distribution for small β values, suggesting a mild underestimation by AlphaFold of the preference for stacked conformations.

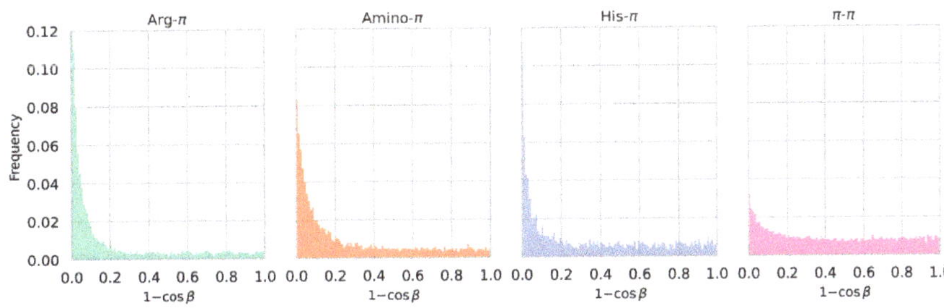

Figure 6. Distribution of $1 - \cos\beta$ values, in which the β angle measures the degree of parallelism between the interacting planar functional groups (see Section 2.3.4), for different types of π interactions detected in the $\mathcal{D}_{\text{monomer}}$ dataset. Parallel or stacked conformations correspond to $\beta = 0°$ and $1 - \cos\beta = 0$; perpendicular or T-shaped conformations, to $\beta = 90°$ and $1 - \cos\beta = 1$. See Supplementary Figure S2 for the corresponding β angle distributions.

3.4. Relationship Between π Interactions and Protein Solubility and Aggregation

Solubility and aggregation are important protein properties that have huge impacts on protein fitness. Molecular mechanisms underlying protein solubility and aggregation have attracted a lot of attention but are still challenging to explain. They are complex properties that arise from the interplay of various forces in both the folded and unfolded or aggregated states. One of the results in this context is that interactions known to decrease solubility and favor aggregation involve delocalized π-electrons [49–52].

Here, we investigated the role of intrachain π-involving interactions in protein solubility. For this purpose, we used the dataset of proteins with known structure and solubility values which was set up in [53]. This set contains curated entries from high-throughput experiments on the *E. coli* proteome [54] and *S. cerevisae* [55] using a cell-free expression system. Each entry in this dataset is a single-chain protein. We refer to this dataset as $\mathcal{D}_{\text{solubility}}$.

We first computed the Pearson correlation between solubility values and the frequency of residues capable of forming π interactions; the results are shown in Table 2. The correlation is statistically significant and most strongly negative for aromatic residues ($\rho = -0.17$), followed by Arg ($\rho = -0.09$). His, Met/Cys, and Asn/Gln do not exhibit any significant solubility preference. In contrast, Lys showed a positive correlation with solubility ($\rho = 0.16$), consistent with the well-known Arg/Lys trend: soluble proteins tend to be depleted in arginine and enriched in lysine [52].

Table 2. Pearson correlation coefficients ρ between protein solubility and (1) the residue frequency F; (2) the frequency F_π of residues involved in a π interaction; (3) the frequency $F_{non-\pi}$ of residues not involved in a π interaction, computed from the $\mathcal{D}_{\text{solubility}}$ dataset. A cross (×) indicates that the result is not statistically significant (p-value > 0.01).

	Arg	Lys	Phe/Tyr/Trp	Asn/Gln	His	Met/Cys
F	−0.09	0.14	−0.17	−0.09 ×	−0.04 ×	0.00 ×
F_π	−0.14	−0.04 ×	−0.23	−0.05 ×	−0.04 ×	−0.15
$F_{non-\pi}$	−0.07 ×	0.15	0.00 ×	−0.09 ×	−0.02 ×	0.04 ×

We further assessed whether the observed correlations differ when considering only residues engaged in π interactions, as identified by PInteract, compared to those that are not. As reported in Table 2, Arg, Met/Cys and aromatic residues involved in intrachain Arg-π, sulfur-π, and π-π interactions, respectively, tend to reduce solubility relative to the same residue types not participating in such interactions. Conversely, Lys exhibits the opposite trend, favoring solubility when it is not engaged in a π interaction. Our results thus indicate that interactions involving delocalized electronic π-systems appear to play a significant role in solubility mechanisms. This is further supported by the negative correlation of approximately -0.17 observed between protein solubility and the frequency of all detected π interactions.

As a further application of PInteract, we investigated π-involving interactions in the stabilization of fibril structures, which is still a matter of debate. Although aromatic residues are not strictly required for amyloid formation, they are known to facilitate and promote the fibril assembly process [56]. Specifically, we analyzed the set of fibril structures provided in [57]; we refer to this set as $\mathcal{D}_{\text{fibril}}$. We observed that the system is dominated by π-π interactions, which account for as much as 91% of all detected π interaction types. We examined the geometric characteristics of these interactions by comparing the distributions of the α and β angles in both fibrilar and monomeric states. Although no significant difference was observed in the α angle distributions, the β angle distributions differ markedly between the two sets, as shown in Figure 7. In particular, fibrilar structures exhibit an almost exclusive preference for stacked conformations, likely driven by the geometric symmetry and packing constraints of extended amyloid aggregates, as illustrated for the Aβ40 fibril [58] in Supplementary Figure S5. In contrast, structures of soluble monomeric proteins display a much broader distribution of β angles (see Figure 6), reflecting a greater conformational variability.

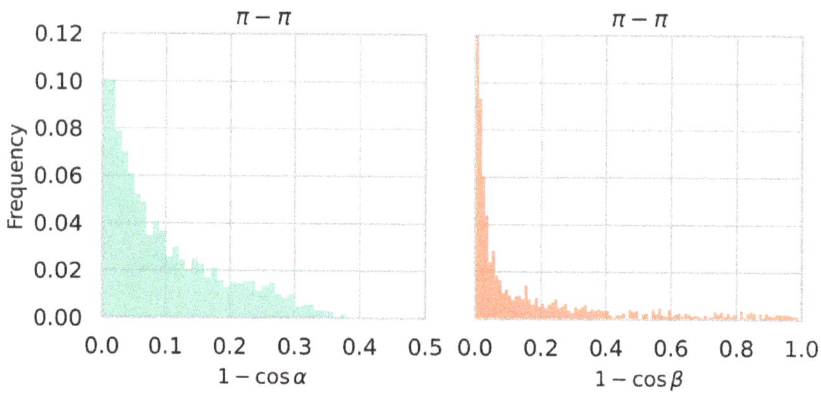

Figure 7. Distributions of $(1 - \cos\alpha)$ and $(1 - \cos\beta)$ values of π-π interactions in the $\mathcal{D}_{\text{fibril}}$ set.

3.5. PInteract: Fast, Scalable, and User-Friendly

PInteract is extremely easy to install and use—it only requires a standard GCC compiler and can be set up simply by cloning our GitHub repository. It is implemented in C, which makes it exceptionally fast and well-suited for large-scale analyses involving millions of structures. PInteract scales efficiently with the size of the dataset: a few thousands of structures can be processed in less than a minute on a standard laptop, and 100,000 structures require only 6 min (for details, see Supplementary Section S5). This level of efficiency makes PInteract particularly valuable for high-throughput structural screening tasks, as we have also highlighted in the previous sections.

4. Conclusions

With the recent breakthroughs in deep learning-based methods for biomolecular structure prediction, an unprecedented number of structures have been generated. We now have access to over a billion predicted protein structures, covering virtually the entire tree of life. This explosion of structural data calls for computational tools capable of efficiently analyzing these structures and extracting meaningful biological insights.

Here, we introduced PInteract, a fast computational tool designed to identify individual π interactions and clusters or chains of π interactions in proteins and in protein-protein, protein-RNA, protein-DNA, and protein-ligand complexes. Moreover, PInteract is the only automatic program that identifies stair motifs at protein-DNA and protein-RNA interfaces, which are recurrent motifs combining specific H-bonds and π interactions. We illustrated PInteract's capabilities across several case studies, demonstrating its effectiveness in detecting π interactions in proteins as well as in protein complexes. We also showed that PInteract can perform large-scale computations on hundreds of thousands of protein structures within minutes, enabling the analysis of the geometric properties of π-involving interactions across large structural databases. Finally, we demonstrated how PInteract can be used to gain meaningful insights into the role of π interactions in protein solubility and aggregation.

In summary, we are confident that PInteract, with its ease of use and high computational efficiency, will serve as a valuable resource for the scientific community, enabling the systematic study of π-involving interactions and shedding light on the wide spectrum of molecular mechanisms in which these interactions play a pivotal role, from protein folding and stability to biomolecular recognition and aggregation.

Supplementary Materials: The following supporting information can be downloaded at: https://www.mdpi.com/article/10.3390/biom15081204/s1.

Author Contributions: Conceptualization, F.P. and M.R.; methodology, M.R.; software, M.R.; validation, D.L., F.P. and M.R.; data curation, D.L.; writing—original draft preparation, D.L.; writing—review and editing, F.P. and M.R. All authors have read and agreed to the published version of the manuscript.

Funding: This research was funded by the FNRS-Belgian Fund for Scientific Research through a PDR project, and by a scholarship from the China Scholarship Council for D.L.

Data Availability Statement: Data available in a publicly accessible repository: https://github.com/3BioCompBio/PInteract, accessed on 19 August 2025.

Conflicts of Interest: The authors declare no conflicts of interest.

References

1. Salonen, L.M.; Ellermann, M.; Diederich, F. Aromatic rings in chemical and biological recognition: Energetics and structures. *Angew. Chem. Int. Ed.* **2011**, *50*, 4808–4842. [CrossRef]
2. Calinsky, R.; Levy, Y. Aromatic Residues in Proteins: Re-Evaluating the Geometry and Energetics of π–π, Cation-π, and CH-π Interactions. *J. Phys. Chem. B* **2024**, *128*, 8687–8700. [CrossRef] [PubMed]
3. Burley, S.; Petsko, G.A. Aromatic-aromatic interaction: A mechanism of protein structure stabilization. *Science* **1985**, *229*, 23–28. [CrossRef] [PubMed]
4. Dougherty, D.A. Cation-π interactions in chemistry and biology: A new view of benzene, Phe, Tyr, and Trp. *Science* **1996**, *271*, 163–168. [CrossRef] [PubMed]
5. Wintjens, R.; Liévin, J.; Rooman, M.; Buisine, E. Contribution of cation-π interactions to the stability of protein-DNA complexes. *J. Mol. Biol.* **2000**, *302*, 393–408. [CrossRef]
6. Dougherty, D.A. The Cation-π Interaction in Chemistry and Biology. *Chem. Rev.* **2025**, *125*, 2793–2808. [CrossRef]
7. Imai, Y.N.; Inoue, Y.; Nakanishi, I.; Kitaura, K. Amide–π interactions between formamide and benzene. *J. Comput. Chem.* **2009**, *30*, 2267–2276. [CrossRef]
8. Cauët, E.; Rooman, M.; Wintjens, R.; Liévin, J.; Biot, C. Histidine-aromatic interactions in proteins and protein- ligand complexes: Quantum chemical study of X-ray and model structures. *J. Chem. Theory Comput.* **2005**, *1*, 472–483. [CrossRef]
9. Shimba, N.; Serber, Z.; Ledwidge, R.; Miller, S.M.; Craik, C.S.; Dötsch, V. Quantitative identification of the protonation state of histidines in vitro and in vivo. *Biochemistry* **2003**, *42*, 9227–9234. [CrossRef]
10. Calinsky, R.; Levy, Y. Histidine in Proteins: PH-Dependent Interplay between π–π, Cation–π, and CH–π Interactions. *J. Chem. Theory Comput.* **2024**, *20*, 6930–6945. [CrossRef]

11. Reid, K.; Lindley, P.; Thornton, J. Sulphur-aromatic interactions in proteins. *FEBS Lett.* **1985**, *190*, 209–213. [CrossRef]
12. Motherwell, W.B.; Moreno, R.B.; Pavlakos, I.; Arendorf, J.R.; Arif, T.; Tizzard, G.J.; Coles, S.J.; Aliev, A.E. Noncovalent interactions of π systems with sulfur: The atomic chameleon of molecular recognition. *Angew. Chem.* **2018**, *130*, 1207–1212. [CrossRef]
13. Nunar, L.; Aliev, A.E. Quantification of the Strength of π-Noncovalent Interactions in Molecular Balances using Density Functional Methods. *Chemistry-Methods* **2023**, *3*, e202200044. [CrossRef]
14. Daeffler, K.N.M.; Lester, H.A.; Dougherty, D.A. Functionally important aromatic–aromatic and sulfur-π interactions in the D2 dopamine receptor. *J. Am. Chem. Soc.* **2012**, *134*, 14890–14896. [CrossRef] [PubMed]
15. Lanzarotti, E.; Biekofsky, R.R.; Estrin, D.A.; Marti, M.A.; Turjanski, A.G. Aromatic–aromatic interactions in proteins: Beyond the dimer. *J. Chem. Inf. Model.* **2011**, *51*, 1623–1633. [CrossRef] [PubMed]
16. Rooman, M.; Liévin, J.; Buisine, E.; Wintjens, R. Cation–π/H-bond stair motifs at protein–DNA interfaces. *J. Mol. Biol.* **2002**, *319*, 67–76. [CrossRef]
17. Biot, C.; Wintjens, R.; Rooman, M. Stair motifs at protein- DNA interfaces: Nonadditivity of H-bond, stacking, and cation-π interactions. *J. Am. Chem. Soc.* **2004**, *126*, 6220–6221. [CrossRef]
18. Kumar, K.; Woo, S.M.; Siu, T.; Cortopassi, W.A.; Duarte, F.; Paton, R.S. Cation–π interactions in protein–ligand binding: Theory and data-mining reveal different roles for lysine and arginine. *Chem. Sci.* **2018**, *9*, 2655–2665. [CrossRef]
19. Kumar, N.; Gaur, A.S.; Sastry, G.N. A perspective on the nature of cation-π interactions. *J. Chem. Sci.* **2021**, *133*, 97. [CrossRef]
20. Shah, D.; Li, J.; Shaikh, A.R.; Rajagopalan, R. Arginine–aromatic interactions and their effects on arginine-induced solubilization of aromatic solutes and suppression of protein aggregation. *Biotechnol. Prog.* **2012**, *28*, 223–231. [CrossRef]
21. Zauhar, R.J.; Colbert, C.L.; Morgan, R.S.; Welsh, W.J. Evidence for a strong sulfur-aromatic interaction derived from crystallographic data. *Biopolymers* **2000**, *53*, 233–248. [CrossRef]
22. Krainer, G.; Welsh, T.J.; Joseph, J.A.; Espinosa, J.R.; Wittmann, S.; De Csilléry, E.; Sridhar, A.; Toprakcioglu, Z.; Gudiškytė, G.; Czekalska, M.A.; et al. Reentrant liquid condensate phase of proteins is stabilized by hydrophobic and non-ionic interactions. *Nat. Commun.* **2021**, *12*, 1085. [CrossRef]
23. Zhao, H.; Tang, L.; Fang, Y.; Liu, C.; Ding, W.; Zang, S.; Chen, Y.; Xu, W.; Yuan, Y.; Fang, D.; et al. Manipulating Cation–π Interactions of Reader Proteins in Living Cells with Genetic Code Expansion. *J. Am. Chem. Soc.* **2023**, *145*, 16406–16416. [CrossRef]
24. Valley, C.C.; Cembran, A.; Perlmutter, J.D.; Lewis, A.K.; Labello, N.P.; Gao, J.; Sachs, J.N. The Methionine-aromatic Motif Plays a Unique Role in Stabilizing Protein Structure. *J. Biol. Chem.* **2012**, *287*, 34979–34991. [CrossRef]
25. Kumar, Y.B.; Kumar, N.; John, L.; Mahanta, H.J.; Vaikundamani, S.; Nagamani, S.; Sastry, G.M.; Sastry, G.N. Analyzing the cation-aromatic interactions in proteins: Cation-aromatic database V2.0. *Proteins Struct. Funct. Bioinform.* **2024**, *92*, 179–191. [CrossRef]
26. Chourasia, M.; Sastry, G.M.; Sastry, G.N. Aromatic–Aromatic Interactions Database, A2ID: An analysis of aromatic π-networks in proteins. *Int. J. Biol. Macromol.* **2011**, *48*, 540–552. [CrossRef] [PubMed]
27. Tina, K.G.; Bhadra, R.; Srinivasan, N. PIC: Protein Interactions Calculator. *Nucleic Acids Res.* **2007**, *35*, W473–W476. [CrossRef] [PubMed]
28. Biot, C.; Buisine, E.; Kwasigroch, J.M.; Wintjens, R.; Rooman, M. Probing the energetic and structural role of amino acid/nucleobase cation-π interactions in protein-ligand complexes. *J. Biol. Chem.* **2002**, *277*, 40816–40822. [CrossRef] [PubMed]
29. Biot, C.; Buisine, E.; Rooman, M. Free-energy calculations of protein- ligand cation-π and amino- π interactions: From vacuum to proteinlike environments. *J. Am. Chem. Soc.* **2003**, *125*, 13988–13994. [CrossRef] [PubMed]
30. Minoux, H.; Chipot, C. Cation-π Interactions in Proteins: Can Simple Models Provide an Accurate Description? *J. Am. Chem. Soc.* **1999**, *121*, 10366–10372. [CrossRef]
31. Gallivan, J.P.; Dougherty, D.A. A Computational Study of Cation-π Interactions vs Salt Bridges in Aqueous Media: Implications for Protein Engineering. *J. Am. Chem. Soc.* **2000**, *122*, 870–874. [CrossRef]
32. McDonald, I.K.; Thornton, J.M. Satisfying hydrogen bonding potential in proteins. *J. Mol. Biol.* **1994**, *238*, 777–793. [CrossRef]
33. Berman, H.M.; Westbrook, J.; Feng, Z.; Gilliland, G.; Bhat, T.N.; Weissig, H.; Shindyalov, I.N.; Bourne, P.E. The protein data bank. *Nucleic Acids Res.* **2000**, *28*, 235–242. [CrossRef]
34. Wang, G.; Dunbrack, R.L., Jr. PISCES: A protein sequence culling server. *Bioinformatics* **2003**, *19*, 1589–1591. [CrossRef]
35. HomodimerDB. Available online: https://seq2fun.dcmb.med.umich.edu/HomodimerDB (accessed on 9 August 2025).
36. Gowthaman, R.; Pierce, B.G. TCR3d: The T cell receptor structural repertoire database. *Bioinformatics* **2019**, *35*, 5323–5325. [CrossRef] [PubMed]
37. Pierce, B.G.; Weng, Z. A flexible docking approach for prediction of T cell receptor–peptide–MHC complexes. *Protein Sci.* **2013**, *22*, 35–46. [CrossRef]
38. Dunbar, J.; Krawczyk, K.; Leem, J.; Baker, T.; Fuchs, A.; Georges, G.; Shi, J.; Deane, C.M. SAbDab: The structural antibody database. *Nucleic Acids Res.* **2014**, *42*, D1140–D1146. [CrossRef] [PubMed]
39. Schneider, C.; Raybould, M.I.; Deane, C.M. SAbDab in the age of biotherapeutics: Updates including SAbDab-nano, the nanobody structure tracker. *Nucleic Acids Res.* **2022**, *50*, D1368–D1372. [CrossRef]

40. Raybould, M.I.; Marks, C.; Lewis, A.P.; Shi, J.; Bujotzek, A.; Taddese, B.; Deane, C.M. Thera-SAbDab: The therapeutic structural antibody database. *Nucleic Acids Res.* **2020**, *48*, D383–D388. [CrossRef] [PubMed]
41. Mitra, R.; Cohen, A.S.; Sagendorf, J.M.; Berman, H.M.; Rohs, R. DNAproDB: An updated database for the automated and interactive analysis of protein–DNA complexes. *Nucleic Acids Res.* **2025**, *53*, D396–D402. [CrossRef]
42. Mitra, R.; Cohen, A.S.; Tang, W.Y.; Hosseini, H.; Hong, Y.; Berman, H.M.; Rohs, R. RNAproDB: A Webserver and Interactive Database for Analyzing Protein–RNA Interactions. *J. Mol. Biol.* **2025**, *437*, 169012. [CrossRef]
43. Dalkas, G.A.; Teheux, F.; Kwasigroch, J.M.; Rooman, M. Cation–π, amino–π, π–π, and H-bond interactions stabilize antigen–antibody interfaces. *Proteins Struct. Funct. Bioinform.* **2014**, *82*, 1734–1746. [CrossRef]
44. Akiba, H.; Tsumoto, K. Thermodynamics of antibody–antigen interaction revealed by mutation analysis of antibody variable regions. *J. Biochem.* **2015**, *158*, 1–13. [CrossRef] [PubMed]
45. Wilson, K.A.; Kellie, J.L.; Wetmore, S.D. DNA–protein π-interactions in nature: Abundance, structure, composition and strength of contacts between aromatic amino acids and DNA nucleobases or deoxyribose sugar. *Nucleic Acids Res.* **2014**, *42*, 6726–6741. [CrossRef]
46. Wu, D.; Sun, J.; Xu, T.; Wang, S.; Li, G.; Li, Y.; Cao, Z. Stacking and energetic contribution of aromatic islands at the binding interface of antibody proteins. *Immunome Res.* **2010**, *6*, S1. [CrossRef]
47. Varadi, M.; Bertoni, D.; Magana, P.; Paramval, U.; Pidruchna, I.; Radhakrishnan, M.; Tsenkov, M.; Nair, S.; Mirdita, M.; Yeo, J.; et al. AlphaFold Protein Structure Database in 2024: Providing structure coverage for over 214 million protein sequences. *Nucleic Acids Res.* **2024**, *52*, D368–D375. [CrossRef] [PubMed]
48. Jumper, J.; Evans, R.; Pritzel, A.; Green, T.; Figurnov, M.; Ronneberger, O.; Tunyasuvunakool, K.; Bates, R.; Žídek, A.; Potapenko, A.; et al. Highly accurate protein structure prediction with AlphaFold. *Nature* **2021**, *596*, 583–589. [CrossRef] [PubMed]
49. Hou, Q.; Bourgeas, R.; Pucci, F.; Rooman, M. Computational analysis of the amino acid interactions that promote or decrease protein solubility. *Sci. Rep.* **2018**, *8*, 14661. [CrossRef]
50. Vernon, R.M.; Chong, P.A.; Tsang, B.; Kim, T.H.; Bah, A.; Farber, P.; Lin, H.; Forman-Kay, J.D. Pi-Pi contacts are an overlooked protein feature relevant to phase separation. *eLife* **2018**, *7*, e31486. [CrossRef]
51. Gazit, E. A possible role for π-stacking in the self-assembly of amyloid fibrils. *FASEB J.* **2002**, *16*, 77–83. [CrossRef]
52. Warwicker, J.; Charonis, S.; Curtis, R.A. Lysine and arginine content of proteins: Computational analysis suggests a new tool for solubility design. *Mol. Pharm.* **2014**, *11*, 294–303. [CrossRef]
53. Hou, Q.; Kwasigroch, J.M.; Rooman, M.; Pucci, F. SOLart: A structure-based method to predict protein solubility and aggregation. *Bioinformatics* **2020**, *36*, 1445–1452. [CrossRef] [PubMed]
54. Niwa, T.; Ying, B.W.; Saito, K.; Jin, W.; Takada, S.; Ueda, T.; Taguchi, H. Bimodal protein solubility distribution revealed by an aggregation analysis of the entire ensemble of Escherichia coli proteins. *Proc. Natl. Acad. Sci. USA* **2009**, *106*, 4201–4206. [CrossRef] [PubMed]
55. Uemura, E.; Niwa, T.; Minami, S.; Takemoto, K.; Fukuchi, S.; Machida, K.; Imataka, H.; Ueda, T.; Ota, M.; Taguchi, H. Large-scale aggregation analysis of eukaryotic proteins reveals an involvement of intrinsically disordered regions in protein folding. *Sci. Rep.* **2018**, *8*, 678. [CrossRef] [PubMed]
56. Stanković, I.M.; Niu, S.; Hall, M.B.; Zarić, S.D. Role of aromatic amino acids in amyloid self-assembly. *Int. J. Biol. Macromol.* **2020**, *156*, 949–959. [CrossRef]
57. van der Kant, R.; Louros, N.; Schymkowitz, J.; Rousseau, F. Thermodynamic analysis of amyloid fibril structures reveals a common framework for stability in amyloid polymorphs. *Structure* **2022**, *30*, 1178–1189. [CrossRef]
58. Kollmer, M.; Close, W.; Funk, L.; Rasmussen, J.; Bsoul, A.; Schierhorn, A.; Schmidt, M.; Sigurdson, C.J.; Jucker, M.; Fändrich, M. Cryo-EM structure and polymorphism of Aβ amyloid fibrils purified from Alzheimer's brain tissue. *Nature Comm.* **2019**, *10*, 4760. [CrossRef]

MDPI AG
Grosspeteranlage 5
4052 Basel
Switzerland
Tel.: +41 61 683 77 34

Biomolecules Editorial Office
E-mail: biomolecules@mdpi.com
www.mdpi.com/journal/biomolecules

www.ingramcontent.com/pod-product-compliance
Lightning Source LLC
Chambersburg PA
CBHW040749250926
51850CB00007BB/1449

* 9 7 8 3 7 2 5 8 5 8 2 5 5 *